ENCYCLOPAEDIA OF
MECHANICAL ENGINEERING

ENCYCLOPAEDIA OF MECHANICAL ENGINEERING

Uday Tusti

ANMOL PUBLICATIONS PVT. LTD.
NEW DELHI - 110 002 (INDIA)

ANMOL PUBLICATIONS PVT. LTD.
H.O.: 4374/4B, Ansari Road, Darya Ganj,
New Delhi-110 002 (India)
Ph.: 23278000, 23261597

B.O.: No. 1015, Ist Main Road, BSK IIIrd Stage
IIIrd Phase, IIIrd Block
Bangalore - 560 085 (India)
Visit us at: www.anmolpublications.com

Encyclopaedia of Mechanical Engineering

First Published, 2007

PRINTED IN INDIA

Printed at Mehra Offset Press, Delhi.

Preface

The Encyclopaedia of Mechanical engineering covers all aspects of Mechanical Engineering and its role in society. This Encyclopaedia covers a range of various Mechanical engineering disciplines such as Product design and manufacturing, Marine engineering and naval architecture, Air conditioning and refrigeration engineering, Vehicle engineering, Energy management, General engineering and Structural engineering.

In recent years there have been dramatic changes in Mechanical engineering on both national scale and international scale. New techniques and products have found global recognition. These changes have had a major impact not only in industry but also on all our lives. Not only Technology is becoming more complex, but even terminology is becoming complex day by day.

This new encyclopaedia integrates all of the diverse approaches that have gone into the making of the modern field of Mechanical engineering. It is unique among available reference works in that it covers the most recent topics of research. Encyclopaedia of Mechanical engineering is intended for all those interested in this field. It will be of immense value to students of all kinds, especially students of different fields of Engineering.

It will also be a major source of reference for professionals and students, who require an authoritative and up-to-date guide to assist them in their work. In preparation of this book, the author has freely consulted large number of books and journals so no authenticity is claimed. Author is especially thankful to Anmol Publications Pvt. Ltd., New Delhi for shaping this book in its final form. Suggestions for further improvement of this book are not only welcome but also greatly appreciated.

Author

A

AB design A single subject research design that contains one baseline (A) and one treatment (B).

AIR 1. an acronym for Air injection reaction 2. The combination of gases that make up the earths atmosphere.

Air bag Properly known as a 'Supplemental Inflatable Restraint System', these are safety devices which inflate in an instant in the unfortunate event of a collision, to provide a cushioning effect which lessens the shock of impact.

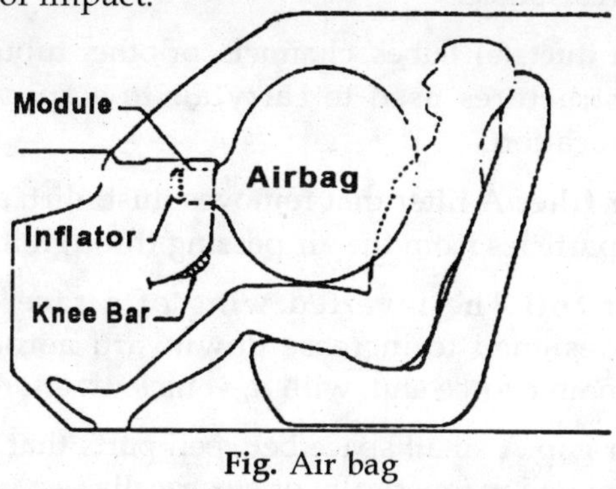

Fig. Air bag

Air bags operate automatically when controlling sensors give the alert, though the presence of these in a car should not give people a false sense of security, which might make them not bother to wear seatbelts. The use of seat and shoulder belts will maximize the effectiveness of air bags.

Air bag igniter A combustible device that converts electric energy into thermal energy to ignite the inflator propellant. The igniter is an integral component of the inflator assembly.

Air bag inflator A term often used for air bag igniter.

Air bag module The air bag and inflator assembly together in a single package. This module is mounted in the centre of the steering wheel.

Air bleed Holes or tubes in the carburettor to allow air to premix with gas flow.

Air box An enclosed chamber to direct air into a carburettor or intake manifold.

Air brake 1. A moveable dynamic spoiler that can be raised against the wind to slow a high speed vehicle.

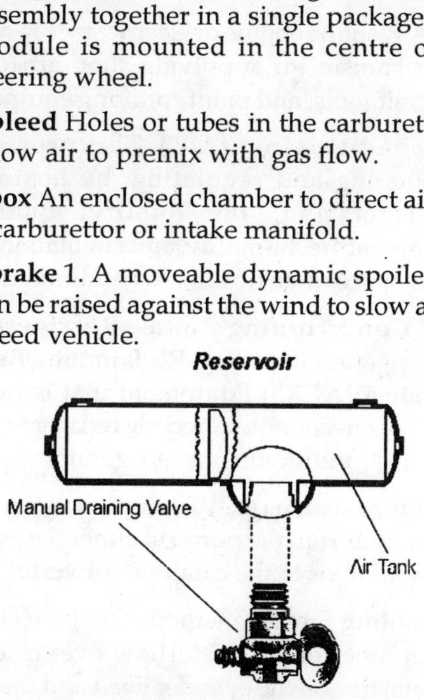

Fig. Air brake

2. The braking system on some heavy trucks that uses compressed air to expand the brake shoes by cam or wedge against the brake drums.

Air carbon arc cutting A carbon arc-cutting process that removes molten metal with a jet of air.

Air charge temperature (ACT) The temperature of incoming air in fuel-injection system.

Air cleaner A device connected to the carburettor in a manner that all incoming air must pass through it. Its purpose is to filter dirt and dust from the air before it passes into the engine.

Air compressor 1. Engine-driven mechanism for supplying high pressure air to truck brake systems. There are basically two types of compressors: those designed to work on in-line engines and those that work on V-type engines. The in-line type is mounted toward the fire wall and is camshaft driven.

2. A gasoline engine or electric -motor-driven mechanism for supplying shop air for the lift, air tools, and paint spraying equipment.

Air conditioning (A/C) The process of adjusting and regulating, by heating or refrigerating, the quality, quantity, temperature, humidity, and circulation of air in a space or enclosure; to condition the air.

Air Conditioning and Refrigeration Refrigerant Recovery, Reclaim, and Recycle System (ACR3) Equipment that is used to recover, reclaim, and recycle refrigerant from an air, conditioning or refrigeration system.

Air control valve (ACV) A component used to route air from the pump to either the exhaust manifold or to the catalytic converter.

Air cooling Simple method of engine cooling that relies on forced airflow over extended metal fins on the cylinder head and the block to maintain proper operating temperature.

Air core coil A method of constructing coils or transformers in which the wire is wound on a hollow form instead of one using magnetic material such as soft iron.

Air cushion restraint system A term used for air bag system.

Air dam 1. A panel across a race car's front end, designed to reduce the air pressure beneath the vehicle for a better ground effect.

2. Panels around the radiator to ensure that all air passes through, not around, the radiator.

Air delivery system The component that contains the air ducts, doors, blower, evaporator core, heater core, and controls that deliver air to the interior via the various outlets.

Air door A door in the duct system that controls the flow of air in the air conditioner and/or heater.

Air drag The resistance of air against an object, such as an automobile, trying to pass through it. Also referred to as aerodynamic resistance and aerodynamic drag, air resistance.

Air ducts(s) Tubes, channels, or other tubular structures used to carry air to a specific location.

Air filter A filter that removes dust, dirt, and particles from the air passing through it.

Air foil The inverted wing of a race car designed to increase downward aerodynamic force and, with it, vehicle traction.

Air gap A small space between parts that are mated magnetically or electrically.

Air hoist A hoisting device using compressed air in a cylinder, acting against a piston, with suitable outside connections, such as a hook.

Air horn 1. A horn that is actuated by compressed air.

2. A tubular passage containing the choke valves in the atmospheric side of a carburettor through which the incoming air must pass.

Air hose Air lines between the tractor and trailer supplying air to the trailer brakes.

Air ingestion Air is coming into the stuffing box because of a negative suction pressure.

Air Injection A system that injects air into the exhaust ports of the engine for combustion of unburned hydrocarbons in the exhaust gases, thus producing "cleaner" exhaust emissions.

Air Injection Reaction One method of reducing harmful exhaust emissions by injecting air into ether the exhaust manifold or the catalytic converter

Air inlet valve A moveable door in the plenum blower assembly that permits the selection of outside air or inside air for both heating and cooling systems.

Air intake and exhaust system The parts on the automobile engine used to get the air into the engine and the exhaust out of the engine, including air cleaner, muffler, tail pipe, and associated duct ing.

Air intake system A system that allows fresh clean air to enter a component such as an air conditioner, passenger compartment, or engine.

Air jack A device that uses compressed air to lift a vehicle. On some Indy and sports GOT cars, the jacks are built into the chassis, permitting the whole car to be raised instantly. This enables the pit crew to change all four tires simultaneously.

Air lift A trade name for a pneumatic helper spring with a Schrader valve that simplifies increasing or decreasing air pressure to compensate for changes in load.

Air line A hose, pipe, or tube through which air passes.

Air lock A compartment separating areas a different environment, especially different air pressures, that is used for entry to and departure from a spacecraft.

Air management system The system of solenoids and valves that control the output of the air pump to the catalytic converter, air cleaner housing, or exhaust manifold.

Air nozzle 1. The tube through which air is delivered to the exhaust gas in an air-injection system.

2. A device used to direct a stream of air into the desired area.

Air outlet valve A movable door in the plenum blower assembly that directs airflow into the heater core or into the duct-work that leads to the evaporator.

Air pressure 1. Atmospheric pressure.

2. The pressure produced by an air pump or compressor in a cylinder.

Air pressure test Applying air pressure to control valve body holes and passages leading to apply devices to check for the proper application

Air pump 1. A belt- or direct-driven vane-type pump that supplies the air needed for most air-injection systems.

2. A term often used for air compressor.

Air resistance The resistance of air against an object, such as an automobile, trying to pass through it. Also referred to as air drag, aerodynamic drag, and aerodynamic resistance.

Air scoop Opening at the front and/or along the side of a vehicle's bodywork to channel cool, ambient air to the radiator, engine, induction system, oil cooler, or brakes.

Air shifting Process that uses air pressure to engage different range combinations in the transmission's auxiliary section without a mechanical linkage to the driver.

Air shock absorber A term often used when referring to an air shock or load-leveling shock absorber.

Air slide release An air-operated release mechanism for positioning a sliding fifth wheel, operated from the cab of a tractor by actuating an air-control valve.

Air spring Also known as air bag. An air-filled device that functions as the spring on axles that utilize air pressure in the suspension system.

Air spring suspension A single or multiaxle suspension relying on air bags for springs and weight distribution of axles.

Air suspension speaker A speaker mounted in a closed box so that the enclosed air acts as a spring against the back side of the speaker cone.

Air suspension system A suspension system that uses contained compressed air for vehicle springing.

Air temperature sensor A unit consisting of an aspirator, bimetallic sensing element, and a vacuum modulator used to sense in-car temperature.

Air/fuel mixture The proportion of air to fuel provided by a carburettor of fuel-injection system.

Air/fuel ratio (A/F, ARF) The relative proportions of air and fuel entering an engine's cylinders as produced by the carburettor or fuel-injection system; the measure of the amount of air and fuel needed for proper combustion. The ideal or Stoichiometric ratio for gasoline is 14.7:1 air to fuel by weight. A higher ratio would contain more air and less fuel, and would be considered a lean mixture. A lower ratio with more fuel and less air would be a rich mixture. The air/fuel ratio is determined by the orifice size of the main jets inside a carburettor, the dwell duration of the mixture control solenoid inside a feedback carburettor, or the orifice opening and fuel pulse duration of a fuel injector.

Airfoil An aerodynamic device designed to improve traction by increasing the downforce on the car. The use of airfoils (also called wings) increases the cornering capability and improves stability at speed, but often at the expense of additional aerodynamic drag.

Alexander Charles Alexander Cesar Jacques Charles (1746 - 1823) A French Scientist after which is named Charles's Law which made the connection that a rise in temperature will cause an expansion in the volume of a gas.

Aliasing Aliasing occurs when a digital oscilloscope does not take enough samples on a particular signal. After the scope reconstructs and displays the waveform you can be fooled by the inaccurate display or it's frequency.

Align bore A stationary or portable boring machine used to correct an engine's out-of-round and/or warped main bearing housings.

Align hone A stationary machine operation that corrects an engine's out-of round and/or warped main bearing housings using a special honing mandrel.

Align ream A machine or hand process to enlarge the inside diameter of bushings to the proper size.

Alignment gap The distance between two adjacent auto-body panels.

Alignment Generally refers to wheel alignment, which is the proper adjustment of the car's front and rear suspension for camber, toe, caster and ride height.

Alignment pin A pin or pins used to align one part with another, such as the pins used to align a cylinder head on an engine block.

Alignment rack A drive-on device used for front end or four wheel alignment

Alignment stud A stud or studs used to align one part with another, such as the studs used to align a cylinder head on an engine block.

Alive (Live) Electrically connected to a source of voltage difference or electrically charged so as to have a voltage different from that of earth; the term may be used in place of 'current-carrying' where the intent is clear, to avoid repetition of the longer term.

Alkali A chemical "base" (the opposite of an acid) Certain types of alkalis (especially potassium hydroxide) have been used as fuel cell electrolytes.

Alkaline Fuel Cell (AFC) A type of Hydrogen/ Oxygen fuel cell in which the electrolyte is concentrated potassium hydroxide (KOH) and hydroxide ions (HO-) are transported from the cathode to the anode.

Alky A performance term used for alcohol, usually methanol, used as a fuel for racing car engines.

Alkyd A coating based on a polyester binder. The polyester binders are chemical combinations of molecules that contain more than one acid or alcohol group. Alkyds are widely used in water-based house paints and automotive primers.

All wheel drive (AWD) This is where all the wheels are utilized as active drive wheels for greater control in challenging road conditions. It works by dividing up the engine torque between the front and rear wheels in varying amounts depending on the needs.

Alligator clip A long-nosed, spring-loaded metal clip with meshing teeth. Used to make temporary electrical connections

Allocated requirements Allocated requirements are requirements that apportion all or part of the performance and functionality of a higher-level requirement on a lower-level element of a system.

Allotropic A term that applies to elements that appear in more than one form, though their atomic composition is the same. For example, the properties of graphite and diamond are the same as the element carbon (C). Since their physical properties are entirely different, both are called an allotropic of carbon.

Alloy A metal containing additions of other metallic or nonmetallic elements to enhance specific properties such as strength and corrosion resistance.

Alloy powder Powder prepared from a homogeneous molten alloy or from the solidification product of such an alloy.

Alloy steel Steel containing specified quantities of alloying elements added to effect changes in mechanical or physical properties.

Alloy wheels This is a term which is an example of how strange a language can sometimes be. It is used in everyday speech to denote any wheels that are made from a combination of metals (an alloy) rather than traditional steel, as most wheels are. However, steel is actually a combination of carbon and iron - and therefore is itself an alloy!.

All-risks policy A policy covering all losses except those specifically excluded in the contract.

All-season tire Tires with special tread designed to improve traction on snow or ice (generally provides 37% higher average snow traction compared with non-all-season tires), while providing acceptable noise levels on smooth road surfaces.

All-terrain cycle A small, three-wheeled, off-road vehicle by Honda.

All-terrain vehicle (ATV) A small, lightweight vehicle designed for recreational off-highway use.

All-wheel drive (AWD) A vehicle drivetrain with a center differential having all four wheels under power at all times.

Alnico A permanent-magnet alloy of aluminium, nickel, and cobalt.

Alpha particle A particle that is a by-product of radioactive decay. It has a positive electrical charge that is twice the negative charge of an electron.

Alpha rockwell hardness Index of the resistance of a plastic to surface penetration by a specified indentor under specified load applied with a Rockwell hardness tester. Higher values indicate higher indentation hardness.

Alpha sintered A type of silicone carbide that does not contain silica.

Alpha version A very early version of a software product.

Altered A drag-racing vehicle with a stock-looking coupe, roadster, or sedan body, but without normal street equipment. The engine and/or cockpit may be moved to the rear for better weight distribution.

Alternate Mode A display mode of operation in which the oscilliscope completes tracing one channel before beginning to trace another channel

Alternating current (AC) 1. The type of electrical current actually produced in an alternator.

2. A current flow that begins at "0" volts with the passage of a magnet and flows in one direction as it's inducted voltage increases to its maximum then decreases to its minimum then the inducted voltage induced by the opposite end of the magnet increases from "0" again however the current flows in the opposite direction.

Alternative fuel An alternative to gasoline or diesel fuel that is not produced in a conventional way from crude oil, for example compressed natural gas (CNG), liquefied petroleum gas(LPG), liquefied natural gas (LNG), ethanol, methanol and hydrogen.

Alternative refrigerant A refrigerant that can be used to replace an existing refrigerant, such as ozone friendly HFC-134a that is used to replace ozone depleting CFC-12

Alternator An electricity-generating device that converts mechanical energy to electrical energy in the form of alternating current. Diodes rectify the alternating current into direct current.

Altimeter A device that measures altitude above the surface of a planet or moon. Spacecraft altimeters work by timing the round trip of radio signals bounced off the surface.

Altitude compensation system An altitude barometric switch and solenoid used to provide better driveability at more than 4,000 feet (1,219 meters) above sea level.

Altitude The distance of a point above sea level. Important to automotive emissions control because the higher the altitude, the fewer oxygen molecules per given volume of air, which alters the effective compression and air/fuel ratios.

Alumel Trade name for an alloy of nickel with up to 5% aluminium, manganese and silicon, used with chrome in K-type thermocouples.

Aluminized coating A metal spray process used to coat engine components subject to high temperatures for long periods of time to increase heat dissipation to the ambient atmosphere.

Aluminized valve A valve with a thin layer of aluminium sprayed on the valve face and, sometimes, on the top of the valve head to provide a thin, hard, corrosion-resistant.

Aluminium (Al) A versatile engineering and construction material that is light in weight, yet some of its alloys have strengths greater than steel. It has high resistance to corrosion and is an excellent conductor of heat and electricity.

Aluminium cylinder block An engine block cast of aluminium or aluminium alloy, usually with cast iron sleeves installed as cylinder bores.

Aluminium-killed steel Steel alloy in which aluminium has been added to "kill" it in the molten stage and refine its grain structure. A process of stopping molten steel from bubbling and combining with oxygen after being poured into ingots.

Amber A type of resin that, when rubbed with a cloth, assumes a strong positive charge.

Amber lens The colour lens used for turn signals and flashers on modern motor vehicles.

Ambient air temperature The temperature of the air that surrounds an object.

Ambient air The air surrounding a given object or system.

Ambient compressor switch An electrical switch that energizes the air conditioner compressor clutch when the outside ambient air temperature is 47° F (8.3°C) or above. Similarly, the switch turns off the compressor when the air temperature drops below 32°F.

Ambient heat/pressure The heat or pressure in the area where the equipment is located.

Ambient sensor A sensor used on computerized automatic temperature-control systems that senses the outside air temperature and uses this information as an input to the system; a thermistor used in automatic temperature-control units to sense ambient temperature.

Ambient switch A switch used to control compressor operation by turning it ON or OFF. The switch is regulated by ambient temperature.

Ambient temperature The average temperature of the environment immediately surrounding the power supply. For forced air-cooled units, the ambient temperature is measured at the air intake.

Ambient temperature sensor A sensor that measures the outside air temperature as it enters the evaporator.

Ammeter An instrument used to determine the amount of amperage (current draw) in a circuit 1. by connecting the ammeter in series with the circuit and measuring the total number of electrons flowing in the circuit or

2. by using an inductive probe that measures the strength of the magnetic field that is created by the current flowing through the wire.

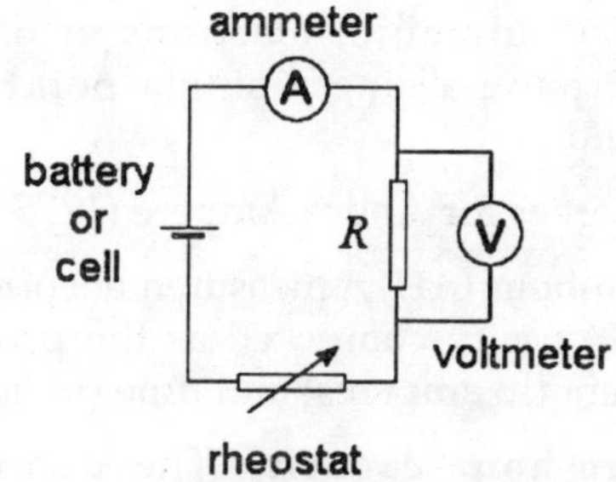

Fig. Ammeter

Ammeter shunt A low-resistance conductor used to increase the range of an ammeter. It is shunted (placed in parallel) across the ammeter movement and carries the majority of the current.

Amortization The gradual reduction of a debt by periodic payments large enough to meet current interest payments and repay the principal.

Amount financed The portion of the purchase price that is actually financed. In addition to the cost of the car, it can include the costs of an extended warranty, credit life insurance, and other items rolled into the payments.

Amp An acronym for ampere.

Amp/hour Amperes per hour; a standard measure for a rate of current flow.

Ampacity The current-carrying capacity of conductors or equipment, expressed in amperes.

Amperage capacity An indication of the length of time a battery can produce an amperage, or the amount of amperage that a battery can produce before being discharged.

Amperage the amount of current, expressed in amperes.

Ampere - turns A measure of the magnetic field produced by a coil of wire with current flowing through it. Equal to amperes of current flow multiplied by the number of turns of wire on the coil.

Ampere (A) 1. A unit of measure for current.

2. 1 Coulomb of electrons or 6.28×10^{18} electrons passing a single point in one second

3. Names for Andre Ampere (1775 - 1836)

Ampere-hour (AH) A measurement of quantity of electricity computed as the product of current (in amperes) and time (in hours).

Ampere-hour capacity The quantity of electricity measured in ampere-hours (Ah) which may be delivered by a cell or battery under specified conditions.

Ampere-hour efficiency The ratio of the number of ampere-hours delivered during the discharge of a secondary cell or battery to the number of ampere-hours necessary to restore the initial state of charge under specified conditions.

Amplification An increase in signal amplitude during its transmission from one point to another

Amplifier A device consisting of one or more vacuum tubs, transistors, or magnetic devices and associated components to increase the strength of the applied signal.

Amplifier, differential An amplifier which has available both an inverting and a non-inverting input, and which amplifies the difference between the two inputs.

Amplifier, inverting An amplifier whose output is 180 degrees out of phase with its input. Such an amplifier can be used with degenerative feedback for stabilization purposes.

Amplifier, non-inverting An amplifier whose output is in phase with its input.

Amplifier, operational A dc amplifier whose gain is sufficiently large that its characteristics and behaviour are substantially determined by its input and feedback elements. Operational amplifiers are widely used for signal processing and computational work.

Amplitude 1. The difference between the highest and lowest voltage displayed

2. The magnitude of an electrical signal above or below some reference, usually measured in voltage above or below "0"

Amplitude modulation (AM) A technique for impressing information upon a carrier by varying the amplitude of the carrier.

Anaerobic sealants A chemical sealant placed on a gasket in an engine to aid in sealing and to position the gasket during installation.

Anaerobic sealer Liquid or gel that bonds two parts together in the absence of air.

Analogue 1. Continuously Variable

2. A signal that varies in amplitude continuously and without interruption.

Analogue computer A computing machine that works on the principle of measuring, as distinct from counting, in which the measurements obtained (as voltages, resistances, etc.) are translated into desired data.

Analogue instrument An instrument having a needle on a dial used for taking measurements, such as temperature and engine RPM.

Analogue signal An electrical signal that varies in amplitude within a given parameter

Analogue-to-digital converter Electrical device used to convert continuous analogue signals to a digital code that represents the amplitude of the original signal sample.

Analysis of variance Analysis of variance is a basic statistical technique for analyzing experimental data. It subdivides the total variation of a data set into meaningful component parts associated with specific sources of variation in order to test a hypothesis on the parameters of a model or to estimate variance components.

Analytical hierarchy process A decision making tool for complex, multi-criteria problems where both qualitative and quantitative aspects of a problem need to be incorporated. AHP clusters the decision elements according to their common characteristics into a hierarchical structure similar to a family tree. It involves building a hierarchy (ranking) of decision elements and then making comparisons between each possible pair in each cluster (as a matrix). This gives a weighting for each element within a cluster (or level of the hierarchy) and also a consistency ratio (useful for checking the consistency of the data). By reducing complex decisions to a series of simple comparisons and rankings, then synthesizing the results, AHP helps arrive at the best decision and also provides a clear rationale for the choice made.

Anaroid tube A thermo-mechanical device in a fuel-injection system that regulates the amount of fuel being injected according to differences in temperature and pressure in the intake manifold.

Anchor 1. A Slang expression for brakes.

2. A mounting point on a vehicle structure for a stressed, non-structural component, such as a seat belt or a seat.

Anchor end That end of a brake shoe that is attached to a fixed point on the backing plate.

Anchor pin A steel pin rigidly attached to the backing plate of drum brakes. Return springs are attached to the anchor pin and to the brake shoes to hold the shoes against the anchor pin in a non-applied position. In an applied position, it prevents the shoes from rotating with the drums.

Ancillary Equipment Consists of the Aperture Plate and Utility Services.

AND gate A digital logic gate that must have all of its inputs high to produce an output

Aneroid bellows An accordion-shaped temperature sensor charged with a small amount of volatile liquid. Temperature change causes the bellows to contract or expand, which, in turn, opens or closes a switch, such as a thermostat.

Angle block A cylinder block that does not have a deck at 90 degrees to the cylinder.

Angle mill A machining operation to mill the deck surface at a shallow angle on the exhaust side of the engine block in order to increase the compression ratio by decreasing the combustion chamber volume.

Angle of attack The angle between the velocity vector and the longitudinal axis of a missile or rocket.

Angle plug head A cylinder head having spark plugs that are angled toward the exhaust valves.

Angstrom A unit for the measurement of wavelength. Equals one hundred millionth of a centimetre (0.003937 millionth of an inch).

Anion A negative ion. Alkali, molten carbonate and solid oxide fuel cells are "anion-mobile" cells.

Annealing A heat-treatment process to reduce hardness or brittleness, relieve stresses, improve machinability, facilitate cold working, or produce a desired microstructure or property.

Annual inventory turns A measure of asset management that is calculated by dividing the value of annual plant shipments at plant cost (for the most recent full year) by the average total inventory value at plant cost. Total inventory includes raw materials, work in process, and finished goods. Plant cost includes material, labour, and plant overhead.

Annular Pertaining to, or having the form of a ring.

Annulus Any type of ring gear, including the ring part of the ring and pinion in the rear end, and the gears in the planetary gear set of an automatic transmission.

Anode The electrode at which oxidation (a loss of electrons) takes place. For fuel cells and other galvanic cells, the anode is the negative terminal; for electrolytic cells (where electrolysis occurs), the anode is the positive terminal.

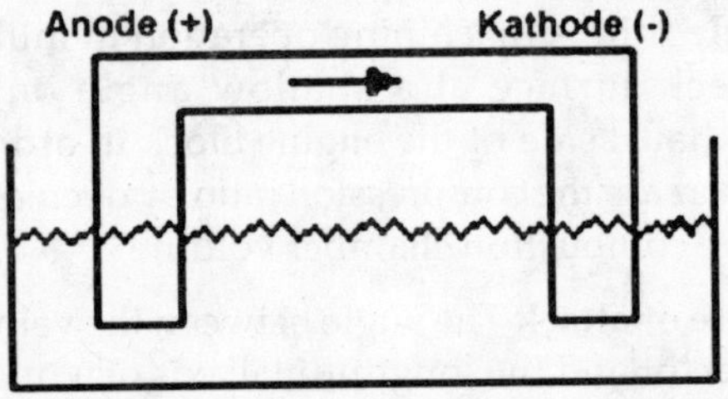

Fig. Anode

Anodize A protective oxide coating to a metal surface using the metal as the anode in an electrical cell and allowing an electrolyte to act upon it.

Anomaly The angular distance between the position of a planet and its last perihelion, or between that of a satellite and its last perigee.

Antenna A wire or other conductive metallic structure used for radiating or receiving electromagnetic signals, such as those for radio, television, or radar.

Antenna trimmer A method used to calibrate the antenna for an AM radio

Anti-Alias Filter An anti-alias (or anti-aliasing) filter allows through the lower frequency components of a signal but stops higher frequencies, in either the signal or noise, from introducing distortion. Anti-alias filters are specified according to the sampling rate of the system and there must be one filter per input signal.

Anti-backfire valve A valve that is used in the air-injection system to prevent backfiring in the exhaust system during deceleration.

Anticipatory failure determination Anticipatory failure determination (afd) is a failure analysis method. Like fmea, it has the objective of identifying and mitigating failures. Rather than asking developers to look for a cause of a failure mode, it reverses the problem by asking developers to view the failure of interest as the intended consequence and try to devise ways to assure that the failure always happens reliably. This viewpoint then facilitates better identifying steps to avoid the failure.

Anti-drive A suspension geometry that resists a vehicle's tendency to drop or dive on the front springs when braking.

Anti-foam agent An additive that reduces foaming caused by the churning action of the crankshaft in the engine oil.

Anti-fouler A device installed on a spark plug in an oil-burning engine to reduce the fouling of the plug.

Anti-freeze A liquid that mixes with the water in a cooling system of a vehicle's engine. Antifreeze keeps the water from freezing in the winter or cold climates, or from overheating in the summer or hot climates.

Anti-friction bearing The term applies to almost any ball-roller or taper-roller bearing.

Anti-icers Chemicals added to gasoline to prevent it from freezing. Typically an alcohol such as ethanol, methanol, or isopropyl alcohol.

Anti-knock additive A compound that may be added to gasoline to increase its octane, thereby decreasing its knock.

Anti-knock index The average of the motor octane number (MON) and the research octane number (RON); a measure of fuel's anti-knock characteristics.

Anti-lock brake system A computer-controlled braking system that senses impending wheel lockup and pulses the brakes many times a second to prevent it. This results in the most important benefit of abs: by preventing wheel lockup, it allows the driver to maintain steering control. Unfortunately, too few drivers are trained to use abs properly. When the system engages, it may vibrate the brake pedal. A driver who doesn't know how to use abs may lift his foot from the pedal when he receives this feedback, effectively disengaging the system

Anti-matter A hypothetical form of matter of which the atoms are composed of anti-particles, as protons, electrons, etc. assumed to carry charges opposite to those associated with ordinary matter. Particles having such properties have been produced in particle accelerators.

Antimony a metal added to non-maintenance-free of hybrid battery grids to add strength.

Anti-oxidant A substance that prevents or slow down oxidation of plastic material that is exposed to air and the elements.

Anti-percolation valve A device used to prevent fuel from evaporating from the fuel bowl of a carburettor while the engine is running. It is connected to the throttle linkage, so it is closed when the throttle is open, and open when the throttle is closed. With the engine off, hot fuel vapours boil out through the vent line and into the charcoal canister.

Antipodal Pertaining to, or located on, the opposite side of the Earth.

Anti-rattle hardware Clips, springs, and washers to prevent disc pads or brake shoes from rattling and vibrating.

Anti-roll bar A suspension component. A steel rod or tube that connects the left and right suspension members to resist roll or swaying of the vehicle. Improves handling.

Anti-rust additive A solution that is added to coolant to retard rust formation inside a cooling system.

Anti-seize compound A thread compound designed to protect threaded connections from damage due to rust or corrosion.

Anti-siphon system A small passage in a carburettor to prevent fuel from siphoning from the float bowl into the engine.

Anti-skid control Accessory for the brake system that operates on rear wheels, or all four wheels, to prevent wheel lockup during braking. Braking pressure is reduced to wheel(s) that are about to lock up and skid, by electronic controls.

Anti-smog device A part or parts designed to reduce or eliminate vehicle exhaust emissions.

Anti-squat Suspension geometry that will resist a vehicle's tendency to drop or squat on the rear springs when accelerating.

Anti-static agents Treatment used during, and/or after the moulding process to minimize static electricity in plastic materials.

Anti-sway bar A suspension component, often called a sway bar, intended to prevent side-to-side body movement in relation to the axles and wheels.

Anti-theft system Deterrent systems designed to scare off would-be car thieves by sounding alarms and/or disabling the ignition system. Common components used in an automobile anti-theft system include an electronic control module, door switches at all doors, trunk key cylinder switch, hood switch, starter inhibitor relay, horn relay and alarm.

Antivirus program A software utility that searches disk drives for viruses and removes any that are found.

Aperture Plate A continuous circular ring that defines the outer clear aperture of M1.

Apex seal A seal used to retain the combustion pressure at all three tips of the rotor in a rotary engine.

Apexes The peaks on the rotor, formed by the meeting of two adjoining rotor faces in a rotary engine.

Aphelion That point in a solar orbit which is farthest from the Sun.

API 1. Application protocol interface.

2. Application programming interface - the standard set of functions provided by a program or operating system to allow for integration of other software. Two programs linked via an API can both be altered and still work together so long as both conform to the API.

API gravity The scale for the density (gravity) of a liquid petroleum product, expressed in API degrees. The lighter the product the higher the number.

API ring The information printed, in a ring, on the top or side of an oil or lubricant container providing the API specifications and ratings of the contents.

A-Pillar In the side view, the foremost roof support of a vehicle, located in most instances between the outer edge of the windshield and the leading edge of the front door upper. Also known as an A-Post.

Apoapsis That point in an orbit which is farthest from the primary.

Apogee That point in a terrestrial orbit which is farthest from the Earth.

Apolune That point in a lunar orbit which is farthest from the Moon.

A-post A term often use for a-pillar.

Apparent power Power value obtained in an ac circuit as the product of current times voltage.

Appearance money The payment of money to a popular driver to encourage competition in a race, so his or her participation can be advertised in advance, thereby trying to attract a larger paying crowd.

Application cable The line or cable that engages the vehicle's emergency brake system.

Applied Voltage The voltage applied to the circuit by the scorce

Apply devices Hydraulically operated bands, multi-disc clutches, and mechanically operated one-way clutches that drive or hold the members of a planetary gearset.

Apply side The side of a piston on which force or pressure is exerted to move the piston to do work.

Apportionment The assignment of goals such as reliability from system to subsystem in such a way that the whole system will meet the required goal.

Appraisal An opinion of the market value of an asset as of a specific date.

Appraiser An insurance company's representative who estimates a vehicle's damage and authorizes payment to the collision repair/refinishing shop.

Approach angle The maximum angle, in degrees, of a line running upward and forward from the front tire contract point to the lowest obstruction under the front of the vehicle.

Approval Written notification by an authorized representative of the acquirer that the developers plans, design, or other aspects of the project appear to be sound and can be used as the basis for further work. Such approval does not shift responsibility from the developer to meet contractual requirements.

Approved data That issue or version of data which has been identified by the developer/supplier as being the master issue or version of data subject to configuration management which has been submitted for acquirer approval, and which has been approved by the acquirer.

Aquaplaning A tire unable to remain in contact with the ground or pavement in wet weather that rides on the water itself. Also known as hydroplaning or, more simply, planing.

Aramid fibres A family of synthetic materials that are stronger than steel but weigh little more than half what an equal volume of fibreglass weighs.

Arbor A tapered metal shaft used to secure a cutting tool or a part being turned on a lathe.

Arbor press A manual- or power-operated press used to force arbors or mandrels into or out of holes and for similar assembly or disassembly operations.

Arc blow In welding, the deflection of an arc from its normal path because of magnetic forces.

Arc braze welding A braze welding process that uses an arc to provide the heat.

Arc cutter A term often used for thermal cutter.

Arc cutting The thermal cutting process that severs or removes metal by melting it with the heat of an arc between an electrode and the work piece.

Arc force The axial force developed by an arc plasma during a welding procedure.

Arc gouging A thermal gouging that uses an arc-cutting process to form a bevel or groove.

Arc length The distance from the tip of the welding electrode to the adjacent surface of the weld pool.

Arc spraying A thermal spraying process using an arc between two consumable electrodes of surfacing materials as a heat source, and a compressed gas to atomize and propel the surfacing material to the substrate.

Arc welding A weld process that produces fusion of work pieces by heating them with an arc with or without the application of pressure and with or without filler metal.

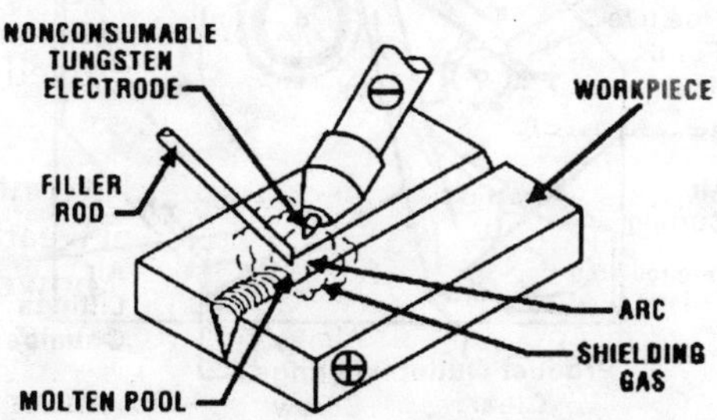

Fig. Arc welding

Architectural principles Architectural principles are statements of preferred architectural direction or practice. Each principle should be stated in such a way that one will know if the architecture has the characteristics expressed by the principle. Principles need to be rationalized, stating why the principle is preferred.

Architecture The physical structure of a computer's internal operations, including its registers, memory, instruction set, input/output structure, and so on.

Archiving Archiving is the intelligent back-up of selected objects (datasets, items etc.) that no longer need to be accessed on a regular basis within the PDM system. These files are removed from the on-line disk storage to a lower cost media such as tape or optical disk.

Arcing 1. The process of grinding or forming drum brake linings to conform to the drum diameter and provide clearance where needed.

2.A term that applies to the spark that occurs in an electrical circuit in an air gap, such as a spark plug.

Arcing time The time elapsing from the severance of the circuit to the final interruption of current flow.

Argument Input parameter to a program.

Argument of periapsis In an orbit, the angular distance between the point of periapsis and the ascending node.

Arianespace A private limited company established for the purpose of producing, financing and marketing the ESA Ariane launch vehicle. Comprises European companies concerned in the rocket's development, CNES and several banks. US agent is Grumman Aerospace.

Arithmetic-logic unit A basic element of the central processing unit (CPU) in a computer. That portion of the CPU where arithmetic and logical operations are performed.

Armature 1. A part moved by magnetism.

2. A part moved through a magnetic field to produce current.

3. The rotating unit inside a dc generator or starter.

Aromatic A type of solvent based on benzene ring molecules. Aromatics are often used as diluents in acrylic lacquers. In gasoline aromatics are commonly used for the BTX group (benzene, toluene, xylene) and are hydrocarbons.

Articulated vehicle Large trucks or buses with two or more wheeled units, so designed for ease of cornering.

Articulating upper coupler A bolster plate kingpin arrangement that is not rigidly attached to the trailer but provides articulation and/or oscillation about an axis parallel to the rear axle of the trailer.

Articulation 1. The action of a chain joint in flexing from the straight to an angle and back to the straight as the joint passes around a sprocket or other path.

2. Vertical movement of the front driving or rear axle, relative to the frame of the vehicle to which they are attached.

Asbestos A fibre mineral that is heat resistant and nonburning. Once used in brake linings, gaskets, and clutch facings, it is no longer used due to health hazards.

Asbestos Information Association (AIA) An association to provide industry-wide information on asbestos and health, and on industry efforts to eliminate existing hazards.

Asbestosis A progressive and disabling lung disease caused by inhaling asbestos fibres over a long period of time.

Ascending node The point at which an orbiting object or spacecraft, travelling from south to north, crosses the plane of the equator.

Ascent module That part of a spacecraft that ascends from the surface of a planet or moon to rendezvous and dock with an orbiting spacecraft.

A-shim A valve spring-adjuster insert with a thickness of 0.060 inch (1.524mm) used to balance spring pressure and to correct installed height.

Asia Pacific Economic Cooperation (APEC) An organization established in 1989 to promote trade and investment in the Pacific Basin. APEC now comprises eighteen countries located in and around the Pacific Ocean: Australia, Brunei, Canada, Chile, China, Hong Kong, Indonesia, Japan, Malaysia, Mexico, New Zealand, Papua New Guinea, the Philippines, Singapore, South Korea, Taiwan, Thailand, and the United States.

Aspect ratio The ratio between the width and sidewall (or height) of the tire. Tires with lower aspect ratios, usually found on sports models, provide superior handling but a harsher ride.

Aspirator A device that uses suction to move air, accomplished by a differential in air pressure; a one-way valve attached to the exhaust system of an engine that admits air during periods of vacuum between exhaust pressure pulses. Used to help oxidize hydrocarbon (HC) and carbon monoxide (CO), and to supply additional air that the catalytic converter may require. Can be used instead of a belt-driven, air-injection pump in some applications.

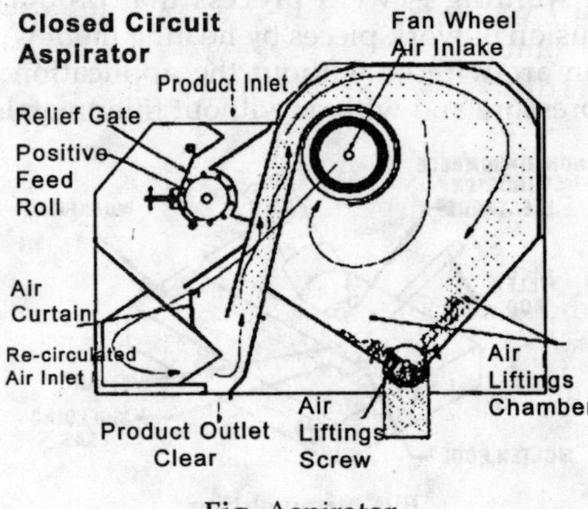

Fig. Aspirator

Aspirator valve A device used to draw out fluids by suction. In this case, a pollution device is used to draw fresh air by suction into the exhaust flow to reduce emissions.

Assembly A number of parts or subassemblies or any combination thereof joined together to perform a specific function and capable of disassembly.

Assembly & packaging A Unaxis division, as well as the term used to describe a production process for encapsulating integrated circuits inside a molded outer casing. Contact pads on the chip are bonded with external through-contacts on the casing, and then the integrated circuit (chip) proceeds to the next manufacturing step.

Assembly line communications link An electrical connector used to check a vehicle's operating system while it is still on the assembly line.

Assembly line data link The information processed for use in assembly-line diagnostics.

Assembly lube A special lubricant used to coat parts that rub or rotate against each other during initial assembly.

Assignable cause Assignable Cause is a source of variation which is not due to chance and, therefore, can be identified and eliminated. An assignable cause is often signalled by an excessive number of data points outside a control limit and/or a non-random pattern within the control limits. Also called "special cause".

Assigned risk plan A risk that must be insured by state law or otherwise.

Associativity A link between two different functions in a cad system that assures that a change made in one area is reflected in all other areas. For example, a change to a solid model will be reflected in its drawing and related cam program. Bi-directional associativity indicates that updates happen in both directions between functions. For example, a change to a drawing will be reflected in its solids model.

Astronaut A person who flies in space, whether as a crew member or passenger.

Astronautics The science and technology of space flight.

Astronomical unit The mean distance of Earth from the Sun, i.e. 92,955,807 miles (149,597,870 km).

Astrophysics Study of the physical and chemical nature of celestial bodies and their environs.

Asymmetrical cam A camshaft having different profiles for the intake and exhaust lobes.

Asymmetrical rear-leafing spring A spring on which the rear axle is not located in its centre.

Asynchronous 1. Not having a constant period or not synchronous.

2. A means of operating electronic circuits without a common timing signal to synchronize events; rather, the events occur at random times.

Asynchronous Fuel Timing Fuel injector firing which is not in direct relationship to the ignition reference pulses

Asynchronous groupware Asynchronous groupware is software used to help people to work in groups, but not requiring those people to be working together at the same time (asynchronous = not coordinating at a single point in time).

A-to-D converter A digital electronic component that converts an electrical signal into discrete binary values

Attenuation The decrease of a propagating physical quantity, such as a radio signal, with increasing distance from the source, or from some obstruction.

Attenuator A circuit, component or device that reduces the signal strength without distorting the signal or reducing its fidelity

Attitude & articulation control subsystem The onboard computer that manages the tasks involved in spacecraft stabilization via its interface equipment. For attitude reference, star trackers, star scanners, solar trackers, sun sensors, and planetary limb trackers are used.

Attitude control The system that turns and maintains a spacecraft in the required direction as indicated by its sensors.

Attribute An entity that describes some aspect of the configuration of a system, subsystem, or component. Certain attributes will be used as command parameters.

Authentication The procedure (essentially approval) used by the approval authority in verifying that specification content is acceptable. Authentication does not imply acceptance or responsibility for the specified item to perform successfully.

Authorization The process of granting or denying access to a network or resource.

Auto control Another term used for automatic control.

Auto ignition Short for automatic ignition.

Autocross A form of automotive competition that is held on a tight, closed course. Vehicle handling and agility is stressed, rather than flat out speed.

Automated clearance sensing (ACS) Commercial vehicle operations (CVO) technology used to help large vehicles negotiate low or limited clearance objects such as bridges and viaducts.

Automatic adjuster A drum-brake mechanism that adjusts the lining clearance as wear occurs. It is commonly actuated on reverse stops or when the parking brake is set.

Automatic chassis lubrication (ACL) A system where the chassis is automatically lubricated at predetermined intervals.

Automatic choke A mechanism that positions the choke valve automatically in accordance with engine temperature.

Automatic control 1. A dial on the instrument panel that is set at a desired temperature level to control the condition of the air automatically.

2. Any system that reacts to a predetermined condition rather than responding to external commands.

3. Also known as auto control.

Automatic door locks A passive system used to lock all doors when the required conditions are met. Many systems lock the doors when the gear selector is placed in drive, the ignition switch is in RUN, and all doors are properly shut.

Automatic headlight dimming Automatically switches the headlights from high beams to low beams under two different conditions when light from oncoming vehicles strikes the photocell-amplifier, or light from the taillights of a vehicle being passed strikes the photocell-amplifier.

Automatic highway advisory radio (AHAR) American traffic information broadcasting system whose transmissions are received through car radios which automatically interrupt other radio reception and tune to the correct station.

Automatic Idle Speed Motor The output control device used by Chrysler to increase or decrease engine idle speed

Automatic ignition A condition where the engine continues to run after the ignition has been shut off. Often called dieseling or running on. Also known as auto ignition.

Automatic level control A shock-absorber system operated by air pressure and provided as an accessory on some vehicles to automatically maintain the correct riding height under various load conditions.

Automatic locking front hubs Found in some four-wheel drive vehicles, this allows the driver to engage, or "lock," the front axle hubs without leaving the vehicle.

Automatic locking retractor Standard on 1995 and later models, this device is built into the shoulder belt retractor and keeps the belt cinched tightly, which is essential for properly securing a child-safety seat.

Automatic on/off with time delay 1. To turn on the headlights automatically when ambient light decreases to a predetermined level.

2. To allow the headlights to remain on for a certain amount of time after the vehicle has been turned off.

Automatic Probe Calibration A procedure to allow a device to adjust internally to compensate for resistance variations between probes

Automatic Shut-Down Relay The relay used by Chrysler to feed ignition voltage to the fuel pump, ignition system and the injectors

Automatic steering effect The tendency of a vehicle to travel straight out of a turn when the steering wheel is released.

Automatic temperature control (ATC) An air-conditioner control system designed to maintain a pre-selected, in-car temperature and humidity level automatically.

Automatic temperature control Automatically controls a vehicle's heating and cooling systems, maintaining a temperature preset by the occupant.

Automatic tensioning The constant tension of a device, maintained at a proper value by some automatic means; to minimize the attention required.

Automatic test equipment / automated test equipment Automated test equipment (ate) built to perform a test or sequence of tests. Ate ranges from simple devices to verify mechanical or electrical continuity to sophisticated computerized systems with automatic sequencing, data processing, and readout. Ate may be stand alone test units or may be built into the operational equipment.

Automatic test pattern generation Automatic test pattern generation is the process that utilizes lists of faults and a model of the circuit to analyze the logical and topical nature of the circuit in order to create test vectors for each fault and, thereby, produce a high-fault-coverage test pattern for a design.

Automatic Traction Control A system that automatically prevents slippage of one of the drive wheels

Automatic transmission A system that varies the power and torque to a drivetrain without the use of a foot-operated clutch.

Automatic transmission cooler A device, often found in the radiator, through which automatic transmission fluid circulates to be cooled by surrounding air or engine coolant.

Automatic transmission fluid (AFT) A red, petroleum-based fluid used to transfer power and control, lubricate, cool, and clean the automatic transmission.

Automatic Transmission Rebuilders Association (ATRA) A trade association for transmission repair shops, technicians, and suppliers of transmission repair equipment, parts, and tools.

Automation 1. Semi-automatic or automatic material handlers, loaders, unloaders, and other labour-saving devices.

2. Automatic cycle control of machines or equipment by tracer, cam, plugboard, numerical control, or computer.

3. The application of machinery and equipment to perform and control semi-automatically or automatically and continuously all operations in a manufacturing plant.

Automobile emissions Certain impurities that may enter the atmosphere during vehicle operation, such as hydrocarbons (HC), carbon monoxide (CO), and oxides of nitrogen (NOx).

Automobile four-wheeled passenger motor vehicle having a seating capacity for not more than 10 people including police cars and racing cars but not including ambulances or hearses.

Automotive air conditioning The process of transferring heat from insider to outside the passenger compartment. The cooled air is also dehumidified, purified, and circulated.

Automotive air pollution Evaporated and unburned fuel, and other undesirable by-products of combustion that escape from a motor vehicle into the atmosphere.

Automotive battery An electro-chemical device that stores and provides electrical energy for the operation of a vehicle.

Automotive Cooling System Institute (ACSI) A Motor and Equipment Manufacturers Association subgroup made up of cooling-system product manufacturers.

Automotive cooling system The many components that operate to absorb and dissipate heat developed in the combustion process, thus maintaining the desired engine-operating temperature.

Automotive emissions Another term for automotive air pollution.

Automotive Friction Material Edge Code (AFMEC) A series of codes on the side of a brake lining material, and coefficient of friction. These codes are for lining identification and comparison; they do not indicate quality.

Automotive power brakes A brake system having a vacuum and atmospheric air-operated power booster or hydraulic power boost to multiply braking force.

Autonomous system A network that is administered by a single set of management rules that are controlled by one person, group or organization. Autonomous systems often use only one routing protocol, although multiple protocols can be used. The core of the Internet is made up many autonomous systems.

Autopilot A system or device that controls a vehicle's flight at a preset course and altitude.

Auto-ranging input An input voltage sensing circuit in the power supply that automatically switches to the appropriate input voltage range (90-132 VAC or 180-264 VAC). Sometimes known as Auto-select input.

Auxiliary A spare; an extra; a back-up system.

Auxiliary air valve A device that allows air to bypass a closed throttle during engine start-up and warm-up.

Auxiliary drum parking brake The incorporation of an auxiliary parking-brake drum inside a rear rotor on some four-wheel drive disc-brake system.

Auxiliary seal A secondary seal mounted outside the seal housing 1. To prevent refrigeration oil from escaping and entering the clutch assembly.

2. To aid in the prevention of the loss of fluid from a system.

Auxiliary section The section of a transmission housing the auxiliary drive gear, main shaft assembly, countershaft,, and synchronizer assembly, where range shifting occurs.

Auxiliary shaft A separate shaft, in an overhead cam engine, that drives devices such as the fuel pump, oil pump, and distributor.

Auxiliary springs 1. A second or third valve spring with a different resonant frequency to cancel out harmonic vibrations that limit engine speed.

2. The spring(s) added to a vehicle, generally in the rear, to support a heavy load.

Auxiliary supply A power source supplying power other than load power as required for the proper functioning of a device.

Auxiliary venturi A small secondary venturi mounted inside the main venturi of a carburettor to provide increased air velocity. May also be called a booster venturi.

Availability The product metric that defines the percentage of time that a product is available and operational for customer use. It is the proportion of total time that an item of equipment is capable of performing its specified functions, normally expressed as a percentage. It can be calculated by dividing the equipment available hours by the total number of hours in any given period.

Available capacity The total capacity, Ah or Wh, that will be obtained from a cell or battery at defined discharge rates and other specified discharge or operating conditions.

Available voltage 1. The amount of voltage available at any point in the circuit

2.The maximum voltage that can be produced in the secondary circuit of a conventional ignition system.

Available-to-promise Automated available-to-promise is achieved by giving order takers access to inventory and capacity information, and in some cases even vendor information, so that they are able to commit to reliable delivery dates while the customer is still on the phone.

Avalanche diode A semiconductor designed to operate in the breakdown region to produce a constant voltage across the diode for regulating through the current though it may vary.

Avalanche point The point at witch a nondestructive PN junction breaks down and creates a reverse bias current flow.

Axial In line along with the axis of centreline of a part or component.

Axial compressor A compressor designed so that the cylinders are arranged parallel to the output shaft.

Axial direction A direction parallel to the geometrical axis of M1.

Axial leads Wires that extend from the ends of an electronic component, such as a capacitor, along the axis of the unit.

Axial load A type of load placed on a bearing that is parallel to the axis of the rotating shaft.

Axial motion Motion that occurs along the axis of a revolving shaft or parallel to the axis of a revolving shaft.

Axial play Axial play in a ball joint means looseness in the same line as the ball joint stud.

Axial support system Consists of 120 support actuators arranged in five circular rings. Support pads at the upper end of each actuator are bonded to the back of M1.

Axiomatic design Axiomatic design recognizes four domains. The needs of the customer are identified in customer domain and are stated in the form of required functionality of a product in functional domain. Design parameters that satisfy the functional requirements are defined in physical domain, and, in process domain, manufacturing variables define how the product will be produced. Solution alternatives are created by mapping the requirements specified in one domain to a set of characteristic parameters in an adjacent domain. The mapping between the customer and functional domains is defined as concept design; the mapping between functional and physical domains is product design; the mapping between the physical and process domains corresponds to process design. The output of each domain evolves from abstract concepts to detailed information in a top-down or hierarchical manner. Two design axioms provide a rational basis for evaluation of proposed solution alternatives and the subsequent selection of the best alternative. The first axiom is the independent axiom, and it states that a good design maintains the independence of the functional requirements. The second axiom is the information axiom and it establishes information content as a relative measure for evaluating and comparing alternative solutions that satisfy the independence axiom.

Axis 1. The centre line of a rotating part.

2. A cross member supporting a vehicle on which one or more wheels are mounter

3. A pair of wheels at either end of a vehicle

Axle A crossmember supporting the vehicle on which one or more wheels are mounted or supported.

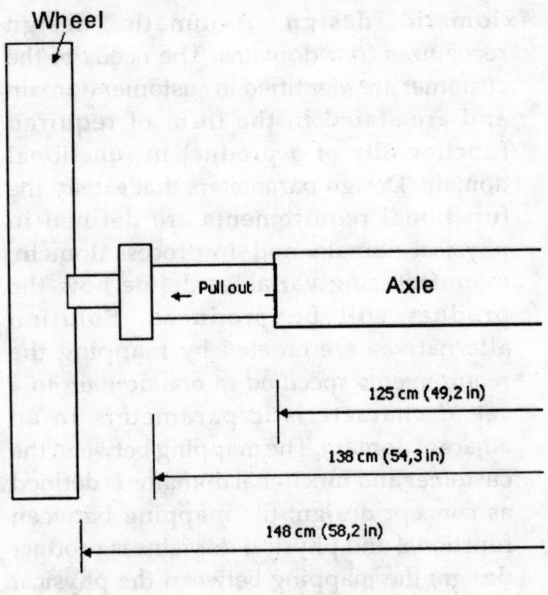

Fig. Axle

Axle bearing A bearing that supports an axle or half shaft in an axle housing.

Axle boot The flexible cover that retains grease and/or oil in a transmission or a constant velocity joint.

Axle carrier assembly A cast-iron framework that can be removed from the rear-axle housing for service and adjustment.

Axle flange The outside end of the axle where the wheel, drum and rotor attaches.

Axle gears Bevel gears that transfer power from the differential pinion gears to the splined axle shafts.

Axle hop The tendency of a live axle housing to rotate with the wheels slightly and then snap back during hard acceleration. This action may be repeated several times, creating a loss of traction until the driver releases the accelerator.

Axle housing Designed in the removable carrier or integral carrier types to house the drive pinion, ring gear, differential, and axle shaft assemblies.

Axle ratio The ratio between the rotational speed (rpm) of the drive shaft and that of the driven wheel; gear reduction through the differential, determined by dividing the number of teeth on the ring gear by the number of teeth on the drive pinion.

Axle seat Suspension component used to support and locate spring on the axle. Also known as a spring chair.

Axle shaft Alloy steel shaft that transfers torque from the differential side gears to the drive wheels. This shaft also supports vehicle weight on most passenger cars.

Axle tamp The tendency of a live axle housing to rotate with the wheels slightly and then snap back during hard acceleration. This action may be repeated several times, creating a loss of traction until the driver releases the accelerator.

Axle windup A term often used for axle tramp.

Axle-shaft end thrust A force exerted on the end of an axle shaft that is most pronounced when the vehicle turns corners and curves.

Axle-shaft tubes Tubes that are attached to the axle housing centre section to surround the axle shaft and bearings.

Azeotrope A mixture of two or more liquids, such as refrigerants, that when mixed in precise proportions, behave like a compound.

Azimuth track assembly The Azimuth Track is the large steel ring upon which the Mount sits. These Azimuth Tracks provide a smooth and flat surface for the hydrostatic axial bearings. They also provide appropriate mounting surfaces for a variety of azimuth Mechanical Systems.

B

B.E.P. The best efficiency point. It is the point where the power coming out of the pump (water horse power) is the closest to the power coming into the pump (brake horse power) from the driver. This is also the point where there is no radial deflection of the shaft cause by unequal hydraulic forces acting on the impeller.

Babbitt An alloy used to line bearings made up of tin, antimony, copper, and other metals.

Baby moons Small chrome-plated wheel covers.

BabySmart Childseat presence orientation detection (CPOD) is a pressure-sensitive foil with integrated antenna and resonators in the child seat to keep the airbag from firing with full capacity if necessary and was developed by Mercedes-Benz.

Back diode A tunnel diode that is selected for its reverse-conduction characteristics.

Back end The second phase in the chip production chain. Backend processes involve slicing the finished wafer into single chips (dicing); affixing those chips to a substrate (die bonding); wiring gold, aluminium or copper electrical connections between the chip and corresponding pads on the substrate (wire bonding); plastic-encapsulating the assembled chip (molding); identifying the molded package (marking); then a final test of the entire semiconductor device.

Back flush The use of a reverse flow of water, with or without a cleaning agent, to clean out the cooling system of a vehicle.

Back motor A mid- or rear-mounted engine.

Back plane The main circuit board of a system, containing edge connectors or sockets so other printed circuit boards can be plugged into it.

Back plate Used in some centrifugal pumps to position the stuffing box and provide an impeller wear surface.

Back pressure 1. Resistance of an exhaust system to the passage of exhaust gases. This can have an adverse effect on performance, fuel economy, and emissions. Excessive back pressure may be caused by the clogged catalytic converter, or a dented or crimped pipe.

2. The excessive pressure build-up in an engine crankcase.

Back pressure EGR Some emissions-control systems use a back-pressure sensor or diaphragm to monitor back pressure so that exhaust gas recirculation flow can be increased when the engine is under maximum load, and producing maximum back pressure.

Back pressure EGR valve A back-pressure-dependent EGR valve.

Back pull out pump A design that allows the wet end of the pump to be left on the piping when the power end and adapter are removed. A.N.S.I. pumps are designed this way.

Back seat The position of a valve stem when turned to the left as far as possible back seating a two-seat service valve.

Back staging Placing a competition vehicle at the start of a drag race behind the usual staging position. Also referred to as shallow staging.

Back to back double seal The rotating seal faces are facing in opposite directions. The worst possible configuration. In the past this term was used to describe a higher barrier fluid pressure between dual mechanical seals.

Backbone frame A chassis structure having one boxed member running down the center. It is usually divided into two parallel members at each end to support the power train and suspension system.

Backbone The major multi-channel link in a network, from which smaller links branch.

Backfire 1. An explosion in the exhaust system of a motor vehicle caused when an unburned air/fuel mixture is ignited, usually upon deceleration.

2. An explosion of the air/fuel mixture in the intake manifold, which is evident at the carburettor or throttle body and may be caused by improper ignition timing, crossed spark plug wires, or an intake valve that is stuck open.

3. The momentary recession of the flame into the welding tip, cutting tip, or flame-spraying gun, followed by immediate reappearance or complete extinction of the flame.

Backfire suppression valve An anti-backfire valve used in the air-injection system of an exhaust emission control.

Backfiring 1. The pre-explosion of an air/fuel mixture so that the explosion passes back around the opened intake valve, through the intake manifold, and through the carburettor.

2. The loud explosion of over-rich exhaust gas in the exhaust manifold that exits through the muffler and tailpipe with a loud popping noise.

Background Multitasking computers are capable of executing several tasks, or programs, at the same time. In some multitasking systems, one of the processes is called the foreground (the one on the monitor) and the others are call processors in the background.

Background noise Extraneous signals that might be confused with the required measurement.

Backing plate Stamped steel plate upon which the wheel cylinder is mounted and the brake shoes are attached; a metal plate that serves as the foundation for the brake shoes and other drum brake hardware.

Backlash The excessive clearance between the meshing teeth of two gears.

Backlight The rear window of a vehicle.

Backlighting A technique used to make flat-panel displays easer to read

Backlite A heatresistant plastic insulator material commonly used in printed circuit boards and transistorized components

Backpressure Restrictions in the exhaust system that slow the exit of exhaust gases from the combustion chamber

Backscattering Reflecting light back in the direction of the source.

Backup A backup is either the process of storing a subset of files on a secure medium or a reference the copied data. Backup's are used for the recovery of the system or lost data. Sometimes the term is used for an Archive.

Bad sector a sector on a computer disk that will not read or write correctly. Usually due to a minor physical flow in the disk.

Badge engineering The act of producing the same car under more than one name.

Baffle 1. A barrier used to reduce noise in an enclosed system, such as the exhaust system.

2. A barrier to prevent splashing of liquid in a tank.

Bail The spring-steel wire loop used to secure a cover, such as on a master cylinder reservoir.

Baja A Spanish term for "lower". The term more generally refers to Mexico's Baja race, which is held in a rugged desert region creating one of the world's toughest off-road courses.

Bakelite A brand name of Union Carbide co. For phenofomaldehyde resin plastic.

Balance 1. To have all rotating parts in a state of equilibrium.

2. To balance all rotating parts statically and dynamically in an engine to ensure maximum performance.

Balance control A control in a stereo amplifier that adjusts the relative output volume from each of the stereo channels.

Balance due The amount currently due, minus previous payments, plus cash advances and purchases.

Balance pipe 1. A pipe that connects the exhaust pipes in a dual exhaust system to equalize the pressures.

2. A pipe that connects the ventures of dual carburettors.

Balanced scorecard A comprehensive performance measurement technique that considers four areas of performance in a balanced way: 1. customer perspective - how customers see us,

2. internal perspective - what we must excel at,

3. innovation & learning - how we continue to improve and create value,

4. financial perspective - how we meet shareholder needs.

Balanced scorecard A comprehensive performance measurement technique that considers four areas of performance in a balanced way: 1. customer perspective - how customers see us,

2. internal perspective - what we must excel at,

3. innovation & learning - how we continue to improve and create value,

4. financial perspective - how we meet shareholder needs.

Balanced seal A design in which the seal face closing area is reduced to lower the closing force, and reduce the heat generation between the faces.

Balanced valve A type of hydraulic valve that produces a pressure change that is proportional to the variations of spring pressure or the movement of a mechanical linkage.

Balancer A heavy crankshaft pulley that aids in overall crankshaft balance as it rotates.

Balancing coil gauge An indicating device, such as a fuel gauge, that contains a pair of coils in the instrument-panel unit.

Balancing The process of proportioning weight or force equally on all sides of an object. Most crankshafts, for example, are balanced both statically and dynamically.

Balk ring A rotating device found in a manual transmission that prevents premature engagement of gears while shifting.

Ball and ramp A common kind of calliper-actuated parking brake.

Ball and socket A term often used for tie rod end.

Ball bearing An anti-friction bearing with an inner and outer race having one or more rows of balls between them.

Ball joint 1. A joint or connection where a ball moves within a socket, allowing a rotary motion while the angle of the rotation axis changes.

2. A suspension component that attaches the steering knuckle to either control arm featuring a ball-and-socket joint to allow pivoting in various directions. Also known as a spherical joint.

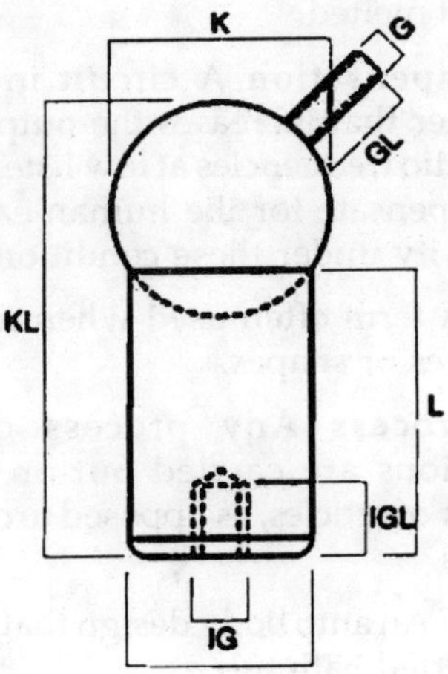

Fig. Ball joint

Ball joint angle The inward tilt of the steering axis from the vertical.

Ball nut In a recirculating ball nut steering gear, the ball nut has internal threads that are meshed to the threads of a worm with continuous rows of ball bearings between the two. The ball bearings are reticulated through two outside loops called ball guides. The sliding ball nut has tapered teeth cut on one face that mate with teeth on the cross-shaft sector.

Ballistics The science that deals with the motion, behaviour, appearance or modification of missiles acted upon by propellants, rifling, wind, gravity, temperature or other modifying conditions of force.

Balloon foot A term used to describe a slow driver; one who tends to back off the throttle early.

Ballute An aerodynamic braking device which is both balloon and parachute.

Band 1. A hydraulically controlled device installed in an automatic transmission around a clutch drum, used to stop or permit drum turning.

2. A manual or hydraulic device installed around a drum to provide a braking action.

Band-Aid A known temporary fix to a problem that may have to be performed one or more times until a permanent fix can be performed. (A permanent fix may consist of scheduled down time requirements, re-engineering, new procedural developments, ect.)

Bandwidth the range of frequencies an oscilloscope can display

Barometer An instrument used for measuring barometric pressure.

Barometric pressure (BARO) The pressure exerted by the weight of the earth's atmosphere, equal to one bar, 99.97 kpa, or 14.5 psi at sea level. Barometric pressure changes with the weather and with altitude. Since it affects the density of the air entering the engine and ultimately the air/fuel ratio, some computerized emissions-control systems use a barometric pressure sensor so that the spark advance and exhaust gas reticulate (EGR) valve flow can be regulated to control emissions more precisely.

Barometric pressure sensor A device that senses barometric pressure and sends an electrical signal to the CPU for optimum engine control.

Barrel (bbl) 1. A term sometimes applied to the cylinders in an engine.

2. A term used to refer to the number of throttle bores in a carburettor.

Barrel faced ring A compression piston ring with a rounded contact face.

Barrel finish The rounded surface on a piston skirt.

Barrel hose A hose constructed with a special liner to prevent refrigerant leakage through its walls. Most air-conditioning systems in vehicles manufactured after 1988 have barrier-type hoses.

Barrel roll A vehicle rollover, sideways.

Barrier 1. A term given to something that stands in the way, separates, keeps apart, or restricts;

2. A term for a diode or transistor junction

3. An obstruction

Barrier charge The potential developed across a semiconductor junction.

Barrier fluid The high pressure fluid that is circulated between two mechanical seals. The fluid should enter the bottom and leave the top to prevent air pockets.

Base 1. The name for the section of a transistor that controls the current flow through the transistor.

2. The unique number of values in a numbering system; decimal has 10, binary has 2, hexadecimal has 16.

Base circle The low portion of each cam on the camshaft, concentric with the journal, which is not part of the lobe.

Base-coat/clear coat A paint system that adds a final clear-coat paint layer over primer and colour coats to provide a deep, "wet-look" shine that resists fading.

Base line A Baseline is a Configuration Identification formally designated and applicable at a specific point in an items life cycle. Baselines, plus approved changes from those baselines, constitute the current configuration identification. A Configuration identification document or a set of such documents formally designated by the acquirer (Customer) at a specific time during a product life cycle.

Base plate temperature The temperature at the hottest spot on the mounting platform of the supply.

Base metal 1. The largest proportion of metal present in an alloy.

2. A metal that readily oxidizes, or that dissolves to form ions.

3. The metal or alloy that is welded, brazed, soldered, or cut.

4. After welding, that part of the metal that was not melted.

Bass compensation A circuit in an audio amplifier that increases the output of low-end audio frequencies at low listening levels to compensate for the human ear's loss of sensitivity under these conditions.

Bastard A term often used when referring to odd sizes or shapes.

Batch process Any process on which operations are carried out on a limited number of articles, as opposed to continuous process.

Bathtub 1. An auto body design that resembles an inverted bath tub.

2. A combustion chamber in an engine with an area that resembles an inverted bath tub with its valves seated at its based.

Bath-tub curve Bath-Tub Curve represents the failure rate of components over the life of the product. Its upward slope at the beginning and end suggests that most components fail either right away (at the beginning of the product life) or towards the end of the expected product life.

Bathymetry Measurement of the depths of features at the bottom of the sea, especially by echo-sounding.

Battery An energy storage device that produces electricity by means of chemical action. It consists of one or more electric cells each of which has all the chemicals and parts needed to produce an electric current.

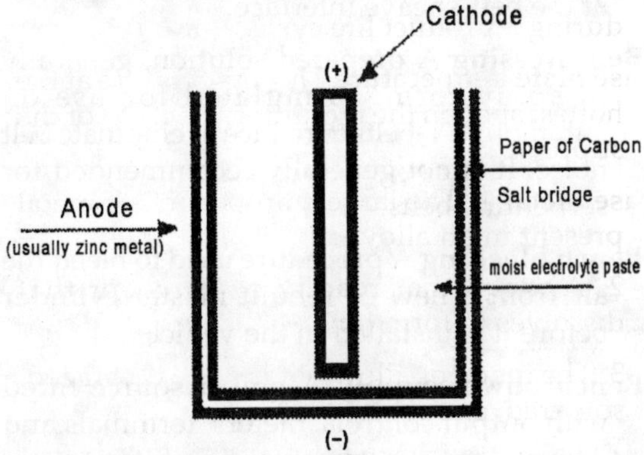

Fig. Battery

Battery backup Auxiliary power that is provided to a computer so volatile memory information is not lost during a power failure, or when otherwise disconnected from its normal power source.

Baud rate The speed at which bits of computer information are transmitted on a serial data stream. Measured in bits per second (bps)

Be Beryllium An element with the atomic number of 4

Bead 1. That part of a tire that contacts the rim of a wheel.

2. A narrow half-round pattern where metal has been welded together.

Bead filler A rubber piece positioned above the bead to reinforce the sidewall and act as a rim extender.

Bead lock A plate that clamps a tire bead to a wheel rim used on circle-dirt-track race cars.

Bead wire A group of circular wire strands moulded into the inner circumference of a tire to anchor the tire on the rim.

Bearing An antifriction device having an inner and outer race with one or more rows of steel balls or rollers.

Bearing block The outside surface of a bearing that seats against the housing bore.

Beater (slang) a car for everyday transportation. Usually not in perfect condition.

Beef A term used for beef up.

Beef up To strengthen and/or reinforce.

Beehive spring A spring that is wound in the shape of a beehive.

Before bottom dead center The position of a piston approaching the bottom of its intake or combustion stroke.

Before top center (BTC) A piston as it is approaching the top of its stroke.

Before top dead center (BTPC) Any position of the piston between bottom dead center and top dead center on its upward stroke.

Behavioural design The design of how an overall system or CSCI will behave, from a user's point of view, in meeting its requirements, ignoring the internal implementation of the system or CSCI. This design contrasts with architectural design, which identifies the internal components of the system or CSCI, and with the detailed design of those components.

Behavioural modeling Behavioural modelling defines a product in terms of required behaviours rather than relationships between geometry elements for mechanical products or relationships between components, gates and registers for electronic products. For example, the use of a hardware description language is a means for describing a behavioural model for an electronic product.

Belief map The belief map is a method of graphically representing relative levels of knowledge and confidence. It is a plot whose horizontal axis represents the evaluator's knowledge of, for example, a concept alternative's ability to meet specified requirements. The vertical axis would represent the evaluator's confidence in the concept's ability to meet those requirements.

Bell housing 1.A bell-shaped housing attached between the engine and the transmission used to house the torque converter or clutch assembly

2. A term often used for clutch housing.

Bellcrank A part that pivots to transfer motion at a right angle (90 degrees).

Belleville spring A diaphragm type spring used to help apply and release a multiple-disc clutch.

Belleville washer A circular disk formed into a conical shape. When a load is applied, the disk tends to flatten, constituting a spring action. May be referred to as a coned-disk spring.

Bell-mouthed drum A well-worn brake drum that is deformed so that its open end has a larger diameter than its closed end.

Bellows Can be manufactured from metal or non metallic materials to eliminate flexing, rolling or sliding elastomers in mechanical seal designs.

Bellows plate A thin, stamped disc. Two are welded together to form a convolution.

Belly pan Body panel(s) covering the bottom of a competition vehicle to reduce the coefficient of drag.

Belly tank A tear shaped, World War II aircraft auxiliary fuel tank used as the body for a lakester.

Belt 1. In a tire, the belt(s), generally steel, restrict ply movement and provide tread stability and resistance to deformation, providing longer tread wear and reducing heat buildup in the tire.

2. A device used to drive the water pump and/or other auxiliary devices, such as the alternator, off the engine.

Belt cover 1. A nylon cover positioned over the belts in a tire that helps to hold the tire together at high speed, and provides longer tire life.

2. A rayon, cotton, or nylon cover to protect the interior of a drive belt from the environment and absorb the ear that occurs at the belt-sheave interface.

Belt dressing A prepared solution, generally in spray form, formulated for use on automotive V-belts to reduce or eliminate belt noise. It is not generally recommended for serpentine belts.

Bench bleeding A procedure used to bleed the air from a new or rebuilt master cylinder before it is installed in the vehicle.

Bench power supply A power source fitted with output controls, meters, terminals and displays for experimental bench-top use in a laboratory.

Bench race To talk about racing, generally just after an event.

Bench seat A full-width seat that can accommodate two or three persons in a vehicle.

Bench seats Full-length seat that can usually seat two or three people.

Bench test The testing of an engine or component, out of the vehicle, for ease of observation and study.

Benchmark A fixed point of reference or a standard for comparison and used to achieve excellence within the manufacturing firm. An outstanding example, appropriate for use as a model.

Benchmarking An improvement process in which a company measures the performance of its products or processes against that of best-in-class products or companies, determines how the product or company achieved their performance level, and uses the information to improve its own performance.

Bend 1. A curve or angle that has been bent.

2. To form a curve or angle by bending.

Bend test Method for measuring ductility of certain materials. There are no standardised terms for reporting bend test results for broad classes of materials; rather, terms associated with bend tests apply to specific forms or types of materials.

Bending sequence The order in which several bends are made so as not to be blocked by a previous bend.

Bending strength Alternate term for flexural strength.

Bending stress A stress, while bending, that involves both tensile and compressive forces that are not equally distributed.

Bending the forming of a material, usually metal, into a particular shape.

Bendix drive A type of starter motor drive that engages, and disengages, the starter and flywheel. Also called an inertia drive not used since the early 1960's

Bendix Folo-Thru drive A starter motor drive engaged by initial rotation of the armature, causing the drive pinion to be twisted outward on a threaded sleeve until it is meshed with the flywheel gear. The gears are disengaged automatically at a predetermined speed of about 400 rpm.

Bendix screw The helix-grooved shaft of a Bendix drive.

Berm 1. A curb-like buildup on the outside of turns on a circular dirt track.

2. The curb-like buildup of dirt along the edges of an unpaved road.

Bernouilli's principle Daniel Bernouilli (1700 - 1782) a Swiss teacher of Physics at the Basel. He advanced our understanding of the physical world in a number of areas, but it is in the kinetic theory of fluids and gasses for which he is most remembered. The Bernouilli's Principle states that as the velocity of a fluid of a gas increases it's pressure decreases.

Bernoulli Daniel An eighteenth-century Swiss scientist, discovered that as the velocity of a fluid increases, its pressure decreases.

Bernoulli's law A moving stream of liquid or gas exerts less sideways pressure than if it were at rest. The result is that things seem to be drawn into the stream, but they are really being pushed in by the higher pressure from outside.

Bernoulli's Principle When a fluid passes through a tube with a narrowing section as it passes the narrow part it's speed increases and it's pressure decreases.

Best Practice Best Practice is a superior method or innovative practice that contributes to the improved performance of an organization, usually recognized as "best" by other peer organizations.

Beta test A test for a computer product prior to its commercial release.

Beta testing Beta testing is the testing a nearly-finished version of a piece of software or hardware, with the goal of finding defects missed by the developers. Often beta testing is carried out by people outside of the developers organization.

Bevel gear A form of spur gear in which the teeth are cut at an angle to form a cone shape, allowing a gear set to transmit power at an angle.

Bezel A trim ring, usually around a gauge, to secure the class cover.

Bezier curve Polynomial used to describe complex curves and surfaces.

Bi Bismuth An element with the atomic number of 83.

Bias 1. A diagonal line of direction.

2. The term applied to the electrical, mechanical, or magnetic force of voltage applied to a device to establish the reference point for operation.

Bias Voltage Voltage applied across a diode

Bias-belted tire A tire that has the ply cords placed diagonally across the tire from bead-to-bead, with alternate ply layers criss-crossing diagonally in opposite directions.

Bias-ply tire Tire construction in which the cords in the body plies of the carcass run from bead to bead at an angle from 26 to 38 degrees instead of 90 degrees as in a radial ply tire.

Big arm 1. The throw of a crankshaft that has been stroked.

2. Sometimes used to identify an engine with a stroked crankshaft.

Bimetal A temperature-sensitive strip made up of two metals having different heat expansion rates.

Bimetal engine A powerplant with block and head made of different metals, such as a cast iron block and an aluminium head.

Bimetal spring A spring made of bimetal, such as a choke spring.

Bimetal strip A temperature regulating or indicating device that works on the principle that two dissimilar metals with unequal expansion rates, welded together, will bend as temperature changes.

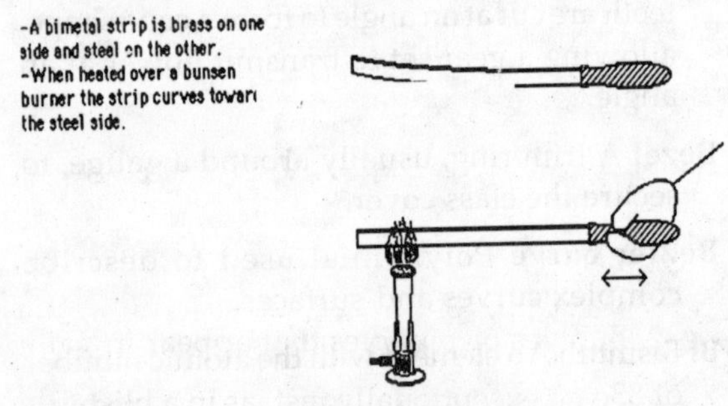

Fig. Bimetal strip

Bimetallic drum A composite brake drum made of cast iron and aluminium.

Bimetallic sensing element Another term used for bimetallic sensor.

Bimetallic sensor A sensing element having a bimetallic strip or coil.

Bimetallic temperature sensor A bimetallic thermostat.

Bimetallic thermostat A thermostat that used a bimetallic strip instead of a bellows or diaphragm for making or breaking electrical contact points.

Binary A system in which numbers are represented as sequences of zeros and ones, which is the basis of digital computing.

Binary digit A character that represents one of the two digits in the number system that has a radix of two.

Binary numeral The binary representation of a number.

Binary star Two stars revolving around a common center of gravity.

Binder 1. The paint material that forms the film, so called because it binds the pigment and additives into a solid, durable film.

2. Adhesive or glue used in brake linings to bond all the other materials together.

Binders A term used for brakes.

Binnacle A console, separate from the dashboard, for switches, instruments, and controls, on or near the steering column.

Biochip An aid to accelerated and deeper understanding of activity in cells, bodily organs, plants, animals, as well as humans. A biochip substrate has a coating with thousands of microscopic reaction sites that might contain molecules of DNA, or proteins, for example. These molecules react exclusively with their complementary reaction partners - the lock-and-key principle. This means samples of cell material can be analyzed for the presence of specific reaction partners. Thousands of biological lock-and-key reactions can take place in parallel on a biochip substrate.

Bipolar plates Conductive plate in a fuel cell stack that acts as an anode for one cell and a cathode for the adjacent cell. The plate may be made of metal or a conductive polymer (which may be a carbon-filled composite). The plate usually incorporates flow channels for the fluid feeds and may also contain conduits for heat transfer.

Bipolar transistor A junction transistor having both majority and minority charge carriers.

Bi-propellant A rocket propellant consisting of two unmixed or uncombined chemicals (fuel and oxidizer) fed separately into the combustion chamber.

Birdcage 1. The plate on an axle housing for attaching suspension components.

2. The chassis space frame with many small pieces of structural tubing.

Black book A guide used to determine the value of a used vehicle.

Black box A term often used for a central processing unit (CPU).

Blanket insurance A property-liability insurance that covers more than one piece of property.

Blanking and punching A mechanical engineering process for stamping out parts from sheet or strip metal, using a punch and die. In blanking, the released profile becomes the workpiece. Punching is the exact reverse, putting holes in a workpiece made of sheet or strip metal.

Bleed 1. To drain fluid.

2. /to remove an air bubble or air lock.

3. To draw air into a system.

Bleed air tanks The process of draining condensation from air tanks to increase air capacity.

Bleed orifice A calibrated orifice in a vacuum system that allows ambient air to enter the system to equalize the vacuum.

Bleeder current A continuous load placed on a power supply by a resistance load that helps improve regulation and safety.

Bleeder jar A glass or transparent plastic container used to detect the escape of air while bleeding brakes.

Bleeder resistor A resistor that allows a small current drain on a power source to discharge filter capacitors or to stabilize an output.

Bleeder screw A small, hollow screw or valve found at drum-brake wheel cylinders, in disc brake callipers, and adjacent to the outlet ports of some master cylinders. It is opened to release pressure and to bleed air and fluid from the hydraulic system.

Bleeding 1. The slow releasing of pressure in the air-conditioning system by recovering some of its liquid or gas.

2. The act of removing air from a hydraulic brake system.

3. A small leaker.

4. When one paint colour shows through another.

Bleeding sequence The order of bleeding brake systems or other components.

Blend air door A door in the duct system that controls temperature by mixing heated and cooled air in correct proportions to achieve the desired effect.

Blend air The control of air quality by blending heated and cooled air to the desired temperature.

Blinky The timing light at the end of a quarter-mile strip.

Blip A quick punch to the throttle to rev the engine momentarily.

Blistering 1. Tire tread separating from the carcass due to high heat.

2. Bubbles or pinholes that appear in paint.

3. To go exceptionally fast, as in a blistering pace.

Block valve A type of thermostatic expansion valve utilizing an internal sensing bulb.

Blow off the doors To defeat a competitor by a wide margin.

Blow off To defeat a competitor.

Blow torch A jet-powered car for drag or lake competition.

Blowby Byproducts of combustion, mostly hydrocarbons, that leak out of the combustion chamber, past the piston and piston rings, into the crankcase during the compression and power strokes. In modern engines, blow by vapours are drawn into the intake through the positive crankcase ventilation (PVC) system and are burned in the engine.

Blower 1. A performance term for a supercharger or turbocharger.

2. A two-stroke, diesel-engine, intake air compressor.

3. The fan motor in a heater-air conditioning system.

Blower belt A reinforced cover with retaining straps on supercharged engines, required by NHRA to prevent parts from scattering should the supercharger explode

Blower circuit All of the electrical components that are required for blower and blower speed control.

Blower motor An electric motor and squirrel cage type of fan moving air inside the vehicle for heating, cooling, and defrosting.

Blower motor relay A relay found in the blower circuit.

Blower relay An electrical device used to control the function or speed of a blower motor.

Blower speed controller A solid-state control device that operates the blower motor and, sometimes, the compressor clutch based on signals from the microprocessor.

Blower switch A dash-mounted device that allows the operator or turn the blower motor ON/OFF and/or to control its speed.

Blown engine 1. An engine that has a supercharger or turbocharger.

2. A seriously damaged engine.

Blown head gasket A broken head gasket that leaks water, oil, or air and reduces engine performance.

Blu Ray Blu Ray is a product that can be described as the successor to DVD. Compared with the DVD system commonly available today which is read and recorded by means of a red laser, the Blu Ray is read by a blue laser. The blue laser has a shorter wavelength than the red laser. This means that the blue laser is capable of reading and storing more information in the same space (i.e. More closely packed). The storage capacity of a Blu Ray varies between about 23 GB and 27 GB.

Blue smoke An exhaust produced when oil has contaminated the air/fuel mixture.

Blue streak The trade name of a Goodyear high-performance tire.

Blueprint To rebuild an engine to original equipment manufacturer (OEM) design specifications.

Bluetooth A computing and telecommunications industry specification that describes how embedded devices such as mobile phones, radios, headsets, and personal digital assistants can easily interconnect with each other and within vehicle devices using a short-range wireless connection.

Bog 1. To lose power and momentarily faulter when coming off the line.

2. To stall or slow due to soft dirt or sand in off-road racing.

3. A mud hole.

Bogey A common way of spelling bogie.

Boggin' Competitive racing of individual off-road vehicles through a long pre-measured mud hole.

Bogie The axle-spring suspension arrangement on the rear of a tandem axle tractor.

Bohr Niels Henrik David Bohr (1885-1962) won the Nobel Prize in Physics for his investigation of the structure of the atom.

Boil tank A very large tank of boiling solution used for cleaning large parts, such as engine blocks.

Boilerplate A metal replica of the flight model (e.g. of a spacecraft) but usually heavier and cruder for test purposes.

Boiling A rapid change in the state of a liquid to a vapour by adding heat, decreasing pressure, or both.

Boiling point The temperature at which a liquid begins to boil.

Boiling the hides Smoking the tires during a drag race.

Bolster plate The flat, load-bearing surface under the front of a semitrailer, including the kingpin, which rests firmly on the fifth wheel when coupled.

Bolt A metal rod, usually with a head at one end and a thread at the other, used to secure parts and assemblies.

Bolt circle diameter The diameter of an imaginary line running through the bolt hole centre.

Bonded lining Brake lining attached to the pad or shoe by high strength, high-temperature adhesive, eliminating a need for rivets.

Bonding Joining together a pair of materials with the aid of adhesives.

Bonding lining The lining attached to a brake shoe with an oven-cured adhesive.

Bondo A tradename for a plastic two-part body filler.

Bone yard A facility that sells used parts for vehicles.

Bonnet 1. A safety helmet.

2. British term for a vehicle hood.

Boost regulator One of several basic families of switching power supply topologies. Energy is stored in an inductor during the pulse then released after the pulse.

Boost The extra power given to a rocket or space vehicle during lift-off, climb or flight, as with a booster rocket.

Boost valve A device that works against a pressure regulator valve to increase mainline pressure when needed

Booster battery A charged battery connected to a discharged battery in order to start the engine.

Booster cables A term used for jumper cables.

Booster seat This child-safety seat is designed for children who are too large for a baby seat, but not big enough to sit safely in the vehicle's seats.

Booster venturi A device in a carburettor that mixes fuel with incoming air.

Boot 1. English term for Trunk (luggage compartment)

2. A sling for a tire

3. A flexible rubber or plastic cover used over the end of a master cylinders, wheel cylinders, transmissions, or constant velocity joint, to keep out water and other matter

4. a computer term to load the first piece of software that starts the computer; usually the BIOS.

Boot disk Term for bootable diskette; a diskette from which you can boot your computer. Usually your computer boots form the hard drive however, if the hard drive is inaccessible you can boot the computer from a bootable diskette usually in the "A" drive.

Boot lid English term for Trunk lid

Borden tube A thermo-mechanical device in the fuel injection system that regulates the amount of fuel being injected according to differences in temperature and pressure in the intake manifold.

Borderline lubrication 1. Poor lubrication as a result of greasy friction.

2. Moving parts coated with a thin film of lubricant.

Bore 1. The diameter of an engine cylinder.

2. To increase the diameter of a cylinder.

Bore align To machine an engine's main bearing journals to assure they are in perfect alignment.

Bore centers The center-to-center distance between two bores.

Bore Diameter of an individual engine cylinder.

Bore in the water To bore a cylinder out far enough to brake through the water passage.

Bored and stroked A combination of an enlarged cylinder bore and a lengthened piston stroke to increase an engine's displacement.

Boss 1. The reinforced extension on a part that holds a mounting pin, bolt, or stud.

2. A slang term for outstanding quality.

Bottleneck Any point at which production is slowed because demand placed on a resource is equal to or more than capacity. Bottlenecks identify machines that are critical to large sections of the production cycle.

Bottom 1. The underside; the lowest part.

2. When a vehicle's chassis hits the lowest point allowed by its suspension system.

3. When the springs are fully compressed.

Bottom dead center (BDC) Piston position at bottom of stoke.

Bottom valve A shock absorber component to control the flow of oil into the reservoir during compression and rebound.

Bounce back A condition that occurs when particles of paint sprayed on a body panel bounce away from the surface.

Bound electron An electron whose orbit is near the nucleus of an atom and is strongly attracted to it.

Bound The opposite of rebound. The inward travel of the piston rod in a shock absorber.

Boundary layer The thin layer of air along the inner walls of an intake port.

Boundary scan A design for testability method that places a scan register at at every pin of every chip on a board for board testing and diagnostics. The test process can control and observe the state of every pin i/o pin without requiring physical access to any of them.

Bourdon tube A curved tube that straightens as the pressure inside it is increased. The tube is attached to a needle on a gauge, that senses the movement of the tube and transmits it as a pressure reading.

Bow shock wave The compressed wave that forms in front of a spacecraft or satellite as it moves rapidly through Earth's atmosphere; more generally, any such wave that forms between an object and a fluid medium.

Bowden cable 1. A wire cable inside a metal or rubber housing used to regulate a valve or control it from a remote place.

2. A small steel cable inside a flexible tube used to transmit mechanical motion from one point to another.

Box stock Standard, conforming to OEM specifications.

Boxer A horizontally opposed engine.

Boxer engine The cylinders are opposite (180 degrees apart) from each other. Also called flat engines, these are relatively flat compared to in-line or v engines.

Breadboard A circuit designed using temporary fabrication techniques. Used in the initial phases of circuit or system design.

Break even after release Break even after release - a metric that measures the time after release of a product for production or sale until the the product has achieved financial breakeven considering the investment in development and other non-recurring expenses.

Break even time Break Even Time - a metric that measures the time from the start of development through production and sales until the product has achieved financial breakeven considering the investment in development and other non-recurring expenses.

Break rule A regulation in drag racing that permits a car, defeated in an elimination, to return to competition if the car that beat it is not able to get to the line for its next round.

Breakdown voltage 1. The voltage level which causes insulation failure.

2. The reverse voltage at which a semiconductor device changes its conductance characteristics.

Breaker A spring-loaded switch in a conventional distributor ignition system that opens (breaks) and closes (makes) the primary circuit.

Breaker cam The rotating part, located near the top of the conventional distributor, that has lobes that cause the breaker points to open and close.

Breaker point gap The space between the fully opened breaker points in a conventional distributor.

Breaker points In a conventional ignition system, and electrical switch that opens to interrupt current flow in the primary circuit. Often called contact points.

Breaker trigger system Any ignition system that utilizes conventional breaker contacts to time and trigger the system.

Breakerless system An electronic ignition system that does not use conventional breaker points to time and trigger the primary voltage but retains the distributor for secondary voltage distribution.

Break-in The operation of a new vehicle at a constant and even speed to assure even initial wear of all engine parts.

Breaking load Load which causes fracture in a tension, compression, flexure or torsion test.

Breakout 1. The point at which conductor(s) are taken out of a multiconductor assembly.

2. The point at which a branch departs from the main wiring harness.

Breakout Box A Test device which connects in series with between the powertrain control module and it's harness and permits measurements of inputs and outputs.

Breakover angle A term often used for ramp angle.

Breathing The ability of an engine to draw in air and exhaust gases. The better the breathing, the better the performance.

B-REP Boundary representation - solids modelling approach based on representing exterior surfaces that define a solid (as opposed to constructive solid geometry).

Bridge circuit Circuit with series-parallel groups of components.

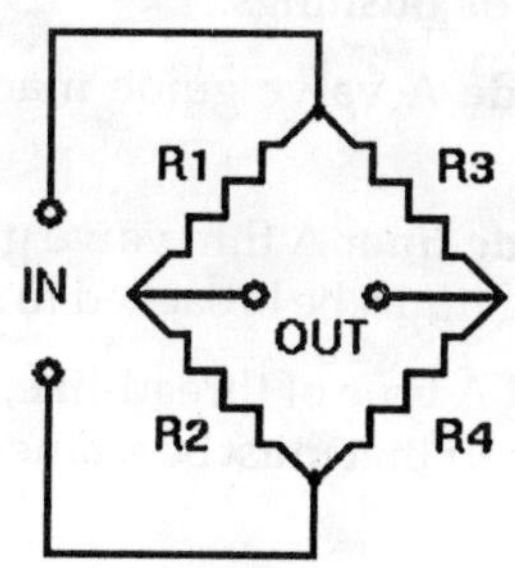

Fig. Bridge circuit

Bridge rectifier A full-wave rectifier in which the diodes are connected in a bridge circuit allowing current flow to the load during both positive and negative alternation of the supply voltage.

Brineling A steel-shim head gasket torqued in place on an engine block.

Brinell test A technique for testing the hardness of a metal.

Brinelling A type of mechanical failure used to describe a dent in metal such as what occurs when a shock load is applied to a bearing. Named after Johann A. Brinell, a Swedish engineer

Brinnell hardness A method of measuring the hardness of metal parts and hard seal faces. Above 350 the standard machining operations of turning, boring, drilling, and tapping become uneconomical.

Brittleness The quality of a material that leads to fracture without appreciable deformation.

Broach A tool used for reshaping or resizing parts.

Brody A controlled, rear-wheel skid often used on dirt tracks for fast cornering.

Broker A person who represents the insured to solicit or negotiate for contracts. A broker may deal with many agents or companies on behalf of his or her client.

Bronze A copper-rich, copper-tin alloy with or without small proportions of other elements such as zinc and phosphorus often used for bearings or bushings.

Bronze guide A valve guide made of bronze alloy.

Bronze guide liner A thin valve guide, easy to install, but must be broached to stay in place.

Bronzewall A type of thread-like, valve-guide repair insert that must be reamed to size after installation.

Browser Short for web browser, a software application used to locate and display web pages.

Brush A block of conducting material, such as carbon, held against an armature commutator or rotor slop ring to form a continuous electrical path.

Bubble gum machine Slang for a blue and/or red police-car roof light.

Bubble memory A method by which information is stored as magnetized dots (bubbles), that rest on a thin film of semiconductor material.

Buck regulator One of several switching power supply topologies where a series of pulses are applied by a switching device to an averaging L-C filter and then applied to a load.

Buckling The bowing or lateral displacement of a compression spring.

Bucks down To have not.

Bucks up bucks down To have and to have not.

Bucks up To have.

Buff An enthusiast.

Buff book A magazine for car enthusiasts.

Buffer A part or circuit used to reduce interaction between two electronic circuits.

Buffer fluid The low pressure fluid that is circulated between dual mechanical seals.

Buffing A surface-finishing process that produces a smooth, lustrous appearance generally free of defined line patterns.

Bug 1. A minor flaw, imperfection, or malfunction.

2. A Volkswagen Beetle

3. a problem in a hardware or software element of a computers system

Bug catcher A scoop-like intake on fuel-injection or blower system.

Buggy A small, lightweight, off-road vehicle.

Bugle A high spot or crown in stretched metal.

Buildup time the time required to form a magnetic field around the primary winding of the coil when current is allowed to flow through it in a conventional ignition system.

Built-in test An internal script intended to check the health of a subsystem, run periodically or on demand.

Built-in-self-test Built-in-self-test - a feature of automatic testing where many test pattern programs are built directly into the circuit generally for go/no-go testing of the assembly or circuit using signature analysis.

Bulb The glass envelope that contains an incandescent lamp or an electronic tube.

Bulk capacitor The energy storage capacitor at the front end of a regulator.

Bulk modulus of elasticity Ratio of stress to change in volume of a material subjected to axial loading. Related to Modulus of Elasticity (E) and Poisson's Ratio (r) by the following equation Bending Strength K = Er 3(1-2r).

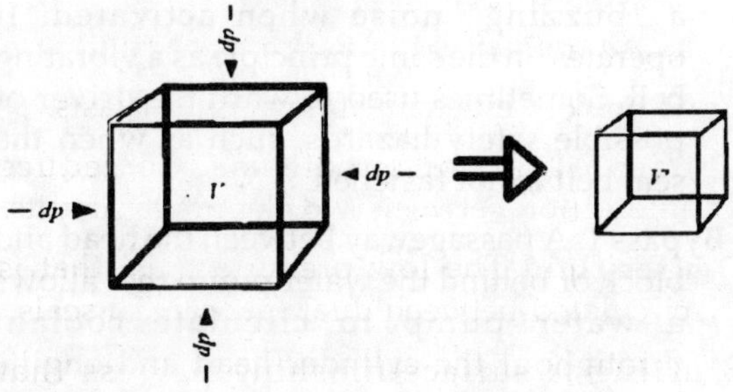

Fig. Bulk modulus of elasticity

Bulk voltage The voltage across a bulk capacitor.

Bulkhead connector A connector for wires or hoses located where they must pass through a partition or lead to another.

Bulkhead English term for Firewall

Bull ring A dirt, oval track, generally one-half mile or less.

Bulletproof Indestructible.

Bullnose The smooth nose of a hood when the ornament has been removed and the holes filled in.

Bump shop An auto body repair facility.

Bunting bronze A trade name for a type of bronze used for bushings.

Bureau of Automotive Repair (BAR) A state agency that regulates the auto service and repair industry in California.

Burette A glass container used to measure liquid in cubic centimeters (cc).

Burn A visible discolouration or sub-surface damage from an excessively high temperature, generally produced by grinding.

Burn in The operation of a newly fabricated device or system prior to application with the intent to stabilize the device, detect defects, and expose infant mortality. In power supplies, a period during which a supply is energized and loaded to peak output, with the intent of finding potentially weak components. Typical burn-in tests can include temperature cycling, input cycling, and/or load cycling

Burn rubber To accelerate at a rate that the traction tires leave black streaks on the pavement.

Burn time Time required for a given amount of air/fuel mixture in the combustion chamber to burn.

Burnout The moment of final oxidation or combustion of fuel in an engine.

Burnt Permanently damaged material caused by heating conditions producing incipient melting or intergranular oxidation.

Burr A feather edge left on metal after being cut or files.

Bus A major part of the structural subsystem of a spacecraft which provides a place to attach components internally and externally, and to house delicate modules requiring a measure of thermal and mechanical stability. The bus also establishes the basic geometry of the spacecraft.

Bus Bar A common electrical connection to which all of the fuses in the fuse box are attached

Bushing A close fitting support device used to restrict flow between two liquids, thermally isolate a hot liquid, support the rotating shaft, break down pressure etc. Commonly made of carbon or Teflon.

Business case Business case refers to the results of market, technical and financial analyses used to justify the feasibility of a new product. Ideally defined just prior to the "go to development" decision (gate), the case defines the product and project, including the project justification and the action or business plan.

Business coupe An inexpensive two-door body type with no rear seat, last available in the mid 1940s.

Business process reengineering Business process reengineering is the analysis and redesign of workflow within and between enterprises. Authors Michael Hammer and James Champy promoted the idea of BPR as the radical redesign and reorganization of an enterprise to lower costs and increase quality of service. They suggested seven principles of reengineering to streamline the work process and thereby achieve significant levels of improvement in quality, time management, and cost: 1. organize around outcomes, not tasks;

2. identify all the processes in an organization and prioritize them in order of redesign urgency;

3. integrate information processing work into the real work that produces the information;

4. geographically dispersed resources as though they were centralized;

5. link parallel activities in the workflow instead of just integrating their results;

6. put the decision point where the work is performed, and build control into the process; and

7. capture information once and at the source.

Bussword A word or phrase that happens to be a popular cliché of a group of people, such as race car drivers.

Butane A colourless gas (C_4H_{10}) that is used as a fuel.

Butt gap The distance between the ends of a piston ring.

Butt weld A type of weld that joins two pieces of metal by fusion.

Butterfly A type of valve used for the choke and throttle valve in a carburettor; a moveable plate that controls the amount of air permitted to enter the carburettor.

Butyl The trade name of a synthetic rubber.

Buying decisions The act of determining whether or not a product purchase or repair, will be made, and/or which product or service will be purchased.

Buzz box A small car with a noisy engine.

Buzzer An electric sound generator that makes a "buzzing" noise when activated. It operates on the same principle as a vibrating bell. Sometimes used to warn the driver of possible safety hazards, such as when the seat belt is not fastened.

Bypass 1. A passageway between the head and block or behind the water pump that allows a water pump to circulate coolant throughout the cylinder head and engine block before the thermostat opens.

2. A valve that is used to regulate pressure or control the quantity of a liquid or gas.

3. A shunt or parallel path around one or more elements in a circuit.

Bypass capacitor A capacitor that provides a low impedance path (usually to ground) to remove unwanted signals from the main signal path.

Bypass control valve A valve used in a bypass system.

Bypass line A line or hose used in a bypass system to transfer liquid or gas.

Bypass line Used to either re-circulate fluid from the pump discharge to the stuffing box, the stuffing box to the pump suction, or the pump discharge to a lower pressure point in the system.

Byte A sequence of binary digits, frequently comprised of eight (8) bits, addressed as a unit.

C

C frame adapter Used to connect and align the pump to the motor with registered fits. (imperial dimensions. Called the D frame adapter in the metric system)

Cable An assembly of two or more wires that may be insulated or bare.

Cable clamp A device used to clamp around a cable to transmit mechanical stain to all elements of the cable.

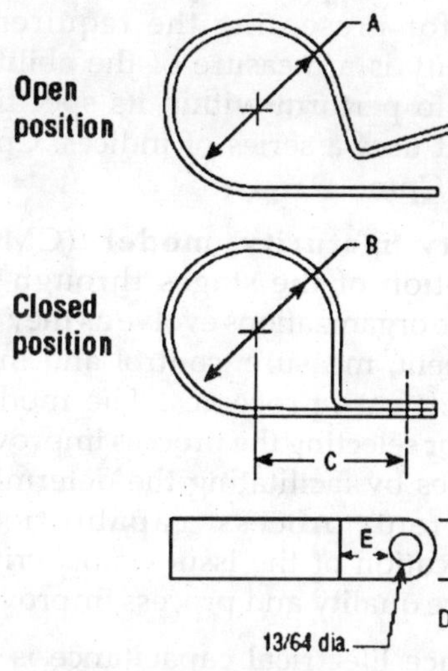

Fig. Cable clamp

Calibrate 1. To check, test, or adjust the initial settings of a unit or system.

2. To compare the reading of some instrument with a fixed standard or another instrument and note or correct any deviation from the correct reading.

Calibration The adjustment of a device or instrument so that output is within a designated tolerance for specific input values.

Calibrator (PROM) An electronic component specifically programmed to met engine operation requirements for each vehicle model. It plugs into the PCM's mother board.

Calliper Non-rotational components of disc brakes that straddle the disc and contain hydraulic components forcing the brake pads against the rotor to slow or stop the vehicle.

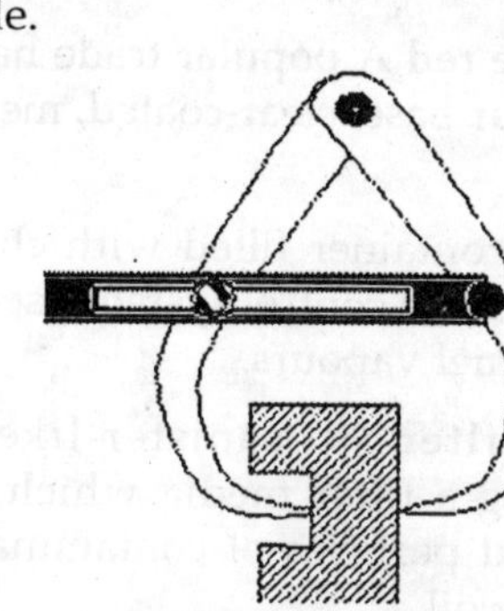

Fig. Calliper

Calliper support The bracket or anchor that holds the brake calliper.

Cam 1. The eccentric element of a one-way roller clutch that carries the profiles through which the rollers transmit torque.

2. An acronym for camshaft.

Cam and kit A specially ground camshaft, complete with a set of compatible camshaft valve train components, including lifters and springs.

Cam angle A term often used for dwell angle.

Cam button A device that keeps a camshaft properly positioned in an engine.

Cam duration The amount of time, measured in crankshaft degrees, that a camshaft holds an exhaust or intake valve open.

Camshaft A machined shaft with lobes that open and close engine-cylinder intake and exhaust valves. As the shaft rotates, the lobes push against valve springs to open the valves and rotate away to close them. Driven by the crankshaft.

Cam-shaped pocket The recess in an overrunning clutch race, large at one end, tapering to small at the other end.

Can tap A device used to pierce, dispense, and seal small cans, such as refrigerant.

Can valve A term often used for can tap.

Candlepower The amount of light energy produced by a bulb is measured in candlepower or in lumens.

Candy-apple red A popular trade name for a gold-colour-base, clear-coated, metallic-red paint job.

Canister A container filled with charcoal in an evaporative control system used to filter and trap fuel vapours.

Canister filter A canister-like device containing a filter media which removes suspended particles of contaminants from air, fuel or oil.

Canister purge solenoid A solenoid valve that admits fuel vapours to the canister for processing.

Canned pump A non seal pump with the shaft, bearings and rotor contained in a can to prevent product leakage. Limited to pumping clean lubricating liquids.

Canted valves A cylinder-head layout where the intake valves are at one angle, while the exhaust valves are at another angle.

Cantilever A lever that is anchored and supported at one end by its fulcrum and provides an opposing force at its opposite end.

Cantilever tire A tire with a tread wider than the rim.

Cap 1. A cover or lid for a container.

2. That part of the distributor that directs electrical energy to the spark plugs.

3. A small metal part, usually of hardened steel, that acts as an interference between the valve-stem end and the rocker arm.

4. The half round, removable part of a connecting rod or main bearing.

Capability A group of related requirements; the word capability may be replaced with 'function', 'subject', 'object' or other term useful for presenting the requirements. Capability is a measure of the ability of a system to perform within its specification limits. It uses a series of indices: Cp, Cpk, Cr, and Cpm.

Capability maturity model (CMM) A description of the stages through which software organizations evolve as they define, implement, measure, control and improve their software processes. The model is a guide for selecting the process improvement strategies by facilitating the determination of current process capabilities and identification of the issues most critical to software quality and process improvement.

Capacitance Electrical capacitance is a term used to measure or describe how much charge can be stored in a capacitor for a given voltage potential difference. Normally measured in farads.

Capacity Maximum production attainable under normal conditions. With regard to normal conditions, the company's operating practices are to be followed with respect to the use of production facilities, overtime, workshifts, holidays, etc.

Capacity requirements planning CRP, used in manufacturing (MRP), is a technique for projecting resource requirements for critical work stations, generated from scheduled receipts and planned order releases.

Cape top The fixed enclosure over the rear seat.

Capillary action The force by which liquid, in contact with a solid, is distributed between closely fitted surfaces of the joint, while being brazed or soldered.

Capillary attraction Another term for capillary action.

Carbide 1. A compound of solid elements, usually metal, with carbon.

2. A mixture of very hard metals, such as tungsten carbide.

Carbon (C) An atom and primary constituent of hydrocarbon fuels. Carbon is routinely left as a black deposit on engine parts, such as pistons, rings, and valves, by the combustion of fuel.

Carbon arc cutting and welding 1. A cutting process that uses a carbon electrode.

2. A welding process that uses an arc between a carbon electrode and the weld pool.

Carbon bushing Used as a restrictive bushing in flushing applications, a thermal barrier in high temperature applications, a disaster bushing in an A.P.I. gland and to support a deflecting shaft in many mechanical seal applications.

Carbonizing A reduction of hydrocarbons resulting in the formation of carbonaceous residue that will interfere with the movement of a mechanical seal. Often called "coking".

Carburetion The action that takes place in a carburettor while converting liquid fuel to a vapour and mixing it with air to form a combustible mixture

Carburettor Device that mixes air with fuel, delivering the mixture into the engine's combustion chambers. Only found on older vehicles. By the mid-1980s, new emissions standards led to the use of fuel-injection systems, which do not require frequent adjustment.

Carburize To treat or combine with carbon.

Carburizing flame A reducing oxyfuel-gas flame in which there is an excess of fuel gas, resulting in a carbon-rich zone extending around and beyond the cone.

Carcass 1. A tire casing to which the rubber tread and sidewall are bonded.

2. The inner part of the tire that holds the air for supporting the vehicle.

3. The empty remains of a stripped vehicle.

Cardan joint A universal joint having two yokes at right angles to each other, with a cruciform-shaped joint in the middle.

Carolina stocker A car built for drag racing, without regard for any recognized rules.

Carousel The large structure that forms the basic building envelope of the enclosure. It consists of a large diameter ring beam at its base, two large arch girders that form the upper structural portion, and all intermediary structural members. The carousel also includes the dual skin system (i.e., inner and outer cladding). The carousel is capable of tracking the sun about the azimuth axis at solar and sidereal rates.

Carrier A part that holds, positions, moves, or transports anther part or parts.

Carrier bearing A bearing that supports the ring-gear carrier in the differential.

Carrier housing Cast-iron rear axle assemble section that contains the working parts of the differential.

Carrier rockets Carrier rockets are the vehicles that transport satellites or human beings (astronauts) into space. Their powerful engines are single-use only, unlike the rocket drives fitted to the space shuttle.

Carson top A removable, non-folding, padded soft top, used on many customized convertibles and roadsters.

Cartridge roll A rolled piece of sandpaper used to blend sharp edges, such as when porting and polishing a head.

Cartridge seal A self contained assembly containing the seal, gland, sleeve, and both stationary and rotating seal faces. Usually needs no installation measurement. Must be used if impeller adjustments are made. Cartridge seals are the standard for A.P.I. seal applications.

Case harden A heat-treating process that hardens the outer surface of metal, while leaving the core soft and ductile.

Casing 1. The outer part of a tire assembly made of fabric or cord to which the rubber is vulcanized.

2. The other housing or shell containing as assembly.

Cast iron A term used for a family of cast ferrous alloys containing at least 2% carbon, plus silicon and sulphur; may or may not contain other alloy elements.

Castellate Formed to resemble a castle battlement.

Castellated nut A nut with six raised portions or notches through which a cotter pin can be inserted to secure the nut.

Caster 1. The angle between the steering-spindle axis and the wheel vertical as viewed from the side.

2. The backward or forward angle of the steering axis viewed from the side of the car.

Caster angle The forward or backward tilt of the steering axis as viewed from the side. If the point of load is ahead of the point of contact, the caster angle is positive. The caster angle tends to keep wheels in a straight line. Proper caster adjustment improves both tire wear and fuel economy.

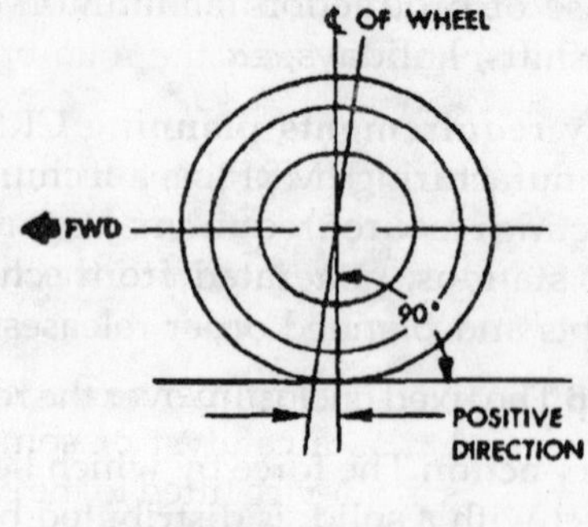

Fig. Caster angle

Casting A process technology that delivers a liquid molten metal into a purpose-built mould. After cooling, the solid metal surface has the shape of the mould cavity.

Casting flash A thin metal exuding at the parting edges of a casting mold, evident when the part is removed from the mold.

Casting number A part number that has been cast into a part during manufacture.

Cast-iron guide A valve guide made of cast iron.

Castor oil A lubricant made from the castor bean.

Casualty insurance Insurance covering the insured's legal liability for damage to other person's property or injuries to them. Other forms of insurance, such as burglary, robbery and worker's compensation, belong in this category.

Catalyst A chemical substance that increases the rate of a reaction without being consumed; after the reaction it can potentially be recovered from the reaction mixture chemically unchanged. The catalyst lowers the activation energy required, allowing the reaction to proceed more quickly or at a lower temperature. In a fuel cell, the catalyst facilitates the reaction of oxygen and hydrogen. It is usually made of platinum powder very thinly coated onto carbon paper or cloth. The catalyst is rough and porous so that the maximum surface area of the platinum can be exposed to the hydrogen or oxygen. The platinum-coated side of the catalyst faces the membrane in the fuel cell.

Catalyst Poisoning The process of impurities binding to a fuel cell's catalyst, lowering the catalyst's ability to facilitate the desired chemical reaction.

Catalytic converter The proper name for what is often known as a catalyst or sometimes just a cat. This is a device fitted to the exhaust and used for making the exhaust gases less harmful to the environment. It accomplishes this task by a process of chemical reactions that change the properties of the emissions but not the converter itself - hence it is a catalyst.

Catalytic decomposition engine A monopropellant engine in which a liquid fuel decomposes into hot gas in the presence of a catalyst. The fuel is most commonly hydrazine.

Catastrophe hazard The risk of loss by a peril to which a large number of insured are subject. Typical examples are hurricanes and tornadoes.

Catch can A container on a race car's radiator or fuel tank to prevent liquid from spilling on the ground during a pit stop.

Catch tank A term often used for catch can.

Catenary effect The curve that a length of chain assumes between its suspension points.

Cathode The electrode at which reduction (a gain of electrons) occurs. For fuel cells and other galvanic cells, the cathode is the positive terminal; for electrolytic cells (where electrolysis occurs), the cathode is the negative terminal.

Cavitate Cavities or bubbles form in the fluid low pressure area and collapse in a higher pressure area of the pump, causing noise, damage and a loss of capacity.

Cell 1. The basic electrochemical unit used to generate or store electrical energy. A cell consists of two electrodes of dissimilar material isolated from one another electronically, in a common ionically conductive electrolyte.

2. An electrochemical system which converts chemical energy into electrical energy and also the reverse for rechargeable units.

Cell control A manufacturing unit consisting of two or more workstations or machines and the materials, transport mechanisms, and storage buffers that interconnect them.

Cell voltage The dc voltage potential between the individual positive and negative terminals of a cell in a battery.

Cellular core A type of radiator core.

Cellular manufacturing A manufacturing approach in which equipment and workstations are arranged to facilitate small-lot, continuous-flow production-often in a U-shaped cell. In a manufacturing "cell," all operations necessary to produce a component or subassembly are performed in close proximity, thus allowing for quick feedback between operators when quality problems and other issues arise. Workers in a manufacturing cell are typically cross-trained and, hence, able to perform multiple tasks as needed.

Cellulose fibre Transparent material made of camphor and guncotton, formed into thread-like filaments.

Cementation A process for introducing elements onto the surfaces of metals by high-temperature diffusion.

Centistroke A metric unit of kinetic viscosity used to indicate the high-temperature operating characteristics of oil.

Central locking system On a vehicle with power door locks, the system locks or unlocks all doors at one time.

Central port injection 1. An early fuel-injection system installed on the 4.3L Chevrolet Vortec V-6, using one throttle-body, injection-style injector to pulse fuel directly to individual nozzles at the intake ports.

2. A fuel metering system consisting of one central fuel injector and separate poppet nozzles

Central Processing Unit Portion of the computer which calculates information

Centrifugal 1. Moving away from the center or axis

2. To develop a force that is progressively away or outward from the axis

Centrifugal advance A mechanical means of advancing spark timing in a conventional distributor with flyweights and springs.

Centrifugal brake drum Combining the strength of steel with the friction characteristics of cast iron by spraying a lining of cast iron on the inside of a steel drum while hot.

Centrifugal clutch A clutch that utilizes a centrifugal force to apply pressure against a friction disc in proportion to the speed of the clutch.

Centrifugal filter A rotating filter that relies on centrifugal force to separate impurities from the fluid, usually oil.

Certification 1. A process, which may be incremental, by which a contractor provides evidence to the acquirer that a product meets contractual or otherwise specified requirements.

2. The approval by a regulatory or standards body that a product meets the applicable requirements or standards.

Chain casing An enclosure, housing, or guard for a chain drive containing the lubricating system and providing protection from contamination.

Chain drive A drive system consisting of a drive, and a driven sprocket and chain, such as a motorcycle or bicycle drive.

Chain guide A device used to support and guide a chain, such as a timing chain, to reduce or prevent whip.

Chain length The distance between the joint centers at each end of a taut chain strand.

Chain pensioner A device used to maintain chain tension, such as a guide rail or a hydraulic pressure piston with a rubbing shoe.

Chain pitch The center-to-center dimension between chain links or joints.

Chaku-Chaku A method of conducting single-piece flow, where the operator proceeds form machine to machine, taking the part from one machine and loading it into the next.

Chamfer face A bevelled surface on a shaft or part that allows for easier assembly.

Change Control The process and procedures that manage how changes are incorporated into a product. Change Control is a part of overall configuration management and uses review and release processes to enforce compliance with enterprise change policies.

Change management Change management is a systematic approach to dealing with change, both from the perspective of an organization and on the individual level. Change management has at least three different aspects including: adapting to change, controlling change, and effecting change. A proactive approach to dealing with change is at the core of all three aspects. For an organization, change management means defining and implementing procedures and/or technologies to deal with changes in the business environment and to profit from changing opportunities.

Change of state The rearrangement of molecular matter as it changes between any of the three physical states solids, liquids, or gases

Change Order A written amendment to a Contract signed by all parties; the only way a change may be made to a Contract or Scope of Work.

Change proposal The formal documentation that is prepared for a proposed change in accordance with the CMP Change Procedure.

Change request The formal documentation that is prepared for a request to change a specification in accordance with the CMP Change Procedure.

Changeover The refitting of equipment to either neutralize the effects of the just completed production or to prepare equipment for production of the next scheduled item, or both.

Changeover time The time required to modify a system or workstation, usually including both tear down time for the existing condition and set-up time for the new condition. Typically associated with a switch to new product.

Channel 1. To lower the body over the chassis.

2. Individual leg of the hydraulic system that relay pressure from the master cylinder to the wheel cylinder.

Chapman strut A type of rear suspension having a telescoping strut that is attached to the chassis at the top and to two links at the bottom, restricting lateral and longitudinal movement.

Characteristic A measured or observed property of a device or substance.

Charcoal canister A container, usually located in the engine compartment, containing activated charcoal. The charcoal absorbs or traps vapours from a vehicle's sealed fuel system, generally when the engine is turned off. This is a basic component of evaporative emissions control systems.

Charge 1. The conversion of electrical energy, provided in the form of a current from an external source, into chemical energy within a cell or battery.

2. The potential energy stored in a capacitive electrical device.

Charge rate The current applied to a secondary cell or battery to restore its capacity. This rate is commonly expressed as a multiple of the rated capacity of the cell or battery. For example, the C/10 charge rate of a 500-Ah cell or battery is expressed as:

Charge temperature sensor A sensor that sends a signal to the computer causing it to vary the temperature of the intake stream.

Charter Charter is a written commitment approved by management stating the scope of authority for a development project or integrated product team.

Chase A process to restore damaged threads.

Chasing threads A manual process using a tap or die to restore threads in a nut or on a bolt.

Chassis dynamometer A drive-on device, used to measure net road horsepower and torque, delivered by the drive wheels.

Chassis dyno A term often used for chassis dynamometer.

Chassis ground in electrical terms, a ground is the desirable or intentional circuit return path. A chassis ground is the desirable return path using the vehicles chassis.

Chassis lubrication 1.An element of preventive maintenance.

2. The procedure of applying the correct type and amount of grease to the chassis lubrication points at recommended intervals.

3. Adding or changing fluids.

4. Miscellaneous services including tire pressure and safety checks.

Chassis The lower structure of a vehicle to which the body and running gear are attached.

Chassis tuning Adjusting the running-gear geometry of a vehicle to compensate for different road conditions.

Chassis waddle A term often used when describing lateral runout or tire waddle.

Cheater slick A tire that is constructed of the same sticky rubber as a racing tire, but has a shallow tread cut into it to make it street legal.

Check ball 1. A device that maintains air or fuel pressure at a predetermined level.

2. A device that permits the flow of fluid or vapour in one direction only.

3. A type of hydraulic valve, consisting of a ball, that seals an orifice when it is seated and can be used to open the orifice.

Check Engine Light (CEL) this light is illuminated when the PCM detects a fault.

Check relay A term often used for check-valve relay or vacuum-check relay.

Check To verify that a component system, or measurement, complies with specifications.

Check valve 1. A valve that permits the passage of a gas or fluid in one direction but not in the other.

2. A device located in the liquid line or inlet to the drier of some systems to prevent refrigerant flow in the opposite direction when the unit is turned off.

Check-in The process of placing or returning a new or modified product information under control within a PDM/PIM system. If a revision is being created, this procedure usually initiates a review/approval process under control of the PDM/PIM system.

Check-out The process of accessing managed product definition information under controlled procedures. Access may be for viewing, reference, for use in another application or task, or for making a change to the information. The PDM/PIM system prevents multiple, simultaneous change activities to ensure product information integrity.

Check-valve relay An electrical switch to control a solenoid-operated check valve.

Cheek The plate-like part of a crankshaft that connects the journals, often serving as a counterweight.

Chem mill A term used for chemical milling.

Chemical energy A form of potential energy released or stored during chemical a reactions.

Chemical fire extinguisher A type of fire extinguisher that used dry chemicals that displace oxygen, thereby extinguishing a fire.

Chemical gasket A liquid or putty-like substance, similar to RTV, used as a substitute for a solid gasket.

Chemical hazards Hazard concerns primarily, cut not exclusively, from solvents, fuels, asbestos, and antifreeze.

Chemical instability An undesired condition that exists when a contaminant causes a fluid, such as refrigerant, to break down into a harmful chemical.

Chemical milling A term often used for acid dip.

Chemical reaction The formation of a new substance when two or more substances are brought together.

Chemiluminescence The emission of light energy, other than by burning, during a chemical reaction.

Cherry 1. In unusually fine shape.

2. As good or better than new.

3. Not used before.

Chicane An artificial series of turns on a straight track, in road racing events, that are marked by pylons or temporary curbs.

Chilled cast iron Cast iron that has been hardened using dry ice.

Chip-on-board Chip-on-board is a component packaging technology in which bare integrated circuits are attached directly to the substrate and interconnected by means of microscopic wires.

Chirp rubber To quickly shift gears during hard acceleration, so the tires momentarily break loose and leave slight streaks of rubber on the pavement.

Chirp the tires A phrase sometimes used for chirp rubber.

Chit box A recreational vehicle, such as a mobile home or trailer.

Chizer A term often given to a Chrysler product.

Chloride stress corrosion Occurs in the 300 series of stainless steel. Caused by a combination of tensile stress, chlorides and heat. No one knows the threshold values.

Chlorinated hydrocarbon solvents A class of chemical compounds that contain various combinations of hydrogen, carbon, and chlorine atoms; best known as a class of cleaning solvents.

Chlorofluorocarbon (CFC) A family of chemicals that includes the automotive air-conditioning refrigerant, R-12. Designated CFC, they have been blamed for a deterioration of the Earth's protective ozone layer and have been phrased out of production by international agreement.

Choke A manually or thermostatically controlled device mounted to a choke shaft having vanes at the mouth of a carburettor that closes when the engine is cold. This increases the gasoline content in the air/fuel mixture that aids in starting when fuel evaporation is low.

Cholorella A genus of unicellular green algae, proposed for converting carbon dioxide into oxygen for use in spacecraft.

Chop 1. To lower the greenhouse of a vehicle.

2. To cut in front of another vehicle in a closed-course race.

Chop channel and enamel The full restyling treatment of a car.

Chop Mode A display mode of operation in which small time segments of each channel are traced sequentially so that more than one waveform can appear on the screen simultaneously

Chopped flywheel To machine the surface of a flywheel to lighten it.

Chrome A simple term for chromium.

Chrome carbide Forms when chrome forms with carbon in the heat affected zone during the welding of stainless steel. The use of low carbon stainless steel is recommended in these applications.

Chute 1. A fast, straight stretch of track on an oval track or road course.

2. A parachute, as used to slow drag and lake cars.

3. The starting position for a dragster.

Circuit 1. A complete path for an electric current from the source, through a resistance, and back to the power source.

2. A race course.

Clad A metal or material covered with another metal by bonding.

Clad metal A metal that is covered with another metal or alloy of different composition, applied to one or both sides by casting, drawing, rolling, surfacing, chemical deposition, or electroplating.

Cladding the application of a surfacing material to impare corrosion and/or heat resistance.

Claim Request for payment of a loss as covered under the terms of an insurance contract.

Clam shell Used to set the pitch (distance between convolutions) while heat treating a metal bellows core.

Clamp A screw-, cam-, or lever-actuated device for temporarily holding parts together.

Clamp circuit A circuit that biases a voltage waveform to a different voltage level.

Clamping Diode A diode that is connected in parallel with a coil to prevent voltage spikes from the coil from reaching other components in the circuit

Class 2 The type of digital data stream used on non-OBD-II diagnostic systems. Class 2 uses two-bit pulse widths and toggles between o (passive) and 7 volts (active)

Climate-control system The non-technical term for the heating, ventilation and air-conditioning system. Most current vehicles have all three - heating, defrost, and ac.

Clinical trial Clinical trial is testing a system in a clinical setting; that is, in a hospital, clinic, doctor's office, etc. User testing and feature testing in such an environment has special limitations, especially because of the potential for unexpected effects on patient care. User testing in clinical settings will often require review by an ethics committee to ensure that patient privacy is not compromised and that no harm will come to patients as a result of testing. As with drug testing, it may not be appropriate to remove a feature from a system for the sake of testing if it appears that the feature is directly benefiting patient care.

Clip A major body repair where the front or rear of a vehicle is replaced with the front or rear of another vehicle of the same mike and model

Clipboard A special file or memory area (buffer) where data is stored temporarily before being copied to another location.

Clock 1. A device for telling the time of day.

2. A term often used for speedometer.

3. A device that generates a basic periodic signal used to control timing of all operations in a synchronous system or computer.

Clock An oscillator producing timing pulses to synchronize various elements of a system. In switching mode power supplies, a clock is used to produce the power pulses that are modulated to control power transfer. In digital interfaces that communicate on a bus (such as the IEEE-488) a clock is used to synchronize the data transfer and commands.

Clock input An input terminal in a circuit where timing or synchronizing signals are applied.

Clock rate The frequency of oscillation of the master clock, or oscillator, in a system.

Clock spring A device, located between the steering column and steering wheel, that conducts electrical signals in an air-bag system to the module, while allowing steering-wheel rotation. This provides electrical continuity in all steering-wheel positions.

Clockspring A winding of electrical conducting tape enclosed within a plastic housing. The clockspring maintains continuity between the steering wheel, switches,the air bag, and the wiring harness in all steering wheel positions.

Close coupled The pump impeller is mounted directly on the motor shaft. There is no separate bearing case.

Close ratios A transmission, usually used in drag racers, with close spacing between the speeds of the gears, allowing for minimum engine rpm reduction when shifting.

Close the door 1. To immediately pull in front of an opponent after overtaking, preventing him/her from re-passing.

2. To move over while entering a curve to block an opponent just as he/she pulls alongside on the inside to pass, preventing him/her from doing so. Also known as close the gate.

Close the gate A term often used for close the door.

Closed course An oval track or road race circuit.

Closed crankcase ventilation (CCV) A system in which crankcase vapours are discharged into the engine intake system (usually via the intake manifold) where they are burned during the combustion process rather than being discharged into the atmosphere.

Closed Loop 1. The basic principle of electronic engine management in which input from an Oxygen sensor allows the engine-control computer to determine and maintain a nearly perfect air/fuel ratio

2. A computer condition in which the air/fuel ratio is being controlled on the basis of various inputs to the computer

3. A continuous circuit from beginning to end and beyond

Closing costs Any expenses added onto the lease that are incidental to the purchase. (for example, title fees and appraisal fees.)

Cloud of points A set of x-y-z coordinates obtained from a 3d scanner or digitizer. The data can be interpreted as a continuous surface and used in a 3d model. This is often used for reverse engineering.

Cloud point 1. The low temperature at which diesel fuel begins to produce wax crystals.

2. The temperature at which wax begins to separate from oil.

Club coupe A two-door, four,- or five- passenger vehicle with limited rear seating.

Clunker 1. A poorly performing car.

2. The family buggy.

Cluster gear A set of three or four gears on a common shaft in a manual transmission.

Clutch 1. A device for connecting and disconnecting the power flow between the engine and standard transmissions, used during starting, shifting, and stopping.

2. A device used to connect two collinear shafts to a driving mechanism such as a motor, engine, or line shaft, and to disconnect them at will.

3. An electromagnetic clutch used to engage and disengage the compressor, to turn the air conditioner on and off.

Coating 1. Paint that is used to provide beauty or weather protection to a vehicle or to its parts.

2. Any material applied to a metal surface to provide protection against the elements.

Coaxial cable A cable consisting of two conductors concentric with, and insulated from, each other.

Coefficient of Drag (CD) This is a number that is a measure of an automobile's resistance to the air it passes through, with lower numbers meaning the car has greater aerodynamic properties. It is one factor that influences the smoothness of ride and better fuel economy.

Coefficient of elasticity An alternate term for modulus of elasticity

Coefficient of friction The measure of the resistance of one surface moving against another. It is obtained by dividing the force required to slide the surfaces across one another by the pressure holding the surfaces togather.

Coefficient of water/oil distribution The ratio of the solubility of a chemical in water compared to its solubility in oil.

Cog A gear, particularly the final drive gear.

Cognitive modelling Cognitive modelling produces a computational model for how people perform tasks and solve problems, based on psychological principles. These models may be outlines of tasks written on paper or computer programs which enable us to predict the time it takes for people to perform tasks, the kinds of errors they make, the decisions they make, or what buttons and menu items they choose. Such models can be used to determine ways of improving the user interface so that a person's task has fewer errors or takes less time and to build into the user interface to make software that reacts more effectively to help people use the system by anticipating their behaviour.

Cognitive walkthroughs Cognitive walkthroughs involve the development of task scenarios from a product specification. Experts then role-play the part of a user working through a set of tasks. Each step of the user's process is evaluated for adherence to established usability principles.

Coherence The property of being coherent, e.g. waves having similar direction, amplitude and phase that are capable of exhibiting interference.

Cohesive strength theoretical stress that causes fracture in tension test if material exhibits no plastic deformation.

Coil 1. A term often used to describe a suspension spring or the component that is formed when several turns of wire are wound on a cylindrical form or on a metal core.

Coil spring A spring-steel bar or rod that is wound into the shape of a coil to provide an up-down springing effect. Found on most vehicle suspensions, these springs are used to support the car's weight, maintain height, and correctly position all other suspension parts, but are little help in supporting side-to-side or lateral movement.

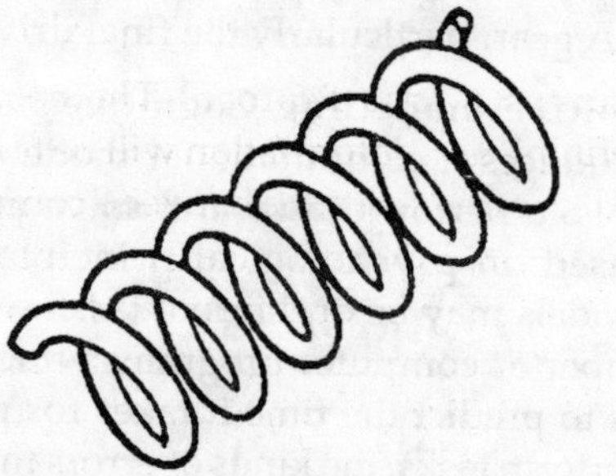

Fig. Coil spring

Collaborative product commerce The Aberdeen group defines collaborative product commerce as "...a class of software and services that uses internet technologies to permit individuals - no matter what role they have in the commercialization of a product, no matter what computer-based tools they use, no matter where they are located geographically or within the supply net - to collaboratively develop, build, and manage products throughout the entire lifecycle. Using a standard browser, an authorized user can review information from an extended-enterprise information system 'view' that operates across a dispersed set of heterogeneous product development resources. These resources typically reside in multiple information repositories and are derived from independently implemented and maintained systems."

Collapsible steering column An energy-absorbing steering column that is designed to collapse if the driver is thrown into it due to a heavy collision.

Collapsible washer A washer that crushes when a nut is tightened against it

Collect A term used for valve keeper.

Collector 1. A device that collects exhaust gases from the primary tubes and channels them into a single exhaust pipe.

2. The tank of a radiator that receives the fluid before it passes through the radiator. Transistor part (Emitter, Collector, Base)

Collector tank The tank that collects coolant from the engine, containing a baffle plate to aid in even distribution of coolant through the core.

Collision coverage Optional insurance that pays for damage to your car caused by collision with another car or object.

Collision shop A specialty paint and body shop that restores a wrecked or damaged vehicle to its pre-accident condition.

Collocation The practice of physically locating multi-function integrated product team members in proximity to one another to enhance communication, coordination and decision-making on a development project. Virtual collocation refers to the use of technology to achieve some of the communication benefits for team members that are geographically dispersed.

Collusion The cooperation between two or more people to secretly defraud another person or company.

Coma The cloud of diffuse material surrounding the nucleus of a comet.

Combination brake system A dual-brake system that uses disc brakes at the front wheels and drum brakes at the rear wheels.

Combination Manifold An assembly which includes both the intake and exhaust manifolds in one casting

Combination valve 1. S hydraulic control valve with two or three valve functions in one valve body.

2. A H-valve, used in some early air-conditioning systems, combining a suction throttling valve and an expansion valve.

Combined emissions-control system An early General Motors transmission-controlled spark system that uses the solenoid valve's plunger as an auxiliary throttle stop.

Combustion A chemical reaction between two or more substances that releases heat, light, and gases.

Combustion Burning, fire produced by the proper combination of fuel, heat, and oxygen. In the engine, the rapid burning of the air-fuel mixture that occurs in the combustion chamber.

Combustion chamber The chamber in a rocket where the fuel and oxidizer are ignited and burned. By common usage the expansion nozzle is included as part of the combustion chamber, particularly for liquid-propelled rocket engines.

Command An instruction commanding a system to start some action. The action may result in some internal parameters being set to particular values. A command may have command parameters that contain the details of the instruction to be obeyed.

Command module The compartment of a spacecraft which contains the crew and main controls, and is used as the reentry vehicle.

Commercialization Commercialization is the process to taking a new product from development to full volume sales. It includes steps such as testing and market validation, production launch and ramp-up, development of marketing programs and materials, supply chain development, sales channel development, training development, training, and service and support development.

Common cause Common cause is a variation that is inherent in the process and cannot be readily identified and controlled.

Common law liability The responsibility of injury or loss imposed upon a person because of his or her actions.

Common point With respect to operationally programmable power supplies one output/ sense terminal is designated "common to which load, reference and external programming signal all return.

Communications port A standard communications interface, such as an IEEE 488 or RS-232, that provides information flow from a processor to a peripheral device, such as a power supply.

Communications Protocol The set of rules defining how information will be transferred . All communications between computers or other electronic devices require that the devices agree on the format of the data. The set of rules defining a format is called a protocol.

Commutating pole An electromagnetic bar inserted between the pole pieces of a generator to offset the cross magnetization of the armature currents.

Commutator the name for the copper segment of the starter or generator that transfers electrical energy to the armature or rotor windings.

Compact A small car that can seat two to four people.

Compact high-pressure tire A term often used for compact spare.

Compact spare A weight- and space-saver tire, especially designed for temporary use.

Compact spare tire A spare tire and wheel that is much smaller than the other tires on the vehicle, and is designed for short-distance driving at relatively low speeds.

Compact tire A term used for compact spare tire.

Companion cylinders Two cylinders in an engine that are at TDC at the same time.

Comparator 1. An instrument for comparing specific critical measurements to a fixed standard

2. A circuit for comparing two input signals which may be ether analogue or digital signals but the output will always be digital

Comparison amplifier A dc amplifier which compares one signal to a stable reference, and amplifies the difference to regulate the power supply power-control elements.

Compensating coil A coil that serves to compensate for the mechanical friction in the moving coil of an electrical meter or gauge.

Compensating port A device used to maintain the proper level of brake fluid in the brake lines.

Compensation The addition of circuit elements to assist in stabilization of a control loop.

Competitive benchmarking Formal programs that compare a plant's practices and performance results against "best-in-class" competitors or against similar operations.

Competitive intelligence Methods and activities for transforming disaggregated public competitor information into relevant and strategic knowledge about competitors' position, size, efforts and trends. The term refers to the broad practice of collecting, analyzing, and communicating the best available information on competitive trends occurring outside one's own company.

Complete circuit An uninterrupted electrical circuit or fluid circuit in which electrical current may travel to and from the battery, or fluid or vapour may circulate continuously through the system.

Completely built up (CBU) The complete building, rebuilding, or modification of a component or vehicle, as in a kit car.

Complex modulus Measure of dynamic mechanical properties of a material, taking into account energy dissipated as heat during deformation and recovery. It is equal to the sum of static modulus of a material and its loss modulus. In the case of shear loading, it is called dynamic modulus.

Compliance Agency certification that a product meets its standards. Compliance range Range of voltage needed to sustain a given constant current throughout a range of load resistance.

Compliance voltage The maximum dc output voltage of a constant current supply.

Component A software entity within a Virtual Instrument with formal interfaces and explicit dependencies; Components are combined to form a Virtual Instrument. Components must be remotely accessible in a distributed environment.

Component assembly A combination of two or more parts or sub-components to form an assembly.

Component Design The activity for the design of specific components including responsibility for material, cost, weight, reliability, durability, function, appearance, serviceability.

Component engineering The application of engineering know-how to the processes of component selection, application, process compatibility and procurement, including analysis of new trends in electronic devices.

Component isolation To isolate a component from the rest of the system or circuit for testing or replacement.

Component location table A table or chart, used with an electrical schematic, that describes or illustrates the actual location of the part being investigated.

Component supplier management Component supplier management is a class of software applications that maintain information about standard components, both purchased and made, to support various functional disciplines such as design, procurement, materials, configuration management, and manufacturing. This system serves as a central repository for component and supplier information to maximize standardization, design retrieval and re-use, and procurement efficiency. CSM systems contain four major elements, part classification and retrieval, component libraries, web component cataloguing, and component/supplier process management.

Components/equipment Reparable assemblies which currently require repair parts support or will require it when introduced into an inventory.

Composit Bulb A headlight assembly that has a replaceable bulb in its housing

Composite drum A drum made of different materials, usually cast iron and steel or aluminium, to reduce weight. The friction surfaces and the hub are cast iron, but supporting parts are made of lighter metal.

Composite headlamps These are improved headlamps for enhanced illumination as compared to the standard sealed beam units. They often, but not always, contain replaceable halogen bulbs with separate acrylic lenses.

Composite headlight A halogen headlight system that uses a replaceable bulb, allowing vehicle manufacturers to produce any style of headlight lens they desire.

Composite materials The bonding of different materials, usually for strength, such as carbon fibre and fibreglass.

Composite rotor A rotor made of different materials, usually cast iron and steel or aluminium, to reduce weight. The friction surfaces and the hub are cast iron, but supporting parts are made of lighter metal.

Composite spring A term used for fibre-composite spring.

Composite technology Composite technology is a form of production in which so-called composite materials are used (generally carbon fibre-reinforced aluminium honeycomb). This technology is called for in applications where the products need to be very light yet very strong - for example in the aerospace industry.

Composite washer A flat washer made of different elements.

Composite When used in the content of mechanical seal faces, it refers to either a non metallic material or a combination of non metallic face inserted into a metal holder.

Composites Structural materials of metal alloys or plastics with built-in strengthening agents, e.g. carbon fibres.

Composition gasket A gasket made of a combination of materials.

Compound A substance composed of tpw or more elements combined chemically

Compound gauge A gauge that registers both above and below atmospheric pressure; used on the low side of an air-conditioning system.

Compound interest Interest computed on the sum of an original principal and accrued interest.

Compound low gear A combination of low gear in the transmission and low range in the transfer case in a four-wheel-drive vehicle with a two-speed transfer case.

Compression 1. The process of squeezing a vapour (gas) into a smaller space.

2. The upward stroke of a piston that compresses the fuel/air mixture into the combustion chamber prior to ignition

Compression braking The slowing of a vehicle utilizing a diesel engine, such as that provided by a Jake brake. It is a misnomer that a gasoline engine will slow the vehicle by compression braking; actually, vacuum causes the braking effect.

Compression Gauge A gauge used in the spark plug hole of the cylinder head to measure cylinder compression pressure

Compression height A distance, as measured from the crown of the piston to the center of the wrist pin.

Compressive strength Maximum stress a material can sustain under crush loading. The compressive strength of a material that fails by shattering fracture can be defined within fairly narrow limits as an independent property. However, the compressive strength of materials that do not shatter in compression must be defined as the amount of stress required to distort the material an arbitrary amount. Compressive strength is calculated by dividing the maximum load by the original cross-sectional area of a specimen in a compression test.

Compressive yield strength Stress which causes a material to exhibit a specified deformation. Usually determined from the stress-strain diagram obtained in a compression test.

Compressor (Air conditioning) The mechanism is an air conditioner that pumps vaporized refrigerant out of the evaporator, compresses it to a relatively high pressure, and then delivers it to the condenser.

Compressor crankshaft seal A term used for compressor shaft seal.

Computer A machine capable of following instructions to alter data in a desirable way and to perform most of these operations without human intervention.

Computer hardware Devices capable of accepting and storing computer data, executing a systematic sequence of operations on computer data, or producing control outputs. Such devices can perform substantial interpretation, computation, communication, control, or other logical functions.

Computer-aided engineering Computer-aided engineering is the use of computers in design, analysis, and manufacturing of a product, process, or project. Sometimes refers more narrowly to the use of computers only in the analysis stage.

Computer-aided manufacturing Computer-aided manufacturing (cam) is the use of the computer description of the part or assembly to drive planning, cutting, forming, assembly and inspection of the item via computerized applications.

Computer-aided process planning Computer-aided process planning uses part data and process rules to generate process plans or work instructions. Variant CAPP is based on group technology classification of parts and part features to search for a predetermined similar process plan that most closely matches the classification. Generative CAPP uses part and feature classification along with rules and knowledge about manufacturing processes associated with features to generate an appropriate process plan.

Concept An idea for a new product or system that is represented in the form of a written description, a sketch, block diagram or simple model. A concept is the earliest representation of a new product or of alternative approaches to designing a new product.

Concept of execution Represents the dynamic relationships of the components. It can include such descriptions as flow of execution control, data flow, dynamically controlled sequencing, state transition diagrams, timing diagrams, priorities among units, handling of interrupts, etc.

Concept testing The process by which a concept statement, sketch or model is presented to customers for their reactions. These reactions can either be used to permit the developer to estimate the sales value of the concept or to make changes to the concept to enhance its potential sales value.

Concept Vehicle A current production vehicle modified for installation of new design concepts for evaluation of environmental functional feasibility.

Conceptual architecture The conceptual architecture represents an appropriate decomposition of the system without delving into the details of interface specification. The conceptual architecture identifies the system components or subsystems, the responsibilities of each component or subsystem, and interconnections between components or subsystems.

Conceptual design review The first major review in the life of an observatory project. It outlines the conceptual approach of the project team to the engineering challenges of the observatory design and development phase.

Conceptual design, studies, and analysis This is a document or document set included in some ATST bid packages describing prior design work performed by or for ATST during the design and development phase of the project.

Concours d'elegance French for "contest of elegance," a showing of luxury cars in a plush setting.

Concurrency The degree to which phases, stages, or activities may be overlapped or done in parallel.

Concurrent engineering A systematic approach to the integrated, concurrent design of products and their related processes, including manufacture and support. This approach is intended to cause the developers, from the outset, to consider all elements of the product life cycle from conception through disposal, including quality, cost, schedule and user requirements.

Condensate Moisture that is removed from air, such as that collected on the surface of an air-conditioning-system evaporator.

Condensation 1. The moisture removed from ambient air.
2. The process of a substance changing state from a vapour (gas) to a liquid.

Condenser A device for storing electrical energy. In a/c application, an air conditioning component used to remove heat from the inside of a vehicle.

Condenser comb A comb-like device used to straighten the fins on the evaporator or condenser.

Condenser temperature The temperature at which compressed gas in the condenser changes to a liquid.

Condensing pressure 1. High side or head pressure, as read from the gauge at the high-side service valve.

2. The pressure present from the discharge side of the compressor to the metering device inlet.

Condition Based Depreciation (CBD) A form of depreciation that directly assesses and measures the run down in service potential of an asset. It is based on an auditable and cost-justified asset renewal plan. The cost of replacing lost service potential over the next 10-30 years (the exact period depends on the agency and the nature of the assets involved) is expressed as an annuity over the period. That annuity is the depreciation estimate. CBD is re-estimated on a continuous basis, based on a rolling future period. It is only used for assets which are essentially renewable rather than replaceable, i.e. infrastructure assets.

Condition Based Maintenance (CBM) Maintenance based on actual condition, obtained from non-invasive test, operating and condition monitoring.

Conditioned air Air that is cool, dry, and clean.

Conductance (G) The ability to conduct current. It is equal to amperes per volt, or the reciprocal of resistance, and is measured in siemens (metric) or mhos (English). G = 1/R

Conduction The transmission of heat or electricity thought, or by means of, a conductor.

Conductor Any material that will conduct an electrical charge or any material with less than three electrons in its valance shell.

Conductor placement A term used for wire placement.

Conduit 1. A tubular raceway, A tubular pathway such as tubing used to protect a wiring system or branch circuit.

2. A flexible metal housing or jacket that houses the parking brake cables to protect them from dirt, rust, abrasion, and other damage.

Cone 1. The conical diaphragm attached to the voice coil of a speaker that produces sound.

2. The conical part of an oxyfuel gas flame adjacent to the tip orifice.

Coned-disk spring A term used for Belleville washer.

Configuration A collection of an item's descriptive and governing characteristics, which can be expressed in a) functional terms, i.e. What performance the item is expected to achieve, and b) physical terms, i.e. What the item should look like and consist of when it is completed.

Configuration control board A board composed of technical and administrative representatives who approve or disapprove proposed engineering changes to an approved baseline.

Configuration control The systematic evaluation, co-ordination, approval or disapproval and dissemination of proposed changes and implementation of all approved changes in the configuration of any item after formal establishment of its configuration Baseline.

Configuration documentation Configuration documentation is the sum of all the documents that define the physical and functional characteristics of a System, subsystem, CSCI, HWCI or designated equipment, for example, specifications, design documents, engineering drawings, source code listings.

Configuration elements Specifications, drawings, source code, etc., that define the configuration of a CSCI.

Conjoint analysis A methodology for exploring and describing subjective customer views of product features. Conjoint analysis avoids direct questioning, e.g., "what do you think of the price of our product?" Instead, the customer is asked what they are willing to pay for a particular product feature. Thus, the real buying situation with consideration of different cost-benefit alternatives is simulated. The resulting analysis show directly the contribution of each product feature to the total product utility. Conjoint analysis can be used to determine to what extent a product's perceived utility changes if some particular product feature is modified.

Connecting link A link in which a removable plate facilitates connecting or disconnecting the ends of a length of chain.

Connecting rod A component used to attach the piston, with pin to the crankshaft rod journal.

Connecting-rod cap The half-round, lower, bolt-on portion of the connecting rod.

Connector A mechanical device used to link conductors.

Consensus Consensus is a group decision resulting from members engaging in full and open discussion and then reaching agreement to live with and openly support the resulting decision.

Console This may refer to the unit found between the front driver and passenger seat that contains the automatic transmission shifter, cupholders and a storage compartment. But it can also refer to the section of the instrument panel that includes the controls for the sound system and climate-control system, particularly if the panel flows down the center of the vehicle and includes the automatic-transmission shifter.

Contextual analysis / contextual inquiry Contextual analysis / contextual inquiry is a structured field evaluation method which uses a combination of methodologies derived from anthropology and journalism. By observing and interviewing users of products in their actual environment and understanding the *context* in which a product is used, better insight is gained into the issues that affect contextual analysis / contextual inquiry is a discovery process that can add insight into the needs of customers.

Contingency A contingency is the planned allotment of time, cost, budget or design margin for unforeseeable elements or risks with a development project.

Contingency design Contingency design is a form of mistake-proofing focusing on the user's experience with the product. The intent is to design in features that help the user avoid mistakes or allow the users to quickly correct input of data or operation of the product. This is accomplished through layout and graphic design, intuitive operation, clear instructions, appropriate markings and warnings, descriptive error messages, avoidance of technical jargon, and simple operation steps.

Contingency money A payment made by accessory and equipment manufacturers to top rated drivers, and some crew members, for using and/or displaying their products in a race.

Continuity The ease or ability of an electrical circuit or component to transfer energy from one point to another. A continuous electrical path.

Continuous code A series of computer diagnostic codes that relate to engine-control functions.

Continuous combustion The constant combustion of an air/fuel mixture.

Continuous duty A demand on an energy-consuming system that requires operation, at a constant load, for an indefinite period of time.

Continuous fuel injection (CFI) A type of fuel-injection system that continuously sprays a stream of fuel into the engine.

Continuous improvement A philosophy of frequent and small changes to production processes developed in Japan. The cumulative results lead to high levels of quality and efficiency.

Continuous process control The use of transducers (sensors) to monitor a process and make automatic changes in operations through the design of appropriate feedback control loops. Such devices have been historically mechanical or electro-mechanical, there is now widespread utilization of computers and centralized control.

Continuous Process Method of producing an article continuously.

Continuous variable transmission (CTV) A stepless transmission that uses a sheave clutch to transmit engine torque.

Continuous-replenishment programs Arrangement with supplier companies in which the supplier monitors the customer's inventory and automatically replaces used materials, eliminating the need for purchase orders and related paperwork.

Contract A mutually binding legal relationship obligating the seller to furnish the supplies or services (including construction) and the buyer to pay for them. It includes all types of commitments that obligate the acquirer to an expenditure of appropriate funds and that, except otherwise authorized, are in writing. In addition to bilateral agreements, contracts include, but are not limited to, awards and notices of awards; job orders or task letter issued under purchase orders, under which the contract becomes effective by written acceptance or performance; and bilateral modifications.

Contract data requirements list That portion of a contract that identifies deliverable data products

Contract documents The sum of a Design and Construction Contract, a Design Requirements Document, and all Modifications issued after execution of the Contract.

Contracting officer's technical representative The COTR will be the single point of contact between a Contractor and the ATST Project.

Contractor An individual, partnership, company, corporation, association or other service, having a contract with an acquirer for the design, development, manufacture, maintenance, modification, or supply of items under the terms of a contract.

Contribution by equal shares Provision in insurance contracts which requires each company to share equally in the loss until the share of each equals the lowest limit of liability under any policy or the full amount of loss is paid.

Control arm A suspension element that has one joint at one end and two joints at the other end, typically on the chassis side. Also known as a wishbone or an a-arm.

Control cable 1. A rigid wire, generally sheathed, used for the remote control of a device, such as a parking brake.

2. An insulated wire used to supply electric current to a motor, controls, or other electrically operated circuits or devices.

Control chart A graphical display of results of a process over time. They are used to determine if a process is in statistical control or in need of adjustment.

Control component A Component that services as the interface to a Virtual Instrument.

Control head The master controls (such as temperature and fan speed) that the vehicle driver uses to select the desired system condition.

Control limits In statistical process control, two horizontal lines are drawn on a control chart denoting the upper control limit and the lower control limit. The sample-means and the ranges from a production lot must be within these limits. If they are so, the process is behaving normally and is said to be under control. If any point lies outside either of the limits, this denotes loss of control - the process must be halted and the reason found.

Control link A term used for control arm.

Control plunger A device that is regulated by the fuel-injection-system airflow sensor to regulated fuel delivery to the injectors.

Control points Holes, points, or flat surfaces that are used to align body parts during assembly or reconstruction.

Control pressure The fuel pressure required for a fuel-injection system to function.

Control rocket A vernier or other rocket used to control the attitude of, or slightly change the speed of, a spacecraft.

Control system The deliberate guidance or manipulation of the elements in a system in order to achieve a prescribed value or performance of a system to complete a defined process.

Control thermostat A temperature-actuated electrical switch used to cycle the compressor clutch on and off thereby controlling the air conditioning system temperature.

Control valve 1.The mechanism, located inside the power steering gearbox or on the steering relay rod, that controls the amount of power assist relayed to the steering linkage via a power piston.

2. A mechanical, pneumatic, or electric valve used to control the flow of coolant into the heater core.

Control valve assembly The block of metal that contains the hydraulic passages and electric solenoids used to direct brake fluid during an ABS event.

Controlled Canister Purge An ECM controlled solenoid valve that permits manifold vacuum to purge the evaporative emissions from the charcoal canister

Controlled-leak governor A governor assembly that reduces the leakage of line pressure as a vehicle's speed increases.

Controller A controller is a computing component that executes software to close servo loops, read external sensors, and interface with higher level software.

Controller area network (CAN) A serial bus system especially suited for networking "intelligent" devices as well as sensors and actuators within a system or sub-system. CAN2.0A is an 11-bit identifier; CAN2.0B is a 29-bit identifier.

Convection A natural circulation of fluid. The hot fluid (lighter) rises and the cool fluid (heavier) sinks.

Convection tank Used to contain fluid between two mechanical seals. An enclosed heater or cooler can be used to control the barrier or buffer fluid temperature. Pressure or level gages can indicate which seal has failed.

Convection-cooled power supply A power supply cooled exclusively from the natural motion of a gas or a liquid over the surfaces of heat dissipating elements.

Conventional cab A truck having a cab behind the engine.

Conventional frame A chassis frame that is separate from the body.

Conventional strut An assembly in which the coil spring that supports the vehicle weight sits on a lower-spring seat, which is part of the strut housing, to maintain vehicle height.

Conventional theory The original theory of electricity that electrical current would flow from the positive potential to the negative potential.

Conventional tire A term used for bias ply tire.

Conventional-frame designs One of several frame designs ie ladder, x-frame, hourglass or perimeter frame

Conversion The substitution of one element or component for another, as in substituting R-134a for R-12 in an automotive air-conditioning system; changing the fuel-type requirements of an engine; or changing ignition systems.

Converter A device that changes the value of a signal or quantity. Examples: DC-DC; a device that delivers dc power when energized from a dc source. Fly-Back; a type of switching power supply circuit.

Converter dolly An axle, frame, drawbar, and fifth-wheel arrangement that converts a semi-trailer into a full trailer.

Converter pressure The operating pressure within the torque converter.

Converter-signal pressure The pressure that signals the torque-converter, lockup clutch engagement.

Convertible Any car with a removable top, either a rag (cloth) top or hard top.

Convertible top A vinyl or canvas soft-top roof system for two- or four-door vehicles that can be manually or automatically lowered and raised.

Convolution Two metal bellows plates welded or formed together. To count the number of convolutions in a metal bellows you count the spaces between the bellows plates.

Cook 1. To overheat, generally to the point of severe or irreparable damage.

2. To perform well as in "s/he's cooking'."

Cool Pack A trade name used to describe an early after-market hang-on or under-dash, after-market air-conditioning system.

Coolant Coolant is a water and anti-freeze mixture that takes heat away from the engine (to stop it overheating) and transports it to the air in the radiator.

Cooling: Forced air Forced air cooling ratings may be specified in either linear feet per minute (LFM), cubic feet per minute (CFM), or pound mass per minute (lb/mm). Linear feet per minute ratings are a more valid measure of the air flow because they specify exactly where the air flows and directly relate to heat transfer. However, CFM values are usually specified. Flow measured in lb/mm is most useful when dealing with varying air densities or high altitude, and is calculated by CFM and density (lb/w). This application note explains the relationship between forced air specifications in linear feet per minute and in cubic feet per minute. This application note applies to all Astec power supplies requiring forced air cooling. To Convert from CFM to LFM and LFM to CFM given the cross sectional area: To convert from CFM to LFM and vice-versa, we need the cross section area of the power supply. This would be the actual area that the air is being blown through. The formula is CFM LFM x cross sectional area. A popular selection is 2.5" x 5.0". This corresponds to 0.0868 square feet. For any fixed cross sectional area, the relationship is linear and can be shown with a graph.

Core 1. The interior of a hollow casing.

2. The coolant passages and fins of a radiator or heater found between the two header tanks.

3. A used part or assembly that is to be returned to the vendor for rebuilding.

Core competencies Core competencies are the essential capabilities that create a firm's sustainable competitive advantage.

Core competency The processes, functions, and activities in a plant or company that are its "life blood"-typically those activities for which the enterprise derives the greatest return for its investments or those that intrinsically align the enterprise with its core market.

Corrosion The act of gnawing, or eating away portions of metals, a chemical action of Oxygen reacting to metals, Iron containing metals corrode in the form of rust.

Corrosive flux A flux composed of inorganic salts and acids, or activated rosin with a residue that chemically attacks the metal.

Corrosive rain A form of pollution produced when sulphur and nitrogen are emitted into the air. Known as acid rain.

Corrugated metal gasket Thin sheets of metal used as a gasket that is shaped into parallel grooves and ridges for added strength.

Cost reduction A formal activity employed to rectify a cost target breach or to reduce the cost of an existing product or design. A cost reduction effort has a specific quantified objective and may affect schedule, performance or support to achieve this objective.

Coudé rotator Also known as the Coudé Platform. The Coudé Rotator is comprised of the structural elements that support Coudé instruments and optical benches underneath the Mount. The Coudé Rotator provides for complete image de-rotation and positioning of the instruments with respect to the incoming light beam that is delivered from the Mount along the azimuth axis of rotation. Included as part of the Coudé Rotator is an Azimuth Track that is identical to that of the Mount Azimuth Track.

Countershaft 1. A shaft used in a V-8 engine to reduce the effects of imbalance.

2. Two shafts used in an I-4 engine to reduce the effects of imbalance.

3. The shaft that supports the cluster-gear set in a manual transmission and rotates in

the opposite direction of the clutch and driveshaft.

Coupling An attachment where one mechanism or part drives another mechanism or part, to follow the movements of the first.

Coupling phase the point in torque-converter operation where the turbine speed is 90% of the impeller speed, and there is no longer any torque multiplication.

Coupling stage A term referring to the torque-converter turbine speed, as it closely approaches the impeller speed; occurs during higher speeds under light loads.

Coupling yoke Two integral or separate Y-shaped components to which the universal-joint bearing cups are attached.

Coupon 1. A piece of metal, of specified size, used for testing.

2. A piece of metal from which a test specimen may be prepared.

Courtesy light Lamps that illuminate the vehicle interior and/or exterior when a door is opened, and are controlled from the headlight and for switches.

Covalent bonding when atoms share valence electrons with other atoms

Covariance The impact of one variable upon others in the same group.

Covered electrode A composite filler metal consisting of a core of bare electrodes or a metal-cored electrode, with a covering sufficient to provide a slag layer on the weld metal.

Crank 1. A crankshaft.

2. To start an engine.

3. To go fast.

Crank kit A reconditioned crankshaft with the appropriate rod and main bearings.

Crank pin The rod journal of a crankshaft.

Crankcase A shaft with one or more cranks, or "throws," that are coupled by connecting rods to the engine's pistons. The combustion process creates reciprocating motion in the rods and pistons which in turn is converted to a rotating motion by the crankshaft.

Crankshaft 1. A revolving part mounted in the lower portion of the engine block.

2. That part of a reciprocating compressor on which the wobble plate or connecting rods are attached.

Crashing Taking action to decrease the total project duration by analyzing a number of alternatives to determine how to get the maximum duration compression for the least cost. Often, it involves reducing the time it takes to complete an activity by adding resources.

Crater A round impression left in a planet or satellite from a meteoroid.

Crazing Fine cracks that may extend in a network on or under the surface of a material, often occurring in the presence of an organic liquid or vapour.

Cream hardener An activating ingredient for a plastic-filler material, such Bondo.

Cream puff A vehicle that is in especially fine condition.

Creature comforts Any options or amenities that improve vehicle driving or riding pleasures.

Creep Deformation that occurs over a period of time when a material is subjected to constant stress at constant temperature. In metals, creep usually occurs only at elevated temperatures. Creep at room temperature is more common in plastic materials and is called cold flow or deformation under load. Data obtained in a creep test usually is presented as a plot of creep vs. time with stress and temperature constant. Slope of the curve is creep rate and end point of the curve is time for rupture. As indicated in the accompanying diagram, the creep of a material can be divided into three stages.

First stage, or primary creep, starts at a rapid rate and slows with time. Second stage (secondary) creep has a relatively uniform rate. Third stage (tertiary) creep has an accelerating creep rate and terminates by failure of material at time for rupture.

Creep recovery Rate of decrease in deformation that occurs when load is removed after prolonged application in a creep test. Constant temperature is maintained to eliminate effects of thermal expansion, and measurements are taken from time load is zero to eliminate elastic effects.

Creep test Method for determining creep or stress relaxation behaviour. To determine creep properties, material is subjected to prolonged constant tension or compression loading at constant temperature. Deformation is recorded at specified time intervals and a creep vs. time diagram is plotted. Slope of curve at any point is creep rate. If failure occurs, it terminates test and time for rupture is recorded. If specimen does not fracture within test period, creep recovery may be measured. To determine stress relaxation of material, specimen is deformed a given amount and decrease in stress over prolonged period of exposure at constant temperature is recorded.

Critical characteristics The characteristics or specifications for a material, part, assembly or product that define those attributes that are essential to the proper fit or functioning of the item to satisfy the intended customer use or need.

Critical design review The final major review in the life of an observatory project. Any and all subsystem, functional and interim reviews must be completed and all issues resolved prior to CDR. The CDR must include verification of compatibility with higher level and interfacing subsystems. The design must be confirmed to be comprehensive, addressing all products and processes.

Cup expander Metal discs formed to fit inside piston cups and to keep the lips of the cups in tight contact with the cylinder walls while the hydraulic system is not pressurized.

Cuppling The method of connecting two circuits together, Circuits connected with a wire are directly coupled (DC) Circuits connected through a capacitor or transformer are indirectly (AC) coupled

Curb height The measurement from the vehicle frame to the road surface.

Curb weight The weight of the vehicle without passengers, driver or cargo, but with all standard features, a full tank of fuel, and all the fluids necessary for proper function.

Curbside The right-hand side of a vehicle, that nearest to the curb.

Curing 1. The hardening of a catalyzed compound, such as fibreglass.

2. The drying of paint.

Curing agent A class of materials used in brake linings to accelerate the chemical reaction of the binders and other materials

Current transformer 1. Instrument Transformer: Intended to have its primary winding connected in series with the conductor carrying the current to be measured or controlled.

2. Metering: Designed for use in the measurement or control of current. Its primary winding may be single turn or bus bar, and is connected in series with the load.

3. Power and Distribution Transformer: Intended to have its primary winding connected in series with the conductor carrying the current to be measured or controlled. (In window-type current transformers, the primary winding is provided by the line conductor and is not an integral part of the transformer.)

Cycle 1. The process of discharging and then recharging a battery.

2. A series of repeated events such as the intake, compression, power, and exhaust strokes of an engine.

3. A complete sequence of a wave pattern that recurs at a regular intervals.

4. Short for motorcycle.

5. The microprocessor action of turning solenoids on and off.

Cylinder 1. A storage tank for gases, such as refrigerant.

2. The round hole(s) inside and engine block that provide space for the reciprocating piston(s).

Cylinder arrangement The way cylinders are placed in an engine, such as in-line, in a row; vee, in two banks or rows at an angle to each other; or flat, pancake.

Cylinder head The removable part of the engine that attaches to the cylinder block directly above the cylinders. The head is cast from aluminium or iron and houses the combustion chambers, the intake and exhaust ports, spark plugs and much or all of the valve train. It has oil and water passages for cooling and lubrication.

D

D An acronym for drive; one of the forward gear positions in an automatic transmission.

D frame adapter Used to connect and align the pump and motor (manufactured in metric dimensions).

D&D Design and Development

Dampening ball joint A term used for non-load-bearing ball joint.

Dampening effect The effect shocks have on reducing suspension-spring oscillations.

Damper 1. English term for Shock Absorber

2. Friction or hydraulic shock absorbers or the equivalent electronic device

3. a device that reduces or eliminates vibration or oscillations

4. a balancer on the front of the crank shaft to reduce or eliminate harmonic vibrations

Damper assembly A device designed to decrease vibrations to the passenger compartment.

Damping The physical touching of a component to arrest vibration.

DAQ Utah's Division of Air Quality

Dark current A small leakage current that flows through a photosensitive device in the absence of light.

Dark factory A term to depict a completely automated factory with no labour.

Dark matter A form of matter which has not been directly observed but whose existence has been deduced by it's gravitational effects.

Darlington pair Two transistors electrically connected to form an amplifier permitting a small amount of current flow to control a larger current flow (an electronic relay). Named for Sidney Darlington, a physicist at Bell Laboratories from 1929 to 1971.

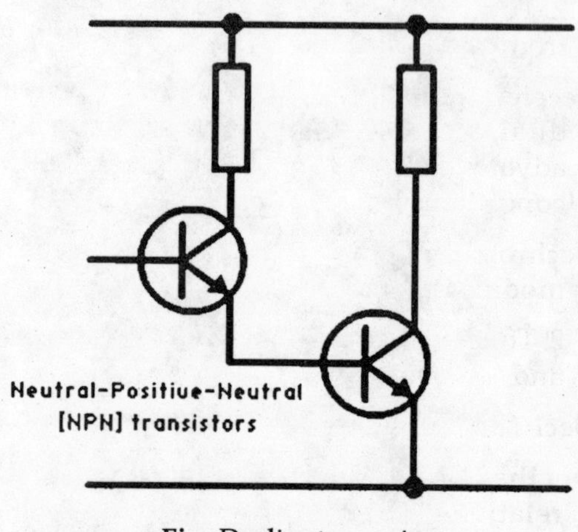

Fig. Darlington pair

Deactivate A series of steps necessary to turn a feature or function off

Dead axle A shaft, such as the rear axle on a front-wheel drive vehicle, that connects wheels to either side of a vehicle but does not provide driving power.

Dead center The extreme top or bottom of the piston stroke.

Dead ending Isolating the stuffing box. No recirculation or flushing lines in or out.

Dead player A vehicle part or assembly that is inoperable.

Dead Short A zero resistance short circuit

Dead spot The momentary loss of power in an engine while increasing its speed.

Decarbonize To remove carbon deposits. Also known as decarburize.

Decarburize Same as decarbonize.

Decay The action of air drag upon an artificial satellite causing it to spiral back into the atmosphere, eventually to disintegrate or burn up.

Decel valve A device that reduces exhaust emissions during vehicle deceleration by keeping rpm up and vacuum down.

Decelerate To decrease speed.

Deceleration 1. A decrease in velocity or speed.

2. To allow the vehicle to coast to idle speed from a high speed.

Deceleration valve A device used withy the dual-diaphragm, vacuum-advance to advance engine timing under deceleration conditions.

Dechrome 1. To strip the vehicle of chrome for modification.

2. To remove the chrome prior to body repair and/or painting.

Deci The prefix representing the number 0.1

Decibel The numerical expression of the relative loudness of two signals, such as sound. The difference in decibels between two signals is ten times the common logarithm of the ratio of their powers.

Decision-support tools PC, client or application server-based systems that use memory-based processing to perform rapid simulations using data drawn from business-transaction processing systems such as enterprise resources planning.

Deck 1. The flat, mating surfaces of an engine block and head.

2. The trunk of a passenger car.

3. To machine an engine block deck flat.

4. To dechrome the deck lid.

Deck lid The trunk door, or cover.

Deck plate A heavy metal plate that is bolted to an engine block during cylinder-honing operations.

Declination One of the coordinates, measured in degrees, used to designate the location of an object on the celestial sphere. Declination is a north-south value similar to latitude on Earth.

Declutching fan An engine-cooling fan system, mounted on the water pump, having a temperature-sensitive device that governs or limits terminal speed.

Decoke A British term for decarbonize.

Decomposition Decomposition is the process of dividing the system into its smallest, coherent, self-contained elements. Decomposition is used in systems engineering, software engineering, process mapping and functional analysis system technique.

Decompression The relief of pressure. Explosive decompression would occur if the cabin of a spacecraft was punctured in space.

Dedicated Set aside for a special purpose.

Deductible The amount of money or percentage of expenses that will be covered by the insured.

Deep cycling A condition where a battery is completely discharged before it is recharged.

Deep staging A drag-racing technique where the driver uses minimum roll out to avoid breakout.

Deep sump An oil pan, used on some drag-race cars, that is deep enough to keep the oil away from the crankshaft at high rpm to slightly increase horsepower.

Deep-dish wheel A wheel with an extremely positive offset.

Deep-rolled fillets The transition radius between a journal and the cheeks of a crankshaft.

Default A value or setting that a device or program automatically selects if you do not specify a subroutine

Default charges Fees charged the lessee as a result of missing payments or otherwise defaulting on the lease. Typical charges include all remaining payments and any additional costs incurred in reclaiming the vehicle. The security deposit may also be lost.

Default setup The setup that exists as long as there are no changes made to the settings.

Defect tracking Defect tracking typically refers to identifying flaws appearing during manufacturing. The goal is not only to spot defects but to track them and identify the source of the problem to prevent them. Were specs incomplete, impractical, or poorly specified? Was the design difficult to produce? Were needed process capabilities lacking?

Defects per million opportunities Defects per million opportunities is a quality measurement where defects are defined as material or process or related, and opportunities for defects are defined as the sum of all parts, lead attachments and/or glue dots as applicable for each part number assembled

Deflection 1. An axial or radial movement away from the normal or standard axis of a part.

2. A bending or distorting motion; usually applied to a brake drum when it is forced out-of-round during brake application.

Deflection angle The angle at which oil is deflected inside a torque converter during operation; the greater the angle, the greater the torque applied to the output shaft.

Deflection rate The number of pounds required to compress a spring one inch (25.4 mm).

Defogger That part of the heater system designed to clear the windshield of fog haze under certain conditions.

Deformable mirror A very thin mirror whose shape can be changed by the force applied by many small pistons behind the mirror.

Deformation energy Energy required to deform a material a specified amount. It is the area under the stress-strain diagram up to a specified strain.

Defrost door A small door within the duct system that diverts a portion of the delivery air to the windshield.

Defrost switch A thermostatic-type switch that senses evaporator temperature and turns the compressor off to prevent frosting or ice formation on the evaporator.

Defroster door A term often used for defrost door.

Defroster That part of the heater system designed to clear heavy frost or light ice from the inside or outside of the windshield.

Deg An acronym for degree.

Degreaser A chemical that breaks oil and grease down.

Degreasing Cleaning parts in a solvent to remove oil and grease.

Degree Part of a circle; one degree is 1/360th of a circle.

Degreeing a cam Using instruments to determine the actual timing of valves opening and closing for a particular camshaft.

Degrees of Freedom The number of unique ways in which a part can move in an alignment system. In static alignment, there are six: one in the direction of and one in rotation about each of the X, Y and Z axes. In dynamic alignment, as of a scanning system, there are up to five more. When aligning a part, each degree of freedom must be constrained sufficiently to avoid unwanted motion but not so much as to create stress. Initially assumed to be of order 1000 for ATST.

Delayed Time Base A time base with a sweep that can start (or be triggered to start) relative to a pre-determined time on the main time bade sweep. Allows the user to see events more clearly and to see events that are not visible solely with the main time base sweep

Departure angle The maximum angle, in degrees, of a line running rearward and upward from the rear tire contact point to the lowest obstruction under the rear of the vehicle.

Dependent demand In manufacturing is demand for an item that results from the production decisions for its parent items. An order for a parent part can only be completed if all of its components are available.

Dependent events Events that occur only after a previous event.

Depolarize To remove or eliminate positive and negative poles from an item.

Deposit A sum of money to hold a deal until the paperwork is complete. If the deal is closed, the deposit is applied to the down payment.

Deposited metal Any filler metal that may have been added during a welding process.

Derived requirements Requirements that are not explicitly stated in the customer requirements, but are inferred a) from contextual requirements (e.g., applicable standards, laws, policies, common practices, and management decisions), or b) from requirements needed to specify a product component. Derived requirements can also arise during analysis and design of components of the product or system.

Descending node The point at which an orbiting object or spacecraft, moving from north to south, crosses the plane of the equator.

Descent engine The rocket used to power a spacecraft as it makes a controlled landing on the surface of a planet or moon.

Descent module That part of a spacecraft that descends from orbit to the surface of a planet or moon.

Desiccant 1. A material, such as silica gel, that absorbs moisture from a gas or liquid.

2. A drying agent used in refrigeration systems to remove excess moisture.

Design and development The first phase of overall observatory construction, this is when the engineering design of the observatory is developed to meet the existing Science Requirements. Followed by a Construction Phase.

Design authority An organization responsible for the detailed design of materiel to approved specifications and authorized to sign certificates of design or certified sealed drawings in accordance with procedures identified in the Program 'Configuration Management Plan'.

Design completeness The design shall be complete from a total system element viewpoint (hardware, facilities, personnel, computer software, procedural data).

Design for assembly Design for Assembly (DFA) refers to the principles of designing assemblies so that they are more manufacturable. DFA principles address general part size and geometry for handling and orientation, features to facilitate insertion, assembly orientation for part insertion and fastening, fastening principles, etc. The objective of DFA is to reduce manufacturing effort and cost related to assembly processes.

Design for environment Design for Environment is a process for the systematic consideration during design of issues associated with environmental safety and health over the entire product life cycle. DFE can be thought of as the migration of traditional pollution prevention concepts upstream into the development phase of products before production and use.

Design for manufacturability 1. Design for Manufacturability (broad definition) is a methodology for designing product's in a way that facilitates the fabrication of the product's components and their assembly into the overall product. In this respect it is synonymous with Design for Manufacturability / Assembly.

2. Design for Manufacturability (narrow definition) is a methodology for designing product's components in a way that facilitates their fabrication.

Design for postponement With a product that offers a lot of different configurations and options, an objective is to delay of postpone the assembly of the unique parts or subassemblies into the product until as late as possible in the final assembly process. This reduces production leadtime (products can be built and stocked up to this point), facilitates a strategy of mass customization, and provides supply chain flexibility. Design for postponement is the design of an assembly in a way that allows the customizable parts of the product to be assembled as late as possible in the assembly process.

Design for reliability Design for reliability is methodology and set of principles to enhance product reliability and reduce overall low life-cycle costs. It is based on early involvement of reliability engineering working with design engineering to enhance reliability by performing steps such as the following: reliability program planning, reliability predictions, parts derating, thermal analysis, failure modes and effects analysis (fmea), fault tree analysis, availability and system modelling, design verification testing, product return rate analysis, fracas, and root cause failure analysis.

Design for serviceability Design for serviceability is a set of principles and a methodology for analyzing product concepts or designs for characteristics and design features which reduce service requirements and frequency, facilitate diagnosis, and minimize the time and effort to disassemble, repair/replace, and reassemble the product as part of the service process.

Design of experiments A statistical methodology for designing, conducting and analyzing experiments or tests to evaluate product and process design parameters or factors that affect the achievement of a product performance characteristic. The response of interest is evaluated under the various conditions to: 1. identify the influential variables among the ones tested,

2. quantify the effects across the range represented by the levels of the variables,

3. gain a better understanding of the nature of the causal system at work in the process, and

4. compare the effects and interactions. These experiments lead to setting parameter or factor levels (values) that can optimize the product performance characteristic under study and minimize the affects of variation. There are several techniques including Taguchi Methods, fractional factorial, and Plackett-Burman. DOE's are often classified in one of three categories: Screening Designs, which are intended to identify which main effects (factors) are the vital few important factors that require further study (usually a fractional factorial); Characterization Designs, which are used to gain some quantitive understanding of

the relationships among the factors, including interactions, on the response variable (usually a factorial); and Optimization Designs, which are used to gain a precise understanding of the mathematical relationships that is sufficient to allow prediction and optimization throughput the experimental region.

Design optimization 1. Design Optimization in the broad sense refers to optimizing a design to meet its functional, environmental and lifecycle requirements at a minimum of cost.

2. Design Optimization in a narrower sense refers to the use of computer-aided engineering applications which analyze a design and, given constraints and objectives, seek to improve or optimize the design to meet the stated objectives within the stated constraints. These applications will typically use an iterative, goal-seeking cycle to seek design optimization.

Design records Design records refers to all technical documentation necessary to define the design, manufacture, packaging, testing, installation, and maintenance of the system and its comprising elements.

Design requirements document Known as Specification Documents. This specifies the "build to", "code to", and "buy to" requirements for products. It includes the Functional Requirements (which are necessary tasks, actions, or activities that must be accomplished by a system) and Performance Requirements (which quantitatively state the extent to which a system function must be executed). It may reference the OCDD and should include a list, at minimum, of identified ICDs. This document should be complete by PDR.

Design reviews Design reviews are formal technical reviews conducted during the development of a product to assure that the requirements, concept, product or process satisfies the requirements of that stage of development, the design is sound, the issues are understood, the risks are being managed, any problems are identified, and needed solutions proposed. Typical design reviews include: requirements review, concept/ preliminary design review, final design review, and a production readiness/launch review.

Design simplicity The concept of design simplicity and standardization shall be evident.

Design structure matrix Design structure matrix - matrix used to represent and analyze task dependencies in a product development project / process.

Desmodromic valves A valve system in which positive cam action, not spring action, is used to open and close the intake and exhaust valves.

Destructive testing A test to determine the point at which a device catastrophically fails.

Developmental configuration The contractor's design and associated technical documentation that defines the evolving configuration of a configuration item under development. It is under the developing contractor's configuration control and describes the design definition and implementation. The developmental configuration for a configuration item consists of the contractors released hardware and software designs and associated technical engineering documentation until establishment of the formal product baseline.

Deviations A specific written authorization, granted prior to the manufacture of a Configuration Item, to depart from a particular performance or design requirement of a specification, drawing or other document for a specific number of units or a specific period of time. A deviation differs from an engineering change in that an approved change requires a corresponding revision of the documentation defining, the affected item, whereas a deviation does not contemplate revision of the applicable specification or drawing.

Device A peripheral which connects to the computer. Mice, keyboards, printers, data acquisition instruments, modules and cards are all devices.

Diagnosibility The ability to identify any faults (or potential faults) in the operation of the product.

Diagnosis 1. A standard procedure that is followed to locate and identify the cause of a malfunction.

2. The art ro act of identifying the cause of a problem or malfunction by observing its signs and symptoms.

Diagnostic Any number of on-board tests run by the diagnostic management system which checks for malfunctions, errors or breakdowns in vehicle systems or components.

Dielectric strength The ability of an insulator to insulate; to resist carrying current without damage.

2. The resistance to electron penetration.

Diesel An internal combustion engine in which the air-fuel mixture is ignited by compression in the cylinder rather than by a spark. Diesel engines use diesel fuel rather than gasoline and tend to be more fuel-efficient and require less maintenance than gasoline engines, but it is more complicated to get them to run cleanly. Also used as a slang term: after turning off the ignition, the engine continues to run for a short period.

Diesel cycle An engine operating cycle where the air is compressed and the fuel is injected at the end of the compression stroke, causing ignition.

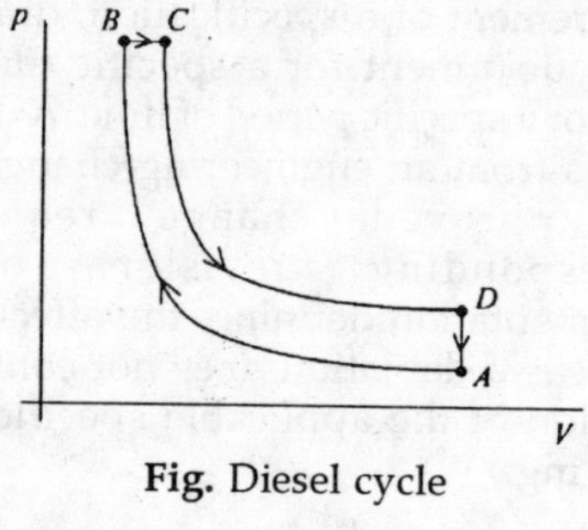

Fig. Diesel cycle

Diesel engine A diesel engine uses heavier weight components than gas engines to handle higher compression ratios. Typically, diesel engines run with greater efficiency and higher torque than similar size gas engines. These attributes lead to better fuel economy and towing performance. Diesel engines do not have spark plugs or carburettors. Instead glow plugs are used to preheat air in the cylinders to ensure easy starts. Once the engine is started, compression heats the fuel in the cylinders for combustion.

Diesel Engine Manufacturers Association (DEMA) A professional association of diesel-engine manufacturers.

Diesel fuel The fuel used by a diesel engine. Usually found in tractor trailers and other trucks.

Diesel injection system The fuel supply side of a diesel engine.

Dieseling A condition in which gasoline continues to fire after the ignition has been shut off. In late-model engines, dieseling , or run-on, is caused by heat and the unusually high manifold pressure that result from retarding the spark at idle. In fuel-injected cars when the engine is turned off, fuel is automatically shut off, eliminating dieseling.

Diff An acronym for differential.

Differential A mechanical gearbox or fluid coupling that allows wheels to rotate at different speeds. Usually located on an axle, it allows the outside wheels to turn faster than the inside wheels during cornering. Four-wheel-drive and all-wheel drive vehicles have two differentials, one for the rear axle and one for the front. All-wheel drive vehicles also may have a third or centre differential on the drive shaft that runs between the front and rear axles.

Differential action An operational situation where one driving wheel can rotate at a slower speed than the opposite driving wheel.

Differential Amplifier One whose output is proportional to the difference between two inputs.

Differential assembly The mechanism that relates to front- or rear -driving axles that permits unequal travel distances and speeds between the vehicle's driving wheels.

Differential case The housing of the differential that contains the side and pinion gears and the pinion shaft, and also serves as a mounting place for the ring gear.

Differential drive gear A large, circular helical gear that is driven by the transaxle pinion gear and shaft and that drives the differential assembly.

Differential global positioning system (DGPS) A differential global positioning system (GPS) land station-based method for improving GPS accuracy or getting around U.S. Department of Defence-imposed selective availability (S/A) encryption.

Differential housing Also known as rear-axle housing. A cast iron assembly that houses the differential unit and the drive axles.

Differential Inputs Using differential inputs can reduce noise picked up by the signal leads. For each input signal there are two signal wires. A third connector allows the signals to be referenced to ground. The measurement is the difference in voltage between the two wires: any voltage common to both wires is removed.

Differential lock A toggle or push-pull type air switch that locks together the rear axles of a tractor so they pull as one for off-the-road operation.

Differential mode noise The component of noise, excluding common-mode noise, that is measured between two lines with respect to a common reference point. The value is the difference of the noise components on the two lines.

Digital 1. Relating to the technology of computers and data communications where all information is encoded as bits of 1's or 0's that represent ON or OFF states.

2. A numerical readout.

Digital audio broadcast (DAB) A digital technology offering considerable advantages over today's FM radio, both to listeners and broadcasters. The most obvious benefit to listeners is DAB's ability to deliver CD-quality stereo sound robustly and reliably, even to mobile receivers, using only a simple rod antenna. FM reception is often distorted or interrupted by multi-path interference in moving vehicles in particular, as reflections from hills and buildings arrive out of phase with the main signal. By contrast, the DAB receiver's powerful processor use these "shadows" to reinforce the main signal.

Digital computer A device that manipulates digital data and performs arithmetic and logic operations on such data.

Digital device A device that has only two states of operation, on and off

Digital electronic fuel injection An early Cadillac electronic fuel-injection system.

Digital instrument A display that indicates an activity, such as rpm, mph, voltage, or fuel supply, with an electronic readout; used in most Indy cars, though some drivers still prefer analogue instruments.

Digitize The process by which an analogue-to digital converter in the horizontal system samples a signal at discrete points in time and converts the signal's voltage at these points into digital values called sample points

Digitized The process of converting an analogue voltage signal to a digital equivalent that the computer can understand.

Dimpling the distortion of an oil-pan rail around the bolt hole.

Ding A small dent or scrape in the body of the vehicle.

Diode A two-element device containing a cathode and an anode that permits flow in one direction and blocks flow from the other.

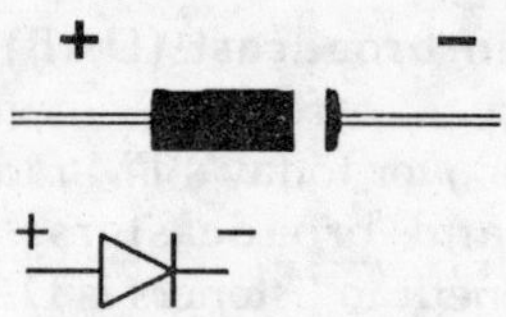

Fig. Diode

Diode trio A group of three diodes grouped together with one output used to provide rectified current to the field and voltage regulator on many internally regulated alternators.

Diolefins An ingredient of gasoline which contribute to gum and lacquer formation in the fuel system.

DIP Dual inline package.

Dip stick A thin, steel strip of metal used to measure the fluid level or quantity in the engine, transmission, or compressor.

Dipole A compact source of magnetic force, with two magnetic poles. A bar magnet, coil or current loop, if their size is small, create a dipole field. The Earth's field, as a first approximation, also resembles that of a dipole.

Dipper A small, metal scoop at the bottom of a connecting rod to scoop up oil from the pan to lubricate the bearing.

Dipper trough A trough aligned under the connecting rod that is fitted with the dipper.

Direct battery power Power available to a circuit in the automobile directly from the battery, without an intermediate switch.

Direct costs Cost that can be specifically identified or traced to an activity, cost object or final cost objective.

Direct current (DC) Flow of electrons in one direction. In text, use lower case: dc.

Direct drive A transmission mode in which the driveshaft and engine crankshaft are at the same speed.

Direct-acting thermostatic air cleaner A component of heated air-intake systems that uses a thermostatic bulb connected to a rod to operate a flapper valve in the air-cleaner snorkel.

Direct-bonded bearing A bearing that is formed by pouring molten babbitt directly into the bearing housing and machining the cooled metal to the desired diameter.

Directed evolution Directed evolution is an advanced methodology used to create scenarios to support the planning and development of future generations of technical systems.

Directional control The ability to steer the automobile while stopping

Directional signals Lights at either side of both the front and rear of the vehicle that flash to indicate an intended turn.

Directional stability A vehicle's ability to maintain a true course of travel despite bumps, crosswinds, uneven road surfaces.

Directional tire A tire having a tread pattern that must be mounted facing forward.

Direct-to-plate thermal imaging A technology in which plates are made directly from digital computer data and in which heat is used rather than light to image printing plates.

Dirt dobber A circle-track race driver that prefers a dirt track over an asphalt track.

Dirt tracking Driving on an asphalt track as if it were a dirt track.

DIS Direct ignition system or distributorless ignition system

DISA Differential air intake control system

Disable 1. To remove or inhibit normal capability.

2. A command that prevents further operation of a peripheral device.

Disassemble To take apart.

Disaster bushing Used in A.P.I. glands to support the shaft in the event of a bearing failure, or to prevent product from rushing to atmosphere after a seal failure. The close clearance (0.025 inch or 0.5 mm.) directs most of the leakage through a drain connection in the seal gland to an appropriate container.

Disc 1. An acronym for discount.

2. A variant of disk.

Disc brakes Shiny metal discs, called brake rotors, are attached to the wheel hub, rotating with the wheel. When the brake pedal is depressed, the brake callipers squeeze the discs to slow the vehicle.

Discharge pressure switch A term used for compressor-discharge pressure switch.

Discharge pressure The high-side pressure of the refrigerant as it is being discharged from the compressor. Also known as high-side pressure.

Discharge recirculation Connecting a line from the discharge side of the pump to the stuffing box. Should be used with a close fitting bushing in the end of the stuffing box to increase the stuffing box pressure. A common application when pumping a fluid close to its vapour point.

Discharge side That portion of the refrigeration system under high pressure, extending from the compressor outlet to the metering-device inlet.

Discharge valve A term used for high-side service valve.

Disconnect A switching device for disconnecting an electrical circuit or load from the power supply.

Discontinuous innovation Discontinuous innovation falls outside of existing markets or market segments, and when successful extends and redefines the market, exposing new possibilities. Discontinuous innovation is characterized by lateral or divergent thinking, by looking outside of defined boundaries, and by discovery of new knowledge related to both market need and technological capability.

Discounted cash flow Discounted cash flow - an analysis technique that determines the present value of a series of positive and negative cash flows using a specified discount factor representing the cost of capital. This can be used to compare investment alternatives such a new product development alternatives.

Discrete components An electrical component, such as a capacitor, that has been fabricated prior to its installation.

Discrete device An individual electrical component, such as a resistor.

Discrete manufacturing The production or assembly of parts and/or finished products that are recognizable as distinct units capable of being identified by serial numbers or other labeling methods-and measurable as numerical quantities rather than by weight or volume.

Discrete-event simulation Material-flow or discrete-event simulators have most often been used by engineers in the design and modification of production systems. The models output statistical estimates of performance, using graphic animation to help create a greater understanding of system dynamics.

Discriminating Sensors Parts of the air bag circuitry that calibrate to close with speed changes that are great enough to warrant air bag deployment

Disengage To disconnect; used to describe the action when a clutch uncouples the power of the driving part from the driven part.

Dish A reflector for radio waves, usually a paraboloid.

Dished piston A piston having a depression in the crown.

Disk 1. A magnetic device for storing information and programs accessible by a computer.

2. The rotor, a revolving piece of metal, against which shoes are applied to provide a braking action.

3. A two-dimensional figure defined by all points enclosed by a circle's diameter.

4. Often referred to as disc.

Disk drive A device that stores, reads, or retrieves data from a magnetic disk and copies it into the computer's memory for use.

Disk runout The amount that a brake disk wobbles during rotation.

Disk-operating system An operating system in which the programs are stored on magnetic disks that keep track of, save, and retrieve files; allocate storage space and manage other control functions.

Disperesant A chemical added to oil that prevents impurities from clinging together and forming lumps that could clog the lubrication system galleys.

Dispersion The spatial property of being scattered about over an area or volume

Displacement current An expression for the effective current flow across a capacitor.

Displacement In an engine, the total volume of air or air-fuel mixture an engine is theoretically capable of drawing into all cylinders during one operating cycle. Generally expressed in litres or cubic inches. Engine displacement is equal to (bore) x (bore) x (stroke) x (number of pistons) x (.785).

Display terminal An output device, such as a CRT, that produces a visual representation of graphic data.

Disposable cylinder A container for the one-time use of packaging, transporting, and dispensing of a fluid, such as refrigerant. It is a violation of Federal Law to refill these cylinders, commonly referred to as "DOT-39s."

Distillation curve The reference to plotting a line connecting the percentages of gasoline that evapourate at various temperatures. The distillation curve is used as an important control for fuel standards such as volatility (vaporization).

Distillation endpoint The temperature at which a fuel is completely vapourized.

Distortion 1. Inaccuracies in size and shape, as in an out-of-round cylinder.

2. An unwanted change in purity of sound, as caused by a weak signal.

3. A term used for tread distortion.

Distributed Control System (DCS) A system customized per facility/organization and designed to meet the total measurement, control and real-time information requirements of today's process plants. A single application to connect to subsystems such as maintenance (CMMS), statistical process control (SPC), and advanced control.

Distribution channel Is the path goods take as their title transfers from producer to consumer. The title transfer for consumer goods is usually accompanied by transfer of the physical goods, as well.

Distribution management Determines optimal quantities of each product to be made at each plant and to be distributed to each warehouse, such that manufacturing and distribution costs are minimized and customer demand is met.

Distribution statement A statement used in marking a technical document to denote the extent of its availability for distribution release, and disclosure without the need for additional approvals and authorizations from the controlling agency.

Distributor Part of the ignition (electrical) system. Delivers electricity from the ignition coil to the distributor cap and the spark plug wires in the correct firing order. (the firing order is that sequence in which each cylinder begins its power stroke.) The spark plugs ignite the fuel and air mixture in each cylinder thousands of times a minute, producing the explosion that pushes the piston down in the cylinder to power the vehicle.

Distributor advance A term used for centrifugal advance, ignition advance, or vacuum advance.

Distributor cam The four-, six-, or eight-sided lobe at the top end of the distributor shaft that rotates to cause the contact points to open and close.

Distributor cap A cover for the conventional ignition-system distributor, having a central terminal that receives secondary voltage from the coil and four, six, or eight peripheral terminals to send this voltage to the spark plugs.

Distributor housing A metal part that contains or provides a mounting for distributor components in a conventional ignition system.

Distributor ignition An ignition system that relies on a conventional distributor for proper operation.

Distributor pipe A pipe to convey fuel from the fuel distributor to the injector.

Distributor plate The plate inside a distributor that is fastened to the housing and does not move.

Diurnal Effect The diurnal effect deals with the heating and cooling of gasses, there expansion and retraction in proportion to there temperature.

Dive The tendency of the front of a vehicle to press down on the front springs during heavy braking.

Divergent nozzle end An exhaust-pipe nozzle end that expands from the inlet to the end.

Divergent-convergent nozzle end An exhaust-pipe nozzle that is expanded then reduced in size at the end.

Diverter valve A vacuum-operated valve in an air-injection system that directs air-pump output to the atmosphere during high-vacuum deceleration to eliminate backfiring.

Division 1. A branch or department of an organization.

2. The mathematic operation to determine how many of one number is contained in another; for example, how many 10s are in 100.

Dobson unit (DU) A measure of ozone density level, named after Gordon Dobson, a British meteorologist, who was the indenter of the measuring device (called a spectrophotometer)

Document management system Allows users to store, search and manipulate documents electronically, and to maintain a library of text and images in a compact space. Most systems also provide a means for passing documents across a network.

Documented procedure A written description of a course of action to be taken to perform a given task.

Dog 1. A poor performing vehicle.

2. To follow another car very closely in oval-track racing.

3. A pin or stub used to mate and/or drive a gear or assembly.

Dog clutch 1. A simple splinted clutch that cannot be slipped; generally found in dirt-track cars without transmissions.

2. The mating collars, flanges, or lugs that can be moved as desired to engage or disengage similar collars, flanges, or lugs in order to transmit rotary motion.

Dog house 1. The housing over an engine or transmission.

2. The front fenders, grille, and hood assembly of a vehicle.

Domain The part of the external world, including users and inmates of the system that effects and is affected by the system.

Dome cranes The array of fixed and moveable jib cranes mounted to the inside of the enclosure carousel. They are used for various handling and maintenance tasks associated with ATST construction and operations.

Domed piston A piston having a raised crown.

Donuts 1. A 360-degree tire burnout.

2. Tires, especially big racing slicks.

Dooley A Chevrolet pick-up truck with dual rear wheels; generally a term given to any dual-wheel vehicle.

Doped solder Solder containing a small amount of an added element to ensure retention of one or more characteristics of the materials on which it is used.

Doping The intentional adding of impurities to pure silicon or germanium to form either P or N type semiconductor materials.

Doppler effect A phenomenon in which waves appear to compress as their source approaches the observer or stretch out as the source recedes from the observer.

Dose A quantity of radiation delivered at a position. In the context of space energetic particle radiation effects, it usually refers to the energy absorbed locally per unit mass as a result of radiation exposure.

Dose equivalent A dose normally applied to biological effects and including scaling factors to account for the more severe effects of certain kinds of radiation.

Dot matrix A type of display or printer output in which each alphanumeric character is made up of an array of small dots.

Dowel pin A round, solid, or hollow pin used to align two or more parts.

Downdraft carburettor A carburettor having a downward airflow.

Downflow radiator A radiator in which the coolant flow is from the tip tank to the bottom tank.

Downforce The downward force of air on a speeding vehicle.

Downlink The radio signal transmitted from a spacecraft to Earth.

Downloading To copy data (usually an entire file) from a main source to a peripheral device.

Downshift The automatic shift from a high gear ratio to a low gear ratio.

Downshift valve An auxiliary shift valve that increases throttle pressure to force a downshift under high driveline loads, also called a kickdown or detent valve.

Downsizing A reduction in the number of employees that occurs due to management decision, not associated with natural attrition.

Downstream A direction indicating the location of a component or fault from it's source.1. A downstream 02 is after the cat.

2. A open in the circuit may be downstream (toward the ground) or up-stream (toward the positive) from the point in the circuit y where you are testing.

Downstream blower A blower arranged in the duct system so as to pull air through the heater and/or air-conditioner core(s).

Downtime 1. The time a system or machine is not available for use, due to failure or routine maintenance.

2. Time lost due to lack of service orders.

Drag coefficient A measure of the aerodynamic sleekness of an object. Drag coefficient is signified by "dc.: the lower the number, the greater the aerodynamic efficiency. The higher the drag coefficient, the more a car's engine must work to keep a given road speed. Also known as "cd" for coefficient of drag."

Drier 1. A tubular device containing desiccant, usually placed in the liquid line to absorb moisture in an air-conditioning system.

2. A catalyst added to paint to speed the time required for curing.

Drift A change in output over a period of time independent of input, environment or load. For power supplies, the change in DC output as a function of time at constant line voltage, load, and ambient temperature. Normally specified for an eight hour period after a half hour warm-up.

Drift oil passage Holes drilled in various parts of the engine for pressurized oil to flow through.

Drift pin A round, tapered, steel pin used to bring plates that are to be bolted or riveted into alignment to permit placing bolts or rivets easily.

Drip pan A shallow pan under the evaporator core to catch the condensation that will be ejected by a drain hose fastened to the drip pan and extending to the outside.

Drive axle An axle or axle shaft that transmits power to the drive wheels.

Drive belt Flexible belt or belts used to connect a drive pulley on the crankshaft to the coolant pump and accessories. Two basic types of drive belts are the serpentine or multiple-ribbed belt and v-belt.

Drivetrain Sometimes known as the powertrain, this is the collective name for all the components that are directly involved in the power production that is needed to get and keep the automobile moving. This includes the engine (obviously) as well as the likes of the clutch, transmission (or gearbox) driveshafts, differential, and wheel axles.

Driving axle An axle capable of transmitting power by way of a differential or other transmission arrangement. Also a term used for drive axle.

Driving lamps Auxiliary lights to provide additional illumination for high-speed driving.

Driving lights A term used for driving lamps.

Drogue A small parachute used to slow and stabilize a spacecraft returning to the atmosphere, usually preceding deployment of a main landing parachute. S

Droop The decrease in amplitude of a flat top square pulse; a form of square pulse distortion.

Drop the hammer 1. To engage the clutch and depress the accelerator suddenly.

2. To suddenly depress the accelerator to the floorboard.

Drop through The undesirable sagging or surface irregularity encountered when brazing or welding; caused by overheating with rapid diffusion between the filler metal and the base metal.

Drop-center rim A wheel that has a smaller diameter centre section, to facilitate tire replacement, designed to prevent the tubeless tire from rolling off the rim when the tire blows out.

Drop-center safety rim A wheel rim with a low center area and raised flanges designed to prevent the tire from accidentally coming off the rim.

Drop-center wheel A conventional wheel that has a space in the center for one bead to fit into while the other bead is being lifted over the rim flange.

Droplight An electrical light on a long extension cord. Also called a trouble light, or repair light.

Dropped axle A lowered beam axle used on older Fords to dump the front end.

Dropping Resistor A resistor inserted in series with a line to reduce the voltage to a desired level.

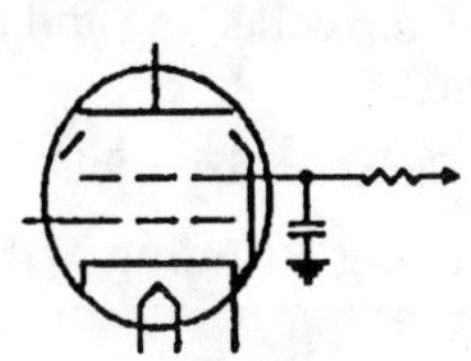

Fig. Dropping resistor

Drop-throttle oversteer A term used for lift-throttle oversteer.

Druid A term used by a dissatisfied contestant, who has problems with the rules, when referring to a race official.

Drum armature A generator or motor armature having its coils wound longitudinally or parallel to its axis.

Drum Brakes A braking system where a metal drum is attached to the wheel, and the rotation of this, and so the wheel, is slowed and halted by curved devices called brake shoes being pressed hard against the brake lining which causes friction on the inside of the drum.

Drum maximum diameter The largest diameter, generally 0.060 inch (1.5 mm) over originally to which a brake drum can be machined or worn before it is considered unsafe.

Drum The part of a brake that rotates with the axle hub and that the brake shoes press against to slow or stop the vehicle.

Drum web The closed side of a brake drum.

Drum-brake fade The loss of braking efficiency due to excessive heat.

Dry 1. The process of changing from a liquid to a solid, as in paint.

2. The process of evaporation, as in water.

Dry deck A condition in which the piston crown is at the level of the block deck.

Dry friction The friction between two dry solids.

Dry fuel A fuel with an additive to prevent water from collecting and freezing in the fuel system.

Dry gas Same as dry fuel.

Dry running Running without fluid at the seal face.

Dry-charged battery A battery that is filled with electrolyte only when put into service.

Dry-disk clutch A clutch in which the friction faces of the friction disk are dry.

Dryer drive A type of starter-motor drive used in heavy-duty applications that provides mechanical meshing means and automatic demeshing.

Dry-film lubricant A petroleum-based chemical used to lubricate operating parts during assembly, prior to engine start up, to provide lubrication until oil is circulated after start up.

Dual camshaft A type of engine that has two camshafts for opening and closing additional valves.

Dual exhaust An exhaust system used on V-type engines with a separate muffler, exhaust, and tail pipe for each cylinder bank.

Dual master cylinder A master cylinder, associated with a dual-brake system, having two sections for displacing fluid under pressure; generally one section for front brakes and one section for rear brakes or, in some applications, one for left brakes and the other for right brakes.

Dual overhead camshaft Two cams in a single-cylinder head.

Dual quads Two four-barrel carburettors.

Dual seal Two seals running in various configurations: back to back, tandem, face to face, or concentric.

Dual trace A feature that allows the oscilloscope to display two separate waveforms at the same time

Dual-brake system A brake system that uses two hydraulic circuits; should one fail, the other remains operational.

Dubble Start Override Prevents the starter motor from being energized if the engine is already running

Ducktail An upswept rear end, often with a spoiler.

Duct A tube or passage used to provide a means to transfer air or liquid from one point or place to another.

Duct tape A heavy duty, fibre-backed tape used for emergency repairs to body parts, in all forms of racing.

Ductility Extent to which a material can sustain plastic deformation without rupture.

Elongation and reduction of area are common indices of ductility.

Duesie 1. Anything truly fine or outstanding.

2. Short for Duesenberg, touted by many to be the finest race car ever built. Also, doozie.

Dummy A mannequin, such as used in crash tests.

Dummy load An electrical test procedure to simulate actual operating conditions.

Dummy shaft A shaft used as an aid in the assembly or disassembly of a parts group.

Dump 1. To lower the front end.

2. To defeat a competitor in a drag race.

3. To damage or destroy a component, such as a transmission.

Dump station A facility where RV's and others clean their holding tanks.

Dump tubes Straight through exhaust headers.

Dumped and tubbed A vehicle that has been lowered in the front and fitted with oversize wheel wells,

Dune buggy A small, light-weight, off-road vehicle with little or no body work.

Duo-servo A drum brake design with increased stopping power due to the servo or self-energizing effect of the brake.

Duplex Simultaneous, independent, two-way transmission in both directions.

Dura Spark ignition An ignition system developed by Ford.

Durability The ability of a product and any of its components to perform the required functions in its intended service environment over its intended service life without unforeseen cost of maintenance and repair.

Duration The time interval between the first and last instants at which the instantaneous amplitude reaches a stated fraction of the peak pulse amplitude.

Dust detector A device for measuring the velocity, mass, charge, flight direction and number of dust particles striking the instrument.

Dust off A term often used for dust.

Dust Particulates which have a direct relation to a specific solar system body and which are usually found close to the surface of this body (e.g. Lunar, Martian or Cometary dust).

Dust shield 1. The upper portion of a shock absorber that surrounds the lower twin-tube unit.

2. A covering that protects an assembly from the elements.

Dust To overtake or defeat a competitor in a race.

DUT Device under test

Dutched A paint job finished with elaborate striping or flame painting.

Duty Cycle 1. The ratio of time on to time off in a recurring event.

2. The operating regime of a cell or battery including factors such as charge and discharge rates, depth of discharge, cycle length a length of time in the standby mode.

Dykes ring A compression piston ring having an "L" cross-sectional shape that provides sealing against the cylinder walls.

Dykum blue A blue dye used to colour metal.

Dynamic analysis Dynamic analysis is the analysis of mechanism's motions that result from forces. Dynamic simulation is more complex than kinematic analysis because the problem needs to be further defined and more data is needed to account for the forces. But dynamic analysis is often required to accurately simulate the actual motion of a mechanical system. Generally, kinematic analysis helps evaluate form, while dynamic analysis assists in analyzing function.

Dynamic balance A term used when wheels are balanced using equipment that spins the tires and wheels as an assembly

Dynamis Response Test A Ford term for a key on/engine running test in which a rapis acceleration is applied to the engine and the engine is checked for appropriate throttle movement, change in RPM, and change in the intake airflow.

Dynamo process The generation of an electric currents by the flow of an electrically conducting fluid through a magnetic field. For instance, the magnetic field originating inside the Earth is believed to come from a dynamo process involving the flow of molten iron in the Earth's hot core. The energy required by the current is obtained from the motion of the flow.

Dynamometer A machine on which a vehicle may be driven, simulating actual driving conditions for emissions and diagnostic purposes.

E

E Cell A cell which consists of a cathode made of silver and an anode made of gold. As current passes through is, a chemical reaction takes place, which causes the silver in the cathode to be depleted to the gold in the anode

Early adopter Early adopter is a person or organization who chooses to purchase or use relatively new technology before it is fully embraced by the mass market. Early adopters are therefore people or organizations who have a stronger need for the technology, a lower reluctance to use it, or the ability to overcome barriers to adopting it.

Early apex Getting to the inside of a turn sooner than usual.

Early fuel evaporation A system that evaporates fuel in the intake manifold with the use of heat from the exhaust, or an electrical element to improve the engine's cold-running characteristics and to reduce emissions.

Early supplier involvement Early supplier involvement is the process of getting the supplier involved early in the development process (when an item is being conceptualized, designed or specified) so that the supplier can make proactive suggestions to improve the design and reduce its cost vs. Providing reactive feedback once the design has been completed.

Early termination charge Additional fee charged when a lease is ended before the agreed-upon term. Often represents a substantial penalty.

Early termination Ending the lease before the contracted time. May be involuntary, due to theft or accident.

Earned premium The portion of the premium which is already paid from an expired policy term.

Earth An electrical connection to the earth frequently using a grid or rod(s).

Earth-sensor A light-sensitive diode which seeks the direction of Earth and then informs the attitude control system of a spacecraft.

Easy exit A system that provides for easy entrance and exit of the vehicle by moving the seat all the way back and down. Some systems may also move the steering wheel up and to full retract.

Eccentric 1. Not round, or concentric

2. An offset, such as a disc, used to convert rotary motion info reciprocation motion, as in a cam lobe.

3. The relationship of two round parts having different centers, a part that contains two round surfaces, not on the same center.

Eccentric bearing A bearing that is thicker at the crown by a few ten thousandths of an inch.

Eccentric Noncircular; elliptical (applied to an orbit).

Eccentric shaft 1. The crankshaft of a Wankle engine.

2. A shaft with eccentric offsets, such as a camshaft.

Eccentricity A condition whereby two or more round parts or holes do not share the same central axis.

Eccentricity of loading Distance between the actual line of action of compressive or tensile loads and the line of action that would produce a uniform stress over the cross section of the specimen.

Econobox A small compact car.

Economizer A device in a carburettor that reduces fuel consumption, particularly under heavy throttle.

Economy of motion Minimum effort required to do a task.

Econorail A dragster with a single, normally aspirated gasoline engine.

Eddy current Currents induced in conducting materials by varying magnetic fields; Sometimes used with dynamiters for fast accurate current changes.

Edge loading A condition that exists when two parts rub together at their edges.

Effective Bits A measure of a digital oscilloscope's ability to accurately reconstruct a sign wave signal's shape. This measurement compares the oscilloscope's actual erroe to that of a theoretical ideal digitizer

Effective diameter In metal bellows terminology it is the calculated diameter where the pressure penetrates between the metal plates. This number is used to determine the hydraulic balance diameter of the seal face.

Effectivity An indicator in a product structure which specifies the versions at which a component part is used. These indicators generally specify a range of either dates, serial numbers, or build lots. Effectivity indicators are typically considered as 'conditions' on the parent-child relationships in a product structure.

Effectivity date Effectivity date is the date from which an intended engineering change is to come into effect (new product configuration) or the past configuration of the product is to go out of effect.

Efficiency 1. The ratio of total output power to total input power, expressed as a percentage, under specified conditions. For power supplies efficiency is generally measured at full load with nominal line conditions.

2. The ratio of the output of a secondary cell or battery on discharge to the input required to restore it to the initial state of charge under specified conditions.

Elapsed time (ET) The time it takes a vehicle to cover a given distance, usually from a standing start, recorded to the thousandths of a second.

Elastic hysteresis Difference between strain energy required to generate a given stress in a material and elastic energy at that stress. It is the energy dissipated as heat in a material in one cycle of dynamic testing. Elastic hysteresis divided by elastic deformation energy is equal to damping capacity.

Elastic limit Greatest stress that can be applied to a material without causing permanent deformation. For metals and other materials that have a significant straight line portion in their stress/strain diagram, elastic limit is approximately equal to proportional limit. For materials that do not exhibit a significant proportional limit, elastic limit is an arbitrary approximation (the apparent elastic limit).

Elastic limit, apparent Arbitrary approximation of the elastic limit of materials that do not have a significant straight line portion on a stress/strain diagram. It is equal to the stress at which the rate of strain is 50% greater than at zero stress. It is the stress at the point of tangency between the stress- Elastic Hysteresis strain curve and the line having a slope, with respect to the stress axis, 50% greater than the slope of the curve at the origin.

Elastic range The stressed part retains its memory and returns to its original shape.

Elasticity Ability of a material to return to its original shape when load causing deformation is removed.

Elastomer A rubber like material that, when compressed and then released will return to 90% of its original shape in less than five seconds.

Electric assist choke A choke containing a small electric heating element to warm the choke spring.

Electric automatic transmission An automatic transmission with computer controlled pressure regulation that gives precise control of shift scheduling and rapid response to changing conditions.

Electrochromatic Mirror This is a rear-view mirror that contains light sensitive properties, so that it can darken to avoid glare from following vehicles' headlamps at night.

Electrochromic mirror A mirror that automatically adjusts the amount of reflectance based on the intensity of glare.

Electrode 1. A component of the electrical circuit that terminates at a gap across which current must arc.

2. The contacts of a semiconductor material that emits, collects or controls the movement of electrons or holes.

3. A rod used in welding.

Electrohydraulic pressure actuator A valve that will provide a continuous adjustment of fuel pressure in certain fuel-injection systems.

Electrohydraulic unit The microprocessor and hydraulic units are combined in one unit.

Electrolysis 1. Charging a material's chemical composition by applying an electromotive force and having an electrical current flow through it.

2. The process where an electric current is passed through an electrolytic solution, causing a chemical reaction.

Electrolyte 1. A substance in which the condition of electricity is accompanied by chemical action.

2. Any substance which, in solution, is separated into ions and is made capable of conducting electric current.

3. The sulphuric acid solution of a lead-acid battery.

4. A chemical compound that conducts ions from one electrode the other inside a fuel cell.

Electrolytic cell A simple battery consisting of the container, two electrodes, and the electrolyte.

Electromagnet A device consisting of a ferromagnetic core and a coil that produces appreciable magnetic effects only when an electric current exists in the coil.

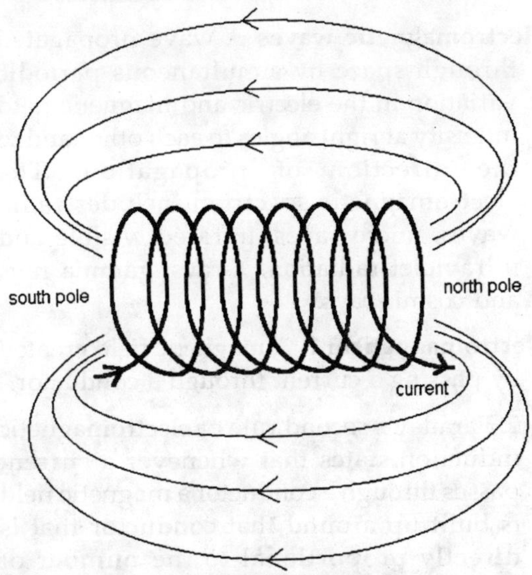

Fig. Electromagnet

Electromagnetic clutch an electrically controlled device used to start and stop the compressor action.

Electromagnetic gauges A type of dash instrument gauge that uses a small electromagnetic coil for gauge needle movement.

Electromagnetic induction 1. The induction of a voltage into a conductor by passing the conductor through a magnetic field or passing the magnetic field across the conductor.

2. Discovered by Michael Faraday in 1831 Faraday's first rule of electromagnetic induction states that whenever a conductor passes through a magnetic field a voltage is induced into that conductor that is directly proportional to the strength of the magnetic field and the speed of it's movement.

Electromagnetic interference (EMI) Any electronic disturbance that interrupts, obstructs, or otherwise impairs the performance of electronic equipment. For power supplies, unwanted energy, generally emitted from switching power supplies, which may be conducted or radiated.

Electromagnetic waves A wave propagated through space by simultaneous periodic variation in the electric and magnetic field intensity at right angles to each other and to the direction of propagation. The electromagnetic spectrum includes radio waves, microwaves, infrared, visible and ultraviolet radiation, X-rays, gamma rays and cosmic rays.

Electromagnetism 1. A magnetic field created by passing a current through a conductor.

2. Faraday's second rule of electromagnetic induction states that whenever a current passes through a conductor a magnetic field is built up around that conductor that is directly proportional to the number of electrons (the amount of amperage) flowing in the conductor.

Electromotive force (emf) Difference of potential produced by sources of electrical energy which can be used to drive currents through external circuits. Unit is the volt.

Electron 1. An element of matter that surrounds the nucleus and helps determine the chemical and or electrical properties of an atom.

2. A negatively charged practical 1/1800 the mass of a proton orbiting around the nucleus in orbital paths called shells.

Electron optics Electronics that apply the behaviour of moving electrons under the influence of electrostatic or electromagnetic forces to devices or equipment.

Electron theory The theory that electrons flows from an area of negative potential to an area of positive potential through a conductor.

Electronic air suspension (EAS) A suspension system having provisions to adjust to road and/or load conditions to ensure a comfortable ride. May also include automatic level control.

Electronic brake control module (EBCM) A system having a monitor at each wheel to sense conditions and feed an electrical impulse into an onboard computer to reduce sideways skidding during rapid braking action.

Electronic Data Interchange (EDI) (EDI) links- Information-system linkages, based on communication protocols and document formats, which permit inter-company computer-to-computer communications. EDI links not only speed communication, but also eliminate re-keying of information and reduces the opportunity to introduce errors. A typical EDI application might speed information exchange between a customer and supplier company for purchase orders, invoices, or other transactions. EDI communications are often facilitated through "electronic mailbox" systems on third-party value-added networks.

Electronic design automation Electronic design automation (eda) consists of hardware and software tools to aid in the design and development of electronic products through design capture, simulation, synthesis, verification, analysis, and testing.

Electronic fuel injection (EFI) There are different types of fuel injection systems, but all have replaced the carburettor for introducing fuel into the engine with much more precise control and timing. This has many advantages in fuel economy terms, as well as providing better engine performance and reliability, with lower exhaust emissions.

Electronic systems design automation Electronic systems design automation is a set of graphical front-end tools that allow designers to use pictures rather than words to describe and analyze their creations. These tools can use hdl's as an interchange format rather than a design medium and allow for higher degrees of abstraction over traditional schematic capture or waveform display programs.

Electronic systems level Electronic systems level is a higher level abstraction for the design of electronic products which will improve design productivity with the design of ever larger and more complex electronic systems. This is the third generation in design methodologies and tools (gate-level, register transfer level, and electronic systems level). Key elements of ESL include behavioural synthesis, integration between the behavioural level and the architecture level, and hardware/software codesign and coverification.

Electroplating A process for depositing metal on a conductive surface that is made by the cathode in an electrolytic bath containing dissolved salts of the metal being deposited.

Electrostatic A kind of electricity where the charges are not moving.

Electrostatic discharge (ESD) The flow of current that results when objects having a static charge come into a close enough proximity to discharge.

Electrostatic field The electrical field of force surrounding an object that has an electrical charge.

Electrostatic shield A shield that protects a device or circuit from electrostatic energy, but not necessarily from electromagnetic energy.

Element 1. A distinct functioning device in an electrical circuit.

2. The negative and positive electrodes together with the separators of single cell. It is used almost exclusively in describing lead-acid cell and batteries.

Elephant foot A valve-adjusting screw having a ball that swivels when it contacts the valve stem.

Elevation The angular measure of the height of an object above the horizon; with azimuth, one of the coordinates defining celestial location and sometimes used in tracking spacecraft.

Eliminations A series of matches between two cars at a time, the winner advancing to the net race.

Eliminator The fastest drag car in its class.

Elliot axle A solid-bar front axle on which the ends span the steering knuckle.

Elliptical spring A term used for leaf spring.

Elongation Measure of the ductility of a material determined in a tension test. It is the increase in gage length (measured after rupture) divided by original gage length. Higher elongation indicates higher ductility. Elongation cannot be used to predict behaviour of materials subjected to sudden or repeated loading.

Embedability The ability of a bearing to permit small dirt particles to sink into the bearing surface, thereby avoiding certain crankshaft scratches.

Embossed gasket A shim-type head gasket having a raised surface.

Embrittlement A condition whereby a part becomes hard and brittle due to excessive flexing.

Emergency brake system A brake system used for stopping a vehicle in the event of a malfunction in the means of operation and control of the service brake system.

Emergency message systems (EMS) Any electronic unit that can control either fuel delivery or ignition timing or both. It is usually used to describe a stand-alone, aftermarket computer.

Emery paper An abrasive-coated paper used for fine finishing. Also called emery cloth

Emi filtering Process or network of circuit elements to reduce electromagnetic interference emitted from or received by an electronic device.

Emission (electrical) The release of free electrons from the surface or a material.

Emission standards Regulatory standards that govern the amount of a given pollutant that can be discharged into the air from a given source.

Emission standards The federal-government-established emission and pollutant standards for all motor vehicles because of environmental damage in the part.

Emissions control A term used for emission controls.

Emissions Unwanted, harmful chemicals and chemical compounds that are released into the atmosphere from a vehicle, especially from the tailpipe, crankcase, and fuel tank including unburned hydrocarbons, carbon monoxides of nitrogen, particulates, and sulphur.

Empathic design Empathic design is based on observation – watching customers/ consumers use products or services. But unlike focus groups, usability laboratories, and other contexts of traditional market research, this observation is conducted in the customer's own environment in the course of normal, everyday routines. This approach enables the researcher to observe and develop information on customer needs that will drive design that is not accessible through other observation-oriented research methods.

Emulation The process by which a device under development and its native software is prototyped before its manufacture.

Emulsifiable chemical Any chemical that will mix with water.

Emulsion A fluid substance containing one liquid disbursed and suspended in another, rather than dissolved.

Emulsion tube A passage in a carburettor where air is mixed with fuel.

EMW Emission maintenance warning

Enable criteria The exact conditions necessary for a diagnostic test or OBD-II monitor to run.

Enable To switch a computer device or facility so it can operate.

Enamel A free-flowing clear or pigmented varnish, treated oil, or other form of organic coating that usually dries to a hard, glossy, or semi gloss finish.

Enamelled wire Wire insulated with a thin, baked-on varnish enamel, used in coils to allow the maximum number of turns in a given space.

Enclosure A housing for an electronic device designed to separate and protect both the internal componentry and the outside environment.

Enclosure base The stationary, approximately cylindrical structure below the azimuth track that transfers all loads from the enclosure to the ground. It includes an approximately circular ring beam as well as the support elements for the utility transfer systems.

Enclosure control system The computer-based system that actively positions and controls the various moving components of the enclosure, including carousel azimuth rotation, shutter operation, sun shade deployment, control of the ventilation gates, and control of the exterior skin temperature via the thermal system sun shade deployment, and control of the ventilation gates.

Encoder An electromagnetic transducer used to produce digital data (code) indicating angular or linear position.

End clearance The distance a shaft moves longitudinally as the crankshaft in an engine, Also known as end play.

End gap The distance between the ends of a piston ring.

End gas The last part of an air/fuel mixture to burn during combustion.

End of lease payment In an open-end lease, the amount the lessee pays at the end of the lease; the difference between the vehicle's residual value, which was used to calculate lease payments, and the actual market value.

End play 1. The total amount of axial play in an automatic transmission. Endplay is typically measured at the input shaft.

2.A term used for end clearance.

End-item A deliverable item which is formally accepted by the acquirer in accordance with requirements of a detail specification.

Endo To flip a vehicle end over end.

End-of-life End-of-life is the term applied to products or components that are being retired from the market because of technology obsolescence or rapidly declining demand.

Endothermic Describes a chemical reaction that absorbs or requires energy (usually in the form of heat).

Endplay The measured amount of axial movement in a shaft

Endurance Alternate term for fatigue limit.

Endurance limit The maximum stress that a metal can withstand without failure, after a specified number of cycles of stress.

Enduro A race of 6 to 24 hours with emphasis on endurance and reliability rather than speed.

Energetic particle Particles that can penetrate outer surfaces of spacecraft. For electrons, this is typically above 100 keV, while for protons and other ions this is above 1 Mev. Neutrons, gamma-rays and X-rays are also considered energetic particles in this context.

Energized Having the electric current or source turned on

Energy content Amount of energy for a given weight of fuel.

Energy density Amount of potential energy in a given measurement of fuel.

Energy The quantity of work a system or substance is capable of doing, usually measured in British thermal units (Btu) or Joules (J).

Eng An acronym for engine.

Engage The mechanical or automatic coupling of two members, like the driving flywheel and pressure plate, to rotate and drive the driven disc of a clutch.

Engagement chatter A shaking, shuddering action that takes place as the driven disc makes contact with the deriving members caused by a rapid grip and slip action.

Engine A device that burns fuel to produce mechanical power; to convert heat energy into mechanical energy.

Engine rebuild To perform an extensive engine repair including machining, re-boring, and honing to factory-stock specifications.

Engine running test A Ford diagnostic term for a section of the EEC-IV self-test that tests the input and control functions under actual operating conditions

Engineering change control Engineering change control is the process and procedures that manage how changes are proposed, reviewed, and approved and incorporated into a product and its associated data items. Change control is a part of an overall configuration management methodology and uses review and release processes to enforce compliance with company change policies.

Engineering change notice / engineering change order Engineering change notice / engineering change order (eco) are formal documents notifying selected persons of proposed, pending, or accomplished changes.

Engineering decision traceability Significant engineering decisions shall be traceable to the system engineering process activities on which they were based.

Engineering GUI The engineering GUI provides control and status information to a user for a particular subsystem. An engineering GUI may be instantiated on a remote computer via a network connection. More than one engineering GUI may be simultaneously opened.

Engineering management The management of the engineering and technical effort to transform a conceptual requirement into an operational system. It includes the system performance parameters and preferred system configuration to satisfy the requirement, the planning and control of technical program tasks, integration of the engineering specialties, and the management of a totally integrated effort of design engineering, computer software engineering, test engineering, safety engineering, security engineering, logistics engineering, production engineering, and specialty engineering (EMC, environmental, etc.,) to meet cost, technical performance and schedule objectives.

Note: 'Engineering Management' is sometimes identified by various institutions and organizations as; 'Program(me) management', 'Quality Management', 'Quality Plan', 'Quality system', 'Project Management', 'Quality assurance', or any combination of the above terms. These are to be considered synonymous with the term 'Engineering Management'.

Engineering plastics Thermoset and thermoplastic materials, such as polycarbonate, ABS, and nylon. Their characteristics and properties enable them to withstand mechanical loads, such as tension, impact, flexing, vibration, and friction, combined with temperature changes, making them suitable for application in structural and load-bearing product design elements.

Engineering specialty integration The timely and appropriate intermeshing of engineering efforts and disciplines such as reliability, maintainability, logistics engineering, human factors, safety, value engineering, computer software, standardization, transportability, etc., to insure their influence on system design.

Engineering stress Load applied to a specimen in a tension or compression test divided by the cross-sectional area of the specimen. The change in cross-sectional area that occurs with increases and decreases in applied load, is disregarded in computing engineering stress. It is also called conventional stress.

Engineering Time A period of time when the equipment is in condition to perform its intended function, but is operated to conduct engineering experiments.

Engine-management system An electronic device that monitors, adjusts, and regulates the ignition and fuel-injection systems to maintain engine control under varying operating conditions.

Enhanced quality function deployment Enhanced quality function deployment is a broader framework that applies a system perspective recognizing the need to decompose more complex products into subsystems and assemblies with supporting deployment matrices and concept selection matrices.

Enleanment The act of reducing fuel delivery to the air-fuel mixture to create a leaner mixture.

Enrichment The act of adding fuel delivery to the air-fuel mixture to create a richer mixture

Enterprise A portion of a company which is related by a common interest in a product or group of products. An enterprise may also logically include a network of subcontractors involved in the common product. PDM systems are often applied across an enterprise.

Enterprise resource planning (ERP) system An extension of MRP II software. ERP systems typically claim the ability to achieve tighter (or "seamless") integration between a greater variety of functional areas, including materials management, supply-chain management, production, sales and marketing, distribution, finance, field service, and human resources. They also provide information linkages to help companies monitor and control activities in geographically dispersed operations; and, in a fully deployed mode, ERP systems facilitate capture of transactional data into data warehouses to support executive decision-making systems.

Epoch An instant in time that is arbitrarily selected as a point of reference, e.g. for a set of orbital elements.

Epoxy A plastic compound that is often used to repair cracks and small holes in metal.

Epoxy cement A durable thermosetting adhesive.

Equal length headers An exhaust system having equal length tubes from each cylinder to the collector.

Equalizer 1. A device in the brake-cable system that prevents one side from being applied before the other. 2. The tube that balances pressure between two exhaust pipes.

Equalizer line A small-bore line used to provide a balance of pressure from one point to another, as in a thermostatic expansion valve.

Equator An imaginary circle around a body which is everywhere equidistant from the poles, defining the boundary between the northern and southern hemispheres.

Equatorial orbit An orbit in the plane of the equator.

Equipment and Tools Institute (ETI) An organization of equipment and tool manufacturers, jobbers, and wholesalers.

Equivalent circuit An electrical circuit that models the fundamental properties of a device or circuit.

Equivalent load An electrical circuit that models the fundamental properties of a load.

Ergonomics Ergonomics is the science of designing products and work to be consistent with the capabilities and limitations of the human body.

Error amplifier An operational amplifier, or differential amplifier, in a control loop that produces an error signal whenever a sensed output differs from a reference voltage.

Error signal The output voltage of an error amplifier produced by the difference between the reference and the input signal times the gain of the amplifier.

Error voltage The output voltage of the error amplifier in a control loop.

Ethnographic studies A qualitative method of researching customer needs based on studying the anthropology or culture of the user. This method involves spending time in the field observing customers and their environment to better understand their lifestyle or culture as a basis for understanding their needs for a new product. A deep understanding of your customer can lead to fundamental insights that impact product design, feature sets, product positioning, marketing communications, advertising execution, etc.

Evacuation A physical process for extracting gas from a vessel. Evacuation techniques fall into two common categories: gas transfer pumps evacuate gas via an inlet and outlet to the outside atmosphere. Gas-binding vacuum pumps operate in self-contained fashion within the vessel, by binding gas molecules to their interior surfaces.

Evaluation The process of determining whether an item or activity meets specified criteria.

Evaporator core The tube and fin assembly located inside the evaporator housing, where refrigerant fluid picks up heat when it changes into a vapour.

Evaporator equalizer valves-in-receiver A term used for valves-in-receiver.

Evaporator fan A fan used to move air across the evaporator and heater core.

Evaporator housing The case that contains the evaporator core, diverter doors, duct outlets, and blower mounting arrangement.

Evaporator pressure-control valve A term used for evaporator control valve.

Evaporator temperature regulator A temperature-regulated device used by early Chrysler Air-Temp systems to control the evaporator pressure.

Exciter Ring A toothed or notched iron or steel disk, which is the moveable part of a wheel speed sensor.

Exclusion Any cause, condition, or property which are not covered by the policy.

Expansion tank 1. A coolant-recovery tank.

2. An auxiliary tank, usually connected to the inlet tank or a radiator, that provides additional storage space for heated coolant.

Expansion tube A metering device used at the inlet of some evaporators to control the flow of liquid refrigerant into the evaporator core.

Expansion valve A term used for thermostatic expansion valve.

Expectancy theory The view that our effort will be greatest when we expect that we can perform the task at hand and that we expect to obtain rewards for our performance.

Experience curve Experience curve (also known as a learning curve) is a mathematical model that relates the cost per unit (or labor time per unit) to the cumulative number of units produced in an exponentially decreasing manner.

Extreme programming Extreme programming (XP) is one of the more popular lightweight, or agile development methods. In general, XP structures the "four basic activities of software development ... Coding, testing, listening, and designing." XP structures coding based on the concepts of pair-programming and test-development. XP structures the testing activity by requiring automated tests that the team runs every day, several times a day. XP structures the listening activity through pair-programming and by requiring that the customer be part of the team and be on-site. Lastly, XP structures the designing activity by encouraging developers to use test-first development: define a test, then code until the test passes, then proceed to the next test. There is no big-design-up-front stage in an XP project.

Extremely hazardous substance (EHS) Any substance or the by-product of any substance that is poisonous or contains carcinogens considered unsafe, in any concentration, for humans, animals, or plant life.

Extrusion Permanent displacement of a portion of the O-ring into a gap, under the action of fluid pressure.

Eye of the impeller The center of the impeller, where the fluid enters.

Eyeball 1. To make a visual estimation or determination.

2. A voltage monitoring circuit

Eyes The light beam that operates an electronic, race-timing system.

Eyewash fountain An emergency water fountain that directs water to the eyes for flushing and cleaning.

F

Fabric cord A cord-like material used for strength and reinforcement in belts and tires.

Face angle The angle of the valve face.

Face combination The materials chosen for the lapped seal faces. Usually a grade of carbon graphite running on a hard face material.

Face feed The application of a filler metal to a joint during brazing and soldering.

Face flatness Measured by an optical flat and a monochromatic light. The measurement is read in helium light bands (.0000116 inches or 0,3 microns).

Face lubrication The fluid or vapour that sometimes exists between lapped mechanical seal faces.

Face pressure The sum of all the loads on the seal face including the spring load, hydraulic load and shaft axial thrust, divided by the area of the seal face. This face load is reduced by friction between the sliding elastomer and the shaft or sleeve.

Face shield A clear-plastic shield that protects the entire face from outside elements.

Face to face seals Two seals running against a common seal face. The barrier fluid pressure is always lower than stuffing box pressure.

Faceing surface The mating surface of a member that is in contact or in close proximity with another member to which it is to be joined.

Facility system Any control system provided as part of ATST beyond those systems specific to a given instrument. This includes the telescope control systems and any facility instrumentation support (filter wheels, polarization units, etc.).

Factor The mathematical evaluation used by NHRA to estimate true horsepower and weight, and assign production cars to specific stock classes.

Factory equipment In used vehicles: the combination of original standard equipment and production options that make up the equipment of a used vehicle. May also be referred to as "base" equipment.

Factory experimental A type of drag-racing car developed by Dodge/Plymouth and Ford/Mercury in the 1960s using their largest engines in light vehicles, a forerunner of the pro-stock class.

Factory recall A vehicle manufacturer's effort to correct a defect discovered after the vehicle has been delivered to the customer, often several years later.

Factory specs Factory specifications; the manufacturer's dimensions, clearances, and tolerances.

Factory team A racing team sponsored and supported by a vehicle manufacturer.

Factory tool A tool designed by a vehicle manufacturer for a specific application.

Fad car A T-bucket with aq bobbed pickup bed.

Fad T A term used for fad car.

Fade A term often used when referring to brake fade.

Fadeaway fenders A front-fender design that flowed back and blended into the front doors of the vehicle; a popular style of the 1940s.

Fahrenheit Temperature scale and unit of temperature (°F). Named for German physicist Gabriel Daniel Fahrenheit (1686-1736) who was the first to use mercury as a thermometric fluid in 1714.

Fail record These records store information about operating conditions when a DTC is stored. The PCM has the ability to store multiple Fail Records, and it also has the ability to update Fail Records.

Fail Safe A computer back-up designed to enable minimal system operation in the event of a sensor or other failure.

Failure A deficiency, defect, nonperformance or nonconformance with specified requirements. An item of equipment has suffered a failure when it is no longer capable of fulfilling one or more of its intended functions. Note that an item does not need to be completely unable to function to have suffered a failure.

Failure analysis Failure analysis is a collection of techniques to determine the root cause of a component or process defect or failure.

Failure analysis Failure Analysis is a collection of techniques to determine the root cause of a component or process defect or failure.

Failure code A term used for trouble code.

Failure in time Failure in time - a reliability measure usually expressed in failures per 10 to the 9th power hours.

Failure modes and effects analysis Failure modes and effects analysis is a procedure in which each potential failure mode in every sub-item of an item is analyzed to determine its effect on other sub-items and on the required function of the item. It is used to identify potential failure modes and their associated causes/mechanisms, consider risks of these failure modes, and identify mitigating actions to reduce the probability or impact of the failure.

Failure modes, Effects and criticality analysis Failure Modes, Effects and Criticality Analysis is a procedure that is performed after a failure mode and effects analysis to classify each potential failure effect according to its severity and probability of occurrence.

Failure Records Failure records data is a GM enhancement of the OBD II freeze frame feature

Failure reporting and corrective action system Failure Reporting and Corrective Action System (FRACAS) is a closed-loop system to capture reports of failure from customers or service technicians in the field or from the factory, analyze these reports, detect trends or problems, and use this analysis to take corrective action in the design, component selection, supplier selection, manufacturing process, or operating manual of the product. Features of a FRACAS system include a database manager, tracking system for document controls, user definable reports which allow selection of data elements and sort options, and search functions.

Failure The total cessation of function of a system or device, such as coil failure.

Fairing A structure whose main function is to streamline and smooth the surface of an aircraft or space vehicle.

Fall time The switching time required for the signal voltage to decrease from logic high to logic low.

Falling Edge The part of a waveform showing the voltage decreasing

False air Air in a fuel-injection system in excess of that required for combustion.

False guide A valve guide used to replace a worn integral-valve guide.

Falt Codes A set of numbers repersenting the result of an onbord diagnostic test.

FAM Feedback actuator motor

Family buggy 1. The family car.

2. A vehicle more than ten years old.

Fan 1. A device having two or more blades attached to the shaft of a motor.

2. A device mounted in the heater/air-conditioner duct that causes air to pas over the heater core and evaporator.

3. A device having four or more blades, mounted on the water pump, that causes air to pass through the radiator and condenser.

4. The spray pattern of a paint spray gun.

5. The incorrect technique of applying paint by waving a spray gun back and forth with the wrist.

Fan belt A flexible V-,or flat poly-grove-type drive belt that transfers power from the crankshaft pulley to the water pump and/ or accessories, such as the alternator.

Fan blade In an engine cooling-system fan, four to six wings on the fan, usually spaced unevenly to reduce vibration and noise.

Fan clutch A device installed between the water-pump pulley and fan of an engine-driven fan that is sensitive to engine speed and underhood temperature.

Fan cooled A method of forced-air cooling used to maintain design temperatures.

Fan hub The mounting surface for the fan.

Fan shroud Plastic or metal housing inside which the fan rotates; on certain vehicles, this allows the fan to pull more air past radiator finned tubing and prevents air recirculation.

Fanning the brakes A term used for brake fanning.

Fanning The use of pressurized air through a spray gun to facilitate drying, a practice not recommended.

FAP Fast Achromatic Polarimeter

FAQT An acronym for Federation of Automotive Qualified Technicians.

FAR Federal Acquisition Regulations

Farad (F) The basic unit for capacitance. A capacitor has a value of one farad when it has stored one coulomb of charge at one volt. Named for an English physicist Michael Faraday (1791-1867)

Faraday One gram equivalent weight of matter is chemically altered at each electrode of a cell for (approximately) each 96,500 international coulombs, or one Faraday, of electricity passed through the electrolyte.

Fast flush The use of a special machine to clean the cooling system by circulating a cleaning solution.

Fast Fourier transfer (FFT) An analysis algorithm - given a finite set of data points, the FFT expresses the data in terms of its component frequencies.

Fast idle The higher speed, 1,100 to 1,500 rpm, at which an engine idles during warming-up, when first started.

Fast overdrive A planetary gear set operating with the planetary carrier as input, the sun gear as output, and the ring gear held.

Fast reverse A planetary gear set operating with the planetary carrier held; the ring gear is input, and the sun gear rotates in the opposite direction.

Fastback An autobody style having a roof line that extends in a single, simple curve from the windshield to the rear bumper.

Fast-idle cam A cam-shaped lever on the carburettor that provides fast-idle action when the engine is cold.

Fast-idle screw A screw in the carburettor linkage to adjust fast-idle speed.

Fast-idle solenoid An electro-mechanical device on the carburettor for adjustment of the fast-idle speed.

FAT a acronym for File Allocation Table. A table that the operating system uses to locate files on a disk.

Fat fenders Bulbous fenders, as on vehicles in the 1930s and 1940s.

Fatigue failure Metal failure due to repeated stress so that the character of the metal is altered and it cracks. This is a condition that frequently causes engine bearing failure due to excessive engine idling or slow engine-idling speed.

Fatigue life Number of cycles of fluctuating stress and strain of a specified nature that a material will sustain before failure occurs. Fatigue life is a function of the magnitude of the fluctuating stress, geometry of the specimen and test conditions.

Fatigue limit Maximum fluctuating stress a material can endure for an infinite number of cycles. It is usually determined from an S-N diagram and is equal to the stress corresponding to the asymptote of the locus of points corresponding to the fatigue life of a number of fatigue test specimens. An alternate term is endurance limit

Fatigue notch factor Ratio of fatigue strength of a specimen with no stress concentration to fatigue strength of a specimen with a notch. Fatigue notch factor is usually lower than the theoretical stress concentration factor because of stress relief due to plastic deformation. An alternate term is strength reduction ratio.

Fatigue Permanent structural change that occurs in a material subjected to fluctuating stress and strain. However, in the case of glass, fatigue is determined by long-term static testing and is analogueous to stress rupture in other materials. In general, fatigue failure can occur with stress levels below the elastic limit.

Fatigue ratio Ratio of fatigue strength or fatigue limit to tensile strength. For many materials, fatigue ratio may be used to estimate fatigue properties from data obtained in tension tests.

Fatigue strength Magnitude of fluctuating stress required to cause failure in a fatigue test specimen after a specified number of cycles of loading. Usually determined directly from the S-N diagram.

Fatigue test A method for determining the behaviour of materials under fluctuating loads. A specified mean load (which may be zero) and an alternating load are applied to a specimen and the number of cycles required to produce failure (fatigue life) is recorded. Generally, the test is repeated with identical specimens and various fluctuating loads. Loads may be applied axially, in torsion, or in flexure. Depending on amplitude of the mean and cyclic load, net stress in the specimen may be in one direction through the loading cycle, or may reverse direction. Data from fatigue testing often are presented in an S-N diagram which is a plot of the number of cycles required to cause failure in a specimen against the amplitude of the cyclical stress developed. The cyclical stress represented may be stress amplitude, maximum stress or minimum stress. Each curve in the diagram represents a constant mean stress. Most fatigue tests are conducted in flexure, rotating beam, or vibratory type machines.

Fatigue The tendency of a material to break under conditions of repeated stressing considerably below its tensile strength.

Fault A crack or break in the crust of a planet along which slippage or movement can take place.

Fault mode input current Input current drawn by a power supply with the output short circuited.

Fault protection Algorithms, which reside in a spacecraft's subsystems, that insure the ability of the spacecraft to both prevent a mishap and to re-establish contact with Earth if a mishap occurs and contact is interrupted.

Fault tolerance The ability of a system to execute tasks regardless of strategic components failure.

Fault tolerant A high level of redundancy to provide continued operation following specified failure

Fault tree analysis Fault tree analysis is a top-down, hierarchical analysis of faults to identify the various fault mechanisms and their cause. It graphically describes the cause and effect relationships that result in major failures. The fault or major failure being analyzed is identified as the "top event." All of the possible causes of the top event are identified in a tree using "or" nodes for independent causes and "and" nodes for multiple causes that must exist concurrently for a failure to occur.

Feasibility Capable of being completed to meet goals. The feasibility of a proposed new product has two dimensions: a) technical feasibility (i.e., will the product work?) And b) financial feasibility (i.e., will the product make an adequate return on investment for the enterprise?)

Featheredge The technique of blending the repair of a damaged area into the undamaged area, maintaining the original surface texture and sheen.

Feathering The technique of modulating the throttle lightly and smoothly for precise, controlled changes in engine speed.

Featherweight leaf spring A fibre composite spring.

Feature 1. Features are elements of the product that provide a distinctive benefit to the customer and are often highlighted in describing the benefits of the product to the customer. Features are the differentiating functionality of a product. This functionality may not be available in other products, or it may not be available with the same quality characteristics.

2. Features are geometric entities that have meaning in the definition and manufacture of a product. Examples of features are through-holes, bosses, bends, chamfers, slots, etc.

Feature car A vehicle displayed at a car show for appearance money and not trophy competition.

Features technology Features technology - a variation of group technology with a focus on coding and classifying based on part features.

Feed forward A control technique whereby the line regulation of a power supply is improved by directly sensing the input voltage.

Feed holes The holes to supply coolant or oil to an engine.

Feed The power supply of hot wire

Feed through A plated-through hole in a printed circuit board which electrically connects a trace a top of the beard with a trace on the bottom side.

Feedback 1. A principle of fuel-system design wherein a signal from an oxygen sensor in the exhaust system is used to give a computer the input it needs to properly regulate the carburettor or fuel-injection system in order to maintain a nearly perfect air/fuel ratio.

2 A signal to a computer that reports on the position of a component, as an EGR valve.

3. The return of a portion of the output (actuator) to the input (computer), used to maintain an output device within predetermined limits

4. the reverse flow of electrical current through a circuit or electrical unit that should not normally be operating, this

feedback current (reverse-bias current flow) is most often caused by a poor ground connection for the sane normally operating circuit

Feedback carburettor A carburettor that controls the air/fuel mixture according to commands from the engine control computer, typically through the operation of a duty solenoid.

Feedback The process of returning part of the output signal of a system to its input.

Feeler Gauge A blade, usually metal of precise predetermined thickness, used to measure the clearance between two parts

Felt dust seal 1. An engine seal made of felt, usually used on the front crankshaft pulley.

2. A compressor-shaft seal made of felt, usually found between the seal face and armature of the clutch.

Felt Natural felts are produced by compressed wool, hair, wool/hair, or synthetic fibres, yielding a wide range of densities and permeabilities of consistent density, pore size, and mesh geometry so that performance is reasonably predictable.

Female The universal designation of a part into which a mating (male) part fits.

Fender cover A protective cover placed on the fender when a mechanic works on an engine, preventing damage to the finish.

Fenderside A narrow-bed pickup truck.

Ferguson Formula (FF) A four-wheel drive system developed in the 1960s, a forerunner of the AWD and 4WD systems in high-performance cars today.

Ferret Satellite using electromagnetic surveillance techniques.

Ferrite A ceramic material that exhibits low loss at high frequencies, and which contains iron oxide mixed (with oxides or carbonates of one or more metals such as manganese, zinc, nickel or magnesium.

Ferromagnetic Material Any material with an Iron-like (Ferrous) magnetic characteristic, such as high permeability to magnetic flux lines.

Ferroresonance The property of a transformer design in which the transformer contains two separate magnetic paths with limited coupling between them. The output contains a resonating tank circuit and draws power from the primary to replace power delivered to the load.

Ferroresonant transformer An AC regulator designed to operate at a given input frequency and which uses a resonant circuit to achieve regulation. Although the regulation is imprecise, the device itself is simple, very rugged and reliable. Typically operating at 50Hz or 60Hz, these "iron-core" (actually laminated silicon steel) transformers incorporate a capacitor in one of the secondaries that resonates with the inductance of the transformer. An excessively high input AC voltage which would otherwise result in excessive secondary voltage will cause the resonant circuit voltage to ring high enough to locally saturate the transformer core. Once saturated, the primary winding magnetic flux no longer links the secondary winding and the secondary voltage is limited.

Fibre optics A data-transmission medium made of tiny threads of plastic (polymethyl methacrylate) that transmit huge amounts of data via light waves, at the speed of light. Also used in some automotive applications for a small light source for cigarette lighter, ashtray, or heater controls.

Field 1. Generally, any area being affected by some physical force such as

2. Electrical force in the space around electrically charged particles.

3. The lines of force in a natural or man-made magnet between its north and south poles.

Field coil The coil, or winding, around the field magnets or pole pieces of a motor or generator. Also known as field winding.

Field density The density of the magnetic field, measured in the number of lines of force per unit area, dependent upon the strength of the field element, the number of turns of wire, and the size and characteristics of the pole piece.

Field effect transistor (FET) Transistor in which the resistance of the current path from source to drain is modulated by applying a transverse electric field between two electrodes.

Field relay a relay that connects the alternator to the battery when alternator voltage is greater than battery voltage and disconnects it when battery voltage is greater than alternator voltage.

Field replaceable unit A collection of hardware or software that is installed or removed from a product as a single serviceable entity. The composition of product fru's is determined by the integrated product team.

Field terminal The electrical terminal connecting the field coil to the voltage regulator.

Field testing Field testing is the testing a product in the actual context in which it will be used, as opposed to laboratory testing; or testing the product in its operating environment.

Field winding A term often used for field coil.

Field-effect transistor (FET) A 3-terminal semiconductor device where current flows from source to drain due to a conducting channel formed by a voltage field between the gate and the source.

Field-frame assembly The round, soft iron frame of a generator or motor into which the field coils are assembled.

Final drive ratio A fluid coupling consists of two fan-like impellers in a sealed, oil-filled housing. The input "fan" churns the oil, and the churning oil, in turn, twirls the output "fan." Such a coupling allows some speed difference between its input and output shafts. The automatic transmission's torque converter is based on the fluid coupling principle.

Final drive ratio The reduction ratio of the transmission gear set furthest from the engine. In other words, the ratio of the number of rotations of the drive shaft for one rotation of a wheel. In general, a low final drive ratio results in better fuel efficiency, and higher final drive ratio results in better performance.

Final drive The pinion, ring, and differential gears that provide power to the drive wheels.

Finished-goods turn rate A measure of asset management that typically is calculated by dividing the value of total annual shipments at plant cost (for the most recent full year) by the average finished-goods inventory value. Plant cost includes material, labour, and plant overhead.

Finished-product first-pass yield The percent of finished-products that meet all quality-related specifications at a final test point. In process industries, yield is often calculated as the percentage of output that meets target-grade specifications (excluding saleable "off-grade" product).

Firewall The metal panel that separates the engine compartment from the passenger compartment. It also often includes sound and heat insulation.

Firing order The order in which the cylinders deliver power strokes.

Firm A business or institution comprising sole proprietorships, partnerships, companies and other forms of organizations.

Firmware The combination of hardware device and computer instructions and/or computer data that resides as read-only software on the hardware device.

First party coverage The compensation for loss or damage from your insurance company rather than the person involved in the accident.

First-aid box A kit made of various first-aid bandages, creams, and wraps for the emergency treating of minor injuries.

Fish eye A crater-like opening in a newly painted surface, caused by water, oil, or a silicone-based material.

Fish to pull wire or cable through a conduit, raceway, or other confine space.

Fish-eye mirror A wide-angle mirror providing a broader view than a standard mirror.

Fishtail To lose traction with the rear wheels, allowing them to uncontrollably slither from side to side.

Fission The release of energy through splitting atoms.

Fit the range of tightness or looseness that results from the application of a specific combination of tolerances in mating parts.

FITS Flexible Image Transport System

Fitting 1. An accessory such as a locknut, bushing, or other part of a wiring system that is intended primarily to perform a mechanical, rather than an electrical function.

2. A hose end that is designed to mate with a mechanical part.

3. The term applied to all plumbing connections on the vehicle.

Five-point seat belt A safety-belt system with two lap belts and two shoulder belts with a single buckle.

Five-window coupe A coupe body having five windows; two in the doors, two in the rear quarter panels, and one in the rear.

Fixed calliper A disc-brake calliper that does not float or slide, rigidly mounted to the steering knuckle, spindle, or control arm, having one or two pistons on each side of the disc.

Fixed costs A cost element that does not vary with changes in the volume of cost drivers or production volumes. The designation of a cost as fixed may vary depending upon the time horizon. In the long term, all costs are variable.

Fixed or one price selling Published fixed price displayed on a new vehicle eliminating need for negotiation.

Fixed rotor A rotor that has the hub and rotor cast as a single part.

Fixed seal A seal for a calliper piston that is installed in a groove in the calliper bore and that does not move with the piston.

Fixed tappet A solid valve lifter or cam follower.

Flame cutting Using a welding torch fitted with a cutting head to cut metal.

Flame front The leading edge of an air/fuel mixture during combustion.

Flame harden To heat-treat metal to increase its surface hardness and wear resistance.

Flame kernel The initial shape of a freshly ignited air/fuel mixture during the first few milliseconds of combustion.

Flame out An ignition failure while a vehicle is in motion.

Flame point The lowest temperature at which a volatile oil will ignite and burn when exposed to a flame.

Flame propagation The expansion of the air/fuel mixture in the chamber as combustion is completed.

Flame spraying A thermal spraying process in which an oxyfuel-gas flame is the source of heat for melting the surfacing material.

Flame suit An aluminized, fire-resistant, driving suit.

Flame travel Distance across the combustion chamber that the flame of the ignited mixture travels.

Flammability limits The flammability range of a gas is defined in terms of its lower flammability limit (LFL) and its upper flammability limit (UFL). Between the two limits is the flammable range in which the gas and air are in the right proportions to burn when ignited. Below the lower flammability limit there is not enough fuel to burn. Above the higher flammability limit there is not enough air to support combustion.

Flange A projecting rim or edge of a part, usually narrow and of approximately constant width for stiffening or fastening.

Flange gasket A gasket that seals the mating surface of a flanged part and base surface.

Flange nut A fastener that incorporates a flange or washer thrust surface.

Flanged bearing 1. A bearing having a flange to affix it position in a bore or on a shaft.

2. A main bearing having a flange to control the end play of a crankshaft.

Flanged sleeve A cylinder sleeve having a flange at the top to allow it to be set at a specific depth in the block.

Flanged weld A weld made on the edges of two or more joint members, at least one of which is flanged.

Flank The flat part of a camshaft lobe.

Flapper valve A ball valve that operates with a vacuum, pressure diaphragm, or motor.

Flare A cone-shaped flange end applied to a piece of tubing to provide a means of fastening to a fitting.

Flared Wheel wells emphasized with raised edges.

Flash 1. Material that is expelled from a flash weld prior to the upset portion of the welding cycle.

2. Excess material found along the parting edges of a cast or forged part.

3. A condition that occurs when the first coat of paint appears to be dull, prior to final drying.

Flash chrome Thin chrome plating on certain engine parts to provide good wear characteristics.

Flash gas Gas resulting from the instantaneous evaporation of refrigerant in a pressure-reducing device, such as an expansion valve.

Flash point The lowest temperature at which a flammable liquid produces sufficient vapour to flash near its surface, but without continuing to burn.

Flash to pass A steering-column-mounted dimmer switch having an additional feature that illuminates the high-beam headlights even with the headlight switch in the OFF or PARK position.

Flashback A recession of the flame back into the mixing chamber of the oxyfuel gas torch or flame-spraying gun.

Flashback arrester A device to limit damage from a flashback by preventing propagation of the flame front beyond the location of the arrester.

Flasher An automatic-reset, circuit-breaker-type switch used in directional signal and emergency signal circuits.

Flashing A rapid change in liquid state from a liquid to a gas.

Flashpoint The lowest temperature under very specific conditions at which a substance will begin to burn.

Flexible machine centre (FMC) Usually an automated system, comprised of CNC machines with robots loading and unloading parts conveyed into and through the system.

Flexible manufacturing system (FMS) A manufacturing process designed so that the production line may be rebalanced often, rapidly matching output to changes in demand.

Flexible member The portion of the seal containing the springs or bellows.

Flexible radiator hose A large-diameter universal hose connecting the radiator to the engine cooling system.

Flexible shaft A shaft with an operating speed higher than its first critical speed

Flexplate A lightweight flywheel with a starter ring gear around its outside diameter, used on engines equipped with an automatic transmission; also serves as the attachment point for the torque converter.

FlexRay A communication system that will support the needs of future in-car control applications.

Flexural modulus of elasticity Alternate term for modulus in bending.

FLI First Light Imager

Flip chip bonding An assembly technology that is a hybrid of die and wire bonding. Chips that have been prepared with tiny solder globules at the wafer stage are attached facedown to a matching substrate.

Flip chip Process for mounting an I.C chip with metalization down and with no bonding wires.

Flip To overturn a vehicle.

Flip top A convertible with a retractable hardtop.

Float 1. Part that floats in the fuel bowl to assist in controlling the gasoline level in the carburettor by operating a float-needle valve.

2. A cruising drive mode in which the throttle setting matches engine speed to road speed, neither accelerating nor decelerating.

Float bowl 1. A section of the carburettor main body that acts as a fuel reservoir.

2. The reservoir from which fuel is metered into the passing air.

Float chamber The fuel reservoir at the bottom of a carburettor.

Float charge A method of maintaining a cell or battery in a charged condition by continuous, long-term, constant-voltage charging, at a level sufficient to balance self-discharge.

Float circuit A circuit that maintains the correct fuel level in the carburettor float bowl.

Float level The float position when the needle valve is against its seat, cutting off the fuel supply.

Float system The system that controls fuel into the carburettor and the fuel level in the float bowl.

Float voltage The voltage required for retaining a charged battery in a fully charged condition. This is also known as float charging.

Floater A term used for full-floating axle.

Floating axle A term used for full- or semi-floating axle.

Floating calliper A disc-brake calliper that has piston(s) on only one side of the disc. The calliper bore moves away from the rotor in order to press the pad on the other side against the disc when the brakes are applied.

Floating drum A brake drum that is not secured to a hub.

Floating output Ungrounded output of a power supply where either output terminal may be referenced to another specified voltage.

Floating piston A piston having a floating piston pin.

Floating piston pin A piston pin that rotates freely within the bore in the connecting rod.

Floating rotor A rotor and hub assembly made of two separate parts.

Floating-calliper disc brakes A brake system in which only one of the two pads are energized and move the calliper so that it is caught between both pads.

Flog To abuse a vehicle by pushing it too hard.

Flood A condition whereby more fuel is in the combustion chamber than can be ignited.

Flooded A condition whereby the air/fuel mixture in a cylinder is too rich to burn.

Floor it To have the accelerator pushed to the floor. Also floorboard it and floored.

Floor pan The panel forming the floor of the interior of the vehicle.

Floorboard The slanted section of the floor pan immediately behind the firewall.

Floored To run at full throttle. Also floorboard it and floor.

Floorplan The high-level physical layout of blocks on a semiconductor device (either chips or boards). Floorplanning tools also typically provide estimations of timing delays.

Floorplanners Floorplanners are EDA software tools that provide an environment where issues such as timing, area, power dissipation, and routeability can be analyzed before a detailed physical layout of a design is completed. These tools provide interactive and/or automatic capability to accurately estimate interconnect resistance/capacitance and predict timing delays before the placement and routing of the design so that users can evaluate and choose the optimum floorplan.

Flopper A funny car with a fibreglass body that flops up in the front to provide access to the engine, chassis, and driver compartment.

Floppy disk A flexible 5.5- or 3.5-inch disk used widely with microcomputers and minicomputers, providing electronic media storage at a relatively low cost.

Flow chart 1. A graphical representation of the sequence of events and or testing that leads to the diagnosis and repair of a system, circuit, or component.

2. In computers, a graphical representation of the sequence of events occurring during program execution.

Flow coating A method of applying paint by passing parts on a conveyor through a chamber where several nozzles direct a shower of coating material over the parts.

Flow Kaizen Radical Improvement, usually applied only once within a *value stream*.

Flowability 1. The ability of molten filler metal to flow.

2. The ability of a fluid or vapour to flow.

Fluid aeration Air bubbles formed in a fluid, giving the appearance of foam.

Fluid Any liquid or gas.

Fluid cooler 1. Small heat-exchanger component in a hydraulic line near the pump to reduce power-steering-fluid temperature.

2. A device inside the radiator to provide cooling for transmission fluid.

3. Any heat exchanger designed to reduce the temperature of a fluid.

Fluid coupling A device in thee power train containing two rotating members, one of which transmits power to the other via fluid.

Fluid pressure The pressure of a fluid that is invariable and uniform in all directions.

Fluid The material assumes the shape of its container. It could be either a liquid or a gas.

Fluidizer oils Oils typically used with deposit control additives to control deposit formation on intake valves. (Removes cauliflower deposits)

Fluorescence The phenomenon of emitting light upon absorbing radiation of an invisible wavelength.

Fluorescent Having the property of giving off light when bombarded by electrons or radiant energy.

Flurocarbon Genetic term for the elastomer called Viton. Viton is a Dupont Dow elastomer product.

Flush bucket An oversized carburettor that feeds more air/fuel mixture than the engine can handle.

Flush Putting an outside liquid into the stuffing box of the pump at a pressure higher than stuffing box pressure. All of this liquid mixes with and dilutes the pumped fluid.

Flushing A term used for various acts of cleaning a system, such as brake flushing.

Flushing agent An approved liquid or gas used to flush a system.

Flushing hydraulic system The procedure for replacing old brake fluid with new fluid.

Flux A material to dissolve and prevent the formation of oxides in molten metal and solid metal surfaces; a wetting agent that facilitates the banding of a filler metal.

Flux density (B) Number of lines of flux per cross-sectional area of a magnetic circuit expressed in Gauss.

Flux Density The number of flux lines per square centermeter measured 1 cenimeter from the center of the magnet

Flux Lines Also called maxwells, the lines of magnetic force

Flux The amount of radiation crossing a surface per unit of time, often expressed in "integral form" as particles per unit area per unit time.

Flux-density The number of lines of force per square inch measured one inch from the center of a magnet.

Flyback converter A power supply switching circuit which normally uses a single transistor. During the first half of the switching cycle the transistor is on and energy is stored in a transformer primary; during the second half of the switching cycle this energy is transferred to the transformer secondary and the load.

Flyboy A drag competitor who races only occasionally as a hobby.

Flyby Space flight past a heavenly body without orbiting.

Flyby spacecraft A spacecraft which follows a continuous trajectory past a target object, never to be captured into an orbit. It must carry instruments that are capable of observing passing targets by compensating for the target's apparent motion.

Flying kilometer The international standard for attempting to set a speed record by entering the measured kilometer afer attaining the highest speed possible.

Flying mile the international standard for attempting to set a speed record by entering the measured mile after attaining the highest speed possible.

Flyweight governor Governor assembly, sensitive to centrifugal force, whose action is controlled by primary and secondary weights.

Flywheel 1. A heavy metal wheel with a starter ring that is mounted at the rear of the crankshaft. It absorbs energy on power stroke(s), return energy on other stroke(s), and transfers power to the clutch or torque converter.

2. Front-most part of a clutch assembly that is bolted to the engine crankshaft with a rear surface to provide a smooth friction area for the disc-front facing to contact during clutch engagement.

Flywheel ring gear A gear, fitted around the flywheel, that is engaged by teeth on the starting-motor drive to crank the engine.

Fm 1. Fermium An element with the atomic number of 100

2. An acronym for Frequency Modulation (Impressing information on a carrier signal by changing its frequency at the information rate.

Focus groups Focus groups are meetings with a group of customers, users, or potential users of a product to explore their needs and obtain their feedback on product ideas and concepts. These meetings are conducted by a facilitator using a prepared script but provide the flexibility to obtain open-ended participant input and feedback. These meetings are often recorded and observed by company for further analysis.

Focused-factory production A plant configuration and organization structure in which equipment and manpower is grouped to create essentially self-contained "mini-businesses," each with a specific product-line or customer focus. A single plant may be divided into several focused-factory units, designed around process flows, each of which has control over such support activities as maintenance, manufacturing engineering, purchasing, scheduling, and customer service.

Fog lamp Auxiliary lamps, often amber, mounted in front of a vehicle to aid visibility during snow, dust or fog conditions.

Fold To bend a material, usually to 180 degrees.

Foldback current limiting A power supply output protection circuit whereby the output current decreases with increasing overload, reaching a minimum at short circuit. This minimizes internal power dissipation under overload conditions. Foldback current limiting is normally used with linear regulators and is not necessary to protect switching regulators. However, many switchers incorporate foldback to protect the load form excessive current.

Follower A term used for lifter.

Following ball joint A term used for non-load-carrying ball joint.

Folothru drive An inertia-starting motor drive, similar to a Bendix Folo-thru drive.

Font A design for a set of characters. A font is the combination of typeface and other qualities, such as size, pitch, and spacing.

Foot (ft) An English measure equal to 12 inches.

Foot in it A driver accelerating rapidly and/or refusing to yield during an attempted pass.

Foot in the carburettor The accelerator pushed to the floor.

Foot Pound The amount of energy or work needed to raise an item weighing one pound a distance of one foot

Foot Supports the wet and power end of the pump and attaches it to the base plate.

Foot valve The foot-operated brake valve that controls air pressure to the service chambers.

Foot-pounds (ft-lb) An English measure for torque.

Footprint 1. The portion of the contact area of a loaded tire with the ground.

2. The bolt pattern of a device.

Footwell The area for passengers' feet in a vehicle.

Force A vector quantity that tends to produce an acceleration of a body in the direction of its application.

Forced induction An intake system that provides a means for the air/fuel mixture to enter the combustion chamber at greater than atmospheric pressure.

Fordor An early designation for a Ford four-door vehicle.

Forecast Prediction of future production or sales in the automotive industry.

Forecast/demand management software Software that provides front-end input to master production scheduling systems and helps to optimize inventory planning. Such software not only takes into account historical demand trends, but also may calculate the impact of planned sales promotions, price reductions, and other factors that cause spikes in demand levels.

Forecasting The work performed to estimate and predict future conditions and events. Of relevance to new product development is the need to forecast future sales of a proposed product to determine its financial feasibility.

Forge To form metal into a desired shape.

Forging A process that transforms solid metal into shapes of varying cross-sectional material thickness, often involving heating.

Form The defined configuration of an item including the geometrically measured configuration, density, and weight or other visual parameters which uniquely characterize an item, component or assembly. For software, form denotes the language, language level and media.

Formal testing The process of conducting testing activities and reporting results in accordance with an approved test plan making provision for customer involvement.

Formal verification The application of rigorous mathematical techniques to prove the functional equivalence of an electronic hardware design with its original specification. Because timing is not included in formal verification, it is used only to verify a design's functional behaviour. It is a collective term used for a number of different tools and methodologies.

Format To prepare a storage medium, usually a disk or hard drive for reading and writing.

Formed metal bellows Manufactured by stretching and compressing the metal bellows material. Not usually used in mechanical seals because of its high spring rate.

Formed-in-place gasket A gel-like material that forms a gasket when clamped between two surfaces.

Forming charge The initial charge applied to a new battery before use.

Forming The process of applying mechanical pressure to a metal component in order to produce the desired three-dimensional form. Thin metal is sheet-formed; solid metal pieces are massive-formed.

Formula libre A type of road racing open to all without any limitation or restrictions on engine size, bodywork, or other design features.

Formula One A single-seat, open-wheeled car for international Grand Prix racing.

Fortuitous event An unforeseen accident.

Forward bias Current flow in normal direction

Forward clutch A clutch that is engaged whenever the vehicle moves forward, controlled by the valve-body forward circuit.

Forward converter A power supply switching circuit that transfers energy to the transformer secondary when the switching transistor is on

Forward rake A car having a lower front than rear, to create an extra down fore at high speeds.

Forward scattering Reflecting light approximately away from the source.

Forward sensor A sensor used in air-bag restraint systems.

Forward shoe A term used for leading shoe.

Forward voltage A voltage applied across a semiconductor junction in order to permit forward current flow through the junction and the device.

Fossil fuels Fuels formed underground from animal and plant matter by chemical and physical change, such as coal, petroleum, and natural gas.

FOTCC an acronym for Fixed orifice tube cycling clutch.

Foul To leave before the green light in drag racing and be disqualified.

Four banger 1. A four-cylinder engine.

2. A vehicle equipped with a four-cylinder engine.

Four by A term used for 4x4.

Four cammer A V-type engine with dual overhead camshafts on each cylinder bank.

Four cycle A term used for four-stroke cycle.

Four on the floor (slang) a four-speed manual transmission.

Four on the floor A four-speed manual transmission with a floor-mounted shift lever.

Four speed A manual transmission with four forward gears.

Four wheel drive (4wd) In a four wheel drive system, a secondary transmission assembly, called a transfer case, is driven from the main transmission. The transfer case distributes power to both axles to drive all four wheels. It is the heart of the four-wheel drive system. Four-wheel drive can be full-time, in which power is delivered to both axles at all times or part-time, where the driver selects two or four wheel drive. Four wheel drive is often combined with independent suspension systems and off-road type tires to enhance driveability on rough, off-road terrain, or on-road driveability in unfavourable driving conditions.

Four wheel independent suspension A type of suspension in which all wheels are mounted to separate suspension members with no rigid axle connecting them. Therefore a disturbance affecting one wheel has no effect on the opposite wheel. Four wheel independent suspension reduces the un-sprung weight, improves ride and handling over rough surfaces and permits room for a larger trunk.

Four wheel On a vehicle equipped with four-wheel anti-lock brakes, all four wheels are equipped with speed sensors. When these sensors determine that the wheels are decelerating so rapidly that lockup may occur, the electro-hydraulic control unit is activated. The electro-hydraulic control unit then modulates the brake pressure in the appropriate brake lines by means of the solenoid-operated valves. This is intended to prevent wheel lockup and help the vehicle maintain directional stability during potentially hazardous braking situations

Four wheeling The off-highway travel in a four-wheel vehicle.

Four-barrel carburettor A carburettor having four ventures.

Four-bolt mains A main bearing cap that is held in place with four bolts.

Four-gas analyzer An exhaust-gas analyzer able to detect and measure exact amounts of hydrocarbons, carbon monoxide, carbon dioxide, and oxygen.

Four-point seat belt A safety-belt system with two shoulder straps, two lap straps, and a single buckle.

Four-speed transmission A transmission providing four forward-gear ratios, one reverse ratio, and neutral, permitting closer matching of engine speed to load requirements than a three-speed transmission.

Four-stroke cycle 1. A cycle of engine operation whereby the combustion occurs in every cylinder on every second revolution of the crankshaft.

2. A complete cycle includes intake or induction, compression, combustion or expansion, and exhaust.

3. Also known as four cycle.

Four-valve head A head design having four valves per cylinder.

Four-wheel drift A controlled four-wheel slide on a paved surface.

Four-wheel steering Vehicle on which all four wheels turn when the driver turns the steering wheel. The rear wheels turn at a smaller angle than the front wheels. This system appeared on a few sports models in the 1980s but was never very popular in North America.

Fractionation A condition when one or more refrigerants of the same blend leak at a faster rate than other refrigerant in that same blend.

Fracture stress True stress generated in a material at fracture

Fracture test Visual test wherein a specimen is fractured and examined for grain size, case depth, etc.

Fracture toughness Ability of a material to resist crack propagation when subjected to shock load as in an impact test.

Frame alignment. The measurement of a vehicle frame to ensure that it is within a manufacturer's specifications.

Frame The substructure of a vehicle supported by the suspension system that supports the bodywork, engine, and power train.

Framework A software infrastructure that provides a common environment for communication and integration of design tools in a design process.

Franchise Establishment that has the right to exercise the powers of a corporation.

Frantz oil filter The tradename of a popular filter that uses a roll of toilet tissue as its filtration medium.

Free electron 1. An electron that is not bound to a particular atom.

2. An electron that is in the valance shell of an atom with less than 4 electrons in that shell.

Free length The uncompressed axial length of a seal.

Free play The distance the brake pedal moves before the master cylinder primary piston moves.

Free wheel diode A diode in a pulse-width modulated switching power supply that provides a conduction path for the counter electromotive force of an output choke.

Freedom of Information Act A federal law giving everyone the right to certain information.

Free-fall The motion of any unpowered body moving in a gravitational field.

Free-flow exhaust An exhaust system with reduced back pressure.

Freeform fabrication Freeform fabrication is the application of rapid prototyping and rapid tooling technologies to the direct manufacture of parts in an on-demand, low volume or a mass customization production environment.

Free-return trajectory Path of a spacecraft that provides for a return to Earth.

Free-standing deck An engine block construction in which the cylinders are cast in place without being tied by the deck to the outer walls.

Free-travel Movement of an operating part before it comes in contact with another part.

Freewheeling A mechanical device in which a driving member imparts motion to a driven member in one direction, but not the other.

Freeze crack A crack in the engine caused by expansion due to coolant freezing.

Freeze frame 1. Operating conditions which are stored in the memory of the PCM at the instant a DTC is stored and the MIL is illuminated. A Freeze Frame is stored for only one DTC and can only be overwritten under certain conditions.

2. A "snapshot" of engine operating conditions taken when a fault occurs. Normally containing such information as engine RPM, throttle position, engine load, vehicle speed, coolant temperature, and loop status.

Freeze plug A term used for core plug.

Freeze protection 1. The controlling of evaporator temperature so that moisture on its surface does not freeze an block the airflow.

2. An additive added to coolant to prevent freeze-up by lowering its freezing temperature.

Freeze To stop a waveform display for closer examination only possable on a digital oscilloscope

Freeze-up 1. Failure of a unit to operate properly due to the formation of ice at the metering device.

2. A term used for seize.

French seam A fabric seam in which the edge of the material is tucked under and sewn on the inner side.

Frenched A body part, normally separate, having been moulded together with another body part.

Freon Registered trademark of E.I. Dupont, for a group of its refrigerants.

Frequency 1. The number of occurrences of an interactive happening per unit of time.

2. The number of complete cycles an alternating electric current, sound wave, or vibrating or rotating object undergoes in one second.

Frequency counter Counts digital pulses over a defined gate time. A typical gate time is between 0.1 and 10 seconds.

Frequency Response A bode point of input to output response of an amplifier of attenuator for sine waves with constant amplitudes a different frequencies over a frequency range

Frequency valve A valve that is used to stabilize the air/fuel mixture on a fuel-injected engine.

Frett or fretting Damage or grooving caused by the removal of the protective oxide that is formed on most corrosion resistant metals. It happens when a softer material (rubber) rubs against a hard shaft or sleeve. A common problem with low cost O.E.M. mechanical seals and bearing grease or lip seals.

Friction ball joint A term used for non-load-carrying ball joint.

Friction bearing A bearing in which there is a sliding contact between the moving surfaces, such as a connecting rod bearing.

Friction disk A flat disc surfaced with a friction material on one or both sides, such as a clutch disk.

Friction facing A hard-molded or woven material that is riveted or bonded to the clutch-driven disc.

Friction horsepower Engine horsepower losses due to friction from such sources as the engine, transmission, and drive train.

Friction material One of several types of material used for friction surfacing, such as metallic friction material or organic friction material.

Friction modifier 1. A class of materials used in brake linings to modify the final coefficient of friction of the linings.

2. Additives used to alter the friction coefficient of a liquid such as ATF.

Friction The resistance to motion of two items in contact with each other.

Friction-modified fluid Automatic transmission fluid that provides smooth automatic shifts; designed to slip.

Front and rear suspension systems The suspension system, with the frame, supplies steering control under all road conditions and maintains proper tracking and directional stability as well as providing proper wheel alignment to minimize tire wear.

Front chute The front straightaway on a circle race track.

Front clip The complete replacement of the front bodywork back to the cowl or A-pillar.

Front control arm Horizontal arms that connect the front wheels to the car and that support the weight of the front of the car.

Front cradle Heavy metal framework that attaches to the chassis and supports the powertrain and allows the attachment of many of the suspension and steering parts.

Front drive A drive system that transmits power through the front wheels.

Front engined A vehicle with the engine in front, ahead of the passenger compartment.

Front idler A pulley used as a means of tightening the drive belt.

Front panel The front surface of a unit, generally containing switches and indicator lights.

Front pump A pump, located at the front of the transmission, driven by the engine through two dogs on the torque-converter housing, to supply fluid whenever the engine is running.

Front seating Closing off the line, leaving the air-conditioner compressor open to the service port fitting, allowing service to the compressor without purging the entire system.

Front steer A steering gear mounted ahead of the front wheel centerline.

Front straight A straight area of a race track, such as between turns four and one at the Indianapolis Speedway.

Front suspension Tp support the weight of the front of the vehicle.

Front to rear brake bias The difference in balance of brake pressure between the front and rear brake cylinders or callipers; higher in front due to weight transfer during heavy breaking.

Frontal area The area, in square feet, of the vehicle's cross section, as viewed from the front.

Front-axle limiting valve A valve that reduces pressure to the front service chambers, thus eliminating front wheel lockup on wet or icy pavements.

Front-body structural components In a perimeter frame design, the front body section is made up of the radiator support, front fender, and front fender apron. These are installed with bolts and form an easily disassembled structure.

Front-end drive A vehicle having its drive wheels located on the front axle.

Front-end geometry The angular relationships involving the front suspension, steering system, and tires.

Front-of-dash components the heating and air-conditioning components that are mounted on the firewall side in the engine compartment.

Front-roll center The center, determined by the front suspension geometry, around which the forward part of a vehicle tends to roll.

Front-suspension system Components that provide support of the vehicle front section, allow wheels to move vertically, and provide adjustments for front wheel alignment. The common parts include upper and lower ball joints, control arms, shaft bushings and shims; sway bar and bushings; strut rod and bushings; coil springs; stabilizers; shock absorbers; and steering knuckle and spindle.

Front-wheel drive (fwd) A drive system where the engine and transaxle components apply the driving force to the front wheels rather than the rear wheels. Benefits of front-wheel drive include: maximized passenger space. Enhanced cargo area. Excellent drive traction; particularly on wet or slippery surfaces, since the drive is through the front wheels, which carry a heavier load.

Fronty In early racing, the Chevrolet Frontenac and Fronty Ford Model T.

Frosting back The appearance of frost on the air-conditioning suction line extending back as far as the compressor.

Froude number In fluid mechanics, a dimensionless number expressing the ratio of nonlinear advection to the pressure gradient acceleration associated with the variation of fluid depth, i.e. of the fluid speed to a measure of the internal wave speed.

Fuel cell 1. A special fuel tank designed for a race car.

2. A device for generating electricity by the chemical combination of fuel and Oxygen.

Fuel Cell Matrix A framework within a fuel cell that supports an electrolyte.

Fuel injection, electronic A computer-controlled method of delivering fuel under pressure. The computer monitors signals from coolant temperatures, manifold vacuum, exhaust oxygen sensor, and engine cranking sensor. It "tells" the injectors to release and adjust the fuel to yield an air/fuel mixture assuring engine operation well matched with emission requirement , optimum fuel economy and overall vehicle performance.

Fuel injector Taking the place of carburettors in the 1980s, the fuel injector is an electrically controlled valve that delivers a precise amount of pressurized fuel into each combustion chamber.

Fulcrum A support, often wedge-shaped, on which a lever pivots when it lifts an object.

Full assist A term used with variable assist steering systems to signify normal power system operation without restrictions.

Full bore Full throttle. Also known as full chat.

Full chat A term used for ful bore.

Full coil suspension A vehicle suspension system in which all four wheels have their own coil spring.

Full field The Maximum AC generator output

Full fielding The method of supplying full battery voltage to the magnetic field of a generator as part of the troubleshooting procedure for the charging system.

Full floater A term used for full-floating axle. Also known as floater.

Full floating Any part that moves and rotates within another part, such as floating piston pin.

Full flow 1. A flow without restrictions.

2. A type of oil filter, having no bypass, through which all of the oil from the oil pump flows.

Full house An engine that has had every normal hot-rodding modification that may be set up for racing or for street use.

Full load An engine delivering its maximum output.

Full race An engine that is built for maximum racing performance.

Full scale output The difference between the minimum output (normally zero) of a data acquisition device and the rated capacity.

Full scan A design for testability methodology that provides complete access to an integrated circuit. EDA tools can insert scan registers automatically during logic synthesis.

Full throttle 1. The wide open throttle position.

2. The accelerator pressed to the floorboard.

Full trailer A trailer that employs a towbar coupled to a swiveling or steerable running gear assembly at the front and does not transfer any of the load to the towing vehicle.

Full wave Rectification The conversion of a complete AC voltage signal to a DC voltage signal

Full wave rectifier A circuit which convert both alternations of an input ac voltage to a pulsating unidirectional dc output voltage.

Full-bridge converter A power switching circuit in which four power switching devices are connected in a bridge configuration to drive a transformer primary.

Full-bridge rectifier A rectifier circuit that employs four diodes per phase.

Full-floating axle An axle that performs only one function, to transfer torque to drive the vehicle; a type of axle popular on trucks.

Full-floating pin A piston pin, held in position by snap rings fitted into grooves in the piston boss, that is allowed to move in both the piston and the connecting rod.

Full-flow filter A type of oil filter designed so that all of the oil from the oil pump flows through it.

Full-load enrichment The injection of additional fuel into an engine during full-load conditions.

Full-metallic brake lining Brake linings made of metal particles that have been fused together into a solid material.

Full-race cam A camshaft that is ground for maximum performance.

Full-size A car that is usually four doors and seats anywhere from two to seven people.

Full-wave rectifier Rectifier circuit that produces a dc output for each half cycle of applied alternating current.

Fully oscillating fifth wheel A fifth-wheel type with fore-aft and side-to-side articulation.

Fume The airborne dispersion of minute particles, a byproduct of heating a solid, that may produce an oxide of the solid.

Function An abstracted description of work that a product must perform to meet customer needs (for value analysis, sometimes stated in a noun-verb format, e.g., "transmit data")

Function analysis Synonymous with value analysis and value engineering. A methodology of focusing on those functions that are valuable to customers and delivering them at the lowest possible cost.

Function analysis system technique Function analysis system technique is a value analysis or function analysis technique to describe a system or product as a series of logically related functions and associate those functions to costs. This technique identifies less important functions that may then be eliminated, thereby reducing costs.

Function cost analysis Function cost analysis is an accounting allocation of cost and importance to product function. It is a tool used to support value engineering or value analysis to identify high cost functions to address.

Function point analysis A top down software development estimating technique which was developed by a.j. albrecht. It entails breaking a project down into 'function points' which are classified by degrees of complexity. Factors are then applied from which time estimates may be developed.

Function The action or actions which an item is designed to perform.

Functional and performance requirements A Functional Requirement is a necessary task, action, or activity that must be accomplished by a system. Performance Requirements state quantitatively the extent to which a system function must be executed. These are a subset of a DRD.

Functional architecture A view of the software architecture that focuses on system behaviour and component interactions

Functional area A distinct group of system performance requirements which, together with all other such groupings forms the next lower-level breakdown of the system on the basis of function.

Functional configuration audit An engineering audit of a configuration item (ci) or system to verify that the performance test results of the item are in accordance with the performance specification of the item.

Functional interfaces Those interfaces that characterize the functional behaviour of specific Components.

Functional requirements Functional requirements capture the intended behaviour of the system or product - what the system will do. This behaviour may be expressed as functions, tasks, or services the system or product is required to perform. Therefore, functional requirements do not include performance characteristics, operating conditions, use cases, and specifications.

Functional test Functional test is a test that identifies functional level faults in printed circuit board assemblies (pcbas), including manufacturing related faults not identified by in-circuit tests (ict), timing related failures, and faults internal to components. Functional test equipment operates at the same frequency the pcba is designed for and may have the capability to margin temperature, voltage and frequency.

Functional worth Functional worth is equal to the least expensive way to perform a given function.

Functionality A term associated with comprehensive component testing in which output from the engine controller can be verified by monitoring specific input signals from other components for expected change.

Fundamental tone The tone produced by the lowest frequency component of an audio or radio frequency (RF) signal.

Funny car This variation of a dragster hides the usual front engine chassis under a lightweight body that looks like a street car.

Furnace brazing A welding process used to repair complex cast-iron castings.

Furnish Contractor shall supply and deliver to the Project site, unload, unpacked, and assembled ready for the intended use or installation, as applicable in each instance.

Fuse 1. A circuit protection device that opens a circuit when the fusible element is severed by heat, due to an over current passing through it.

2. To join two pieces of metal by bonding them together.

Fuse block A box-like enclosure that contains the fuses and circuit breakers for the electrical circuits of a vehicle.

Fuse link A term used for fusible link.

Fuse Safety protective device that permanently opens an electric circuit when overloaded.

Fuse wire 1. Wire made of an alloy that melts at a low temperature.

2. A term used for fusible link.

Fusible link A type of fuse that will melt and open the protected circuit in the event of a short circuit, Most fusible links are actually wires four gauge sizes smaller than the wire of the circuit being protected, covered with a Hypolan insulation

Fusion welding Any welding process that uses fusion of the base metal to make the weld.

Fussion The release of nuclear energy through the uniting of atoms.

Fuzzy front end The is the process for determining customer needs or market opportunities, generating ideas for new products, conducting necessary research on the needs, developing product concepts, and evaluating product concepts up to the point that a decision is made to proceed with development. This process is described as the fuzzy front end because it is the least defined and most unstructured part of product development.

Fuzzy logic A method used to model linguistic expressions that have nonbinary truth values. It has been used with PID algorithms in process control, especially where process relationships are nonlinear.

G

Gain Accuracy An indication of how accurately the vertical system attenuates or amplifies a signal, usually represented as a percentage error

Gain Any increase in the current, voltage or power level of a signal.

2. The ratio of output to input signal level of an amplifier.

Gain margin The gain of a control loop at the frequency for which there is 360 degrees of phase shift around the control loop.

Galaxy A very large system of stars, gas and dust isolated from its neighbours by an immensity of space; an "island universe".

Gallery A passageway inside a wall or casing, such as for oil circulation.

Galling 1. The friction-induced roughness of two metal surfaces in direct sliding contact.

2. The welding up and tearing sown of metals due to extreme high temperature and/or inadequate lubrication.

Galloping hinge A connecting rod.

Galvanic action the action of electropositive and electronegative metals due to the wasting away of the positive metal. This is most noted when iron (Fe) and brass, copper (Cu), or copper (Cu) and zinc (Zn), are in contact in the presence of acid-diluted water.

Galvanic battery A D/C voltage source generated by the chemical action of an electrolyte.

Galvanic reaction An electrical reaction caused by heat and water passing through metal pipes or parts at high speeds.

Galvanic series A list of metals with those on the top of the list being attacked by those lower down in the list. The father apart on the list, the faster the attack.

Galvanize An electro mechanical process of coating zinc (Zn) to ferrous metals to reduce or prevent rust.

Galvanized steel A specially zinc-coated steel used on many major painted panels and in key unpainted areas of a vehicle to help prevent rust and corrosion.

Galvanizing Electrochemical processing of metallic surfaces. A coat of dissimilar metal is electrolytically deposited in baths of metallic salt solution; the passage of electric current decomposes the dissolved metallic compounds.

Galvanometer An instrument used for measuring a small electrical current.

Gamma rays Very short, highly-penetrative electromagnetic radiation with a shorter wavelength than X-rays; produced in general by emission from atomic nuclei.

Gantt chart Gantt Chart is a diagram used in project management, where the x axis is time and the y axis shows tasks to be performed to complete the project. Each task is displayed as a horizontal bar spanning the time period during which it is expected to take place. Arrows may be drawn from one task to another to indicate dependencies (when one task can't be begun until another is completed). The Gantt chart was developed by Charles Gantt in 1917.

Gap The space between two adjacent parts or surfaces.

Garbage in, garbage out GIGO A famous computer axiom meaning that if invalid data is entered into a system, the resulting output will also be invalid.

Gas 1. A state of matter

2.A vapour.

3. A short term for gasoline.

Gas bypass line A term used for hot-gas bypass line.

Gas bypass valve A term used for hot-gas bypass valve.

Gas cylinder A container used for the transportation and storage of compressed gas.

Gas defrosting The use of hot, high-pressure gas in the evaporator to remove frost or ice buildup.

Gas diffusion Mixing of two gases caused by random molecular motions. Gases diffuse very quickly; liquids diffuse much more slowly, and solids diffuse at very slow (but often measurable) rates. Molecular collisions make diffusion slower in liquids and solids.

Gas fade Brake fade due to hot gases and dust that reduce friction between the drum or rotor during prolonged hard braking.

Gas filled shock absorbers A nitrogen gas chamber is used to pressurize the shock absorber in place of the traditional air/oil combination. Gas filled shock absorbers provide more stable damping in a variety of conditions and thus improves ride and road contact.

Gas Fuel gas, such as natural gas, undiluted liquefied petroleum gases (vapour phase only), liquefied petroleum gas-air mixtures, or mixtures of these gases.

Gas generator A chamber in which propellant is burned to produce high pressure gas that is then used to drive a turbine, e.g. turbopump.

Gas giant A large planet composed mostly of gas, e.g. the Jovian planets.

Gas guzzler A vehicle having excessive fuel consumption.

Gas hog A term often used for gas guzzler.

Gas ports A series of holes drilled from the crown of a piston to the top ring groove, providing combustion pressure to force the ring against the cylinder wall, ensuring a more effective seal.

Gas regulator A device used to control the delivery of gas at a substantially constant pressure.

Gas shock A term used for gas-filled shock absorber.

Gas-charged shocks Also called gas-filled shocks. They are shock absorbers filled with a low-pressure gas to smooth the vehicle's ride during up-and-down movement.

Gas-filled shock absorber A shock that uses nitrogen gas, at 25 times atmospheric pressure, to pressurize the fluid in the shock to reduce or prevent aeration or foaming. Also known as gas shock.

Gasifier section The part of a gas-turbine engine that draws in air, mixes it with fuel, and burns the mixture.

Gasket 1. A metal ring used on some spark plugs to seal pressure between the plug shell and the cylinder head.

2. A piece of thin compressible material such as cork, rubber, or soft metal, placed between two mating surfaces to form a seal.

Gasket cement A liquid gel or paste adhesive material used to bond gaskets to their mating surfaces.

Gasket shellac A liquid form of gasket cement.

Gasket Used between two static surfaces to provide a seal. Made from a variety of deformable materials.

Gas-metal arc welding (GMAW) An arc-welding process that uses an arc between a continuous-filler metal electrode and the weld pool.

Gasohol In the United States the term gasohol refers to gasoline which contains 10 percent ethanol. This term was used in the late 1970's and early 1980's but has been replaced by terms such as Super Unleaded Plus Ethanol or Unleaded Plus.

Gasoline consumption A term used for fuel consumption.

Gasoline container Specially approved OSHA containers used to safely hold and store gasoline.

Gassing The conversion of battery water into hydrogen and oxygen gas.

Gas-transfer velocity The speed at which the air-fuel mixture spreads out in the combustion chamber during the power phase.

Gas-tungsten arc cutting (GTAC) An arc-cutting process that uses a single tungsten electrode with gas shielding.

Gas-tungsten arc welding (GTAW) An arc-welding process that uses an arc between a non-consumable tungsten electrode and the weld pool.

Gas-turbine engine An internal-combustion engine in which the shaft is spun by combustion gases flowing against curved turbine blades located around the shaft.

Gate 1. A device or element that has the ability to block or pass a signal.

2. A device having one output channel and two or more input channels that performs a logic function.

3. A control electrode in a semiconductor device such as a triac, or FET.

Gate count A metric for the size of an ASIC design, usually expressed in terms of the equivalent number of basic 2-input NAND gates used. A gate count can be roughly converted to a transistor count by multiplying by a factor of four.

Gatekeeper Gatekeepers are the members of management that conduct the stage-gate or phase-gate reviews that are part of a stage-gate process in new product development. The gatekeepers are usually members of a formal group known as product committee or similar name that are charged with portfolio management and pipeline management.

Gateway A device that allows consumer products to interface with the communication system in vehicles, while protecting the vehicle's system from defective devices or inappropriate messages.

GATV Gemini-Agena Target Vehicle.

Gauge 1. An instrument used for measuring, such as for compression.

2. A dashboard-mounted component used for visual indication of engine and system conditions, such as oil pressure.

Gauge pressure A pressure indicated on a scale with atmospheric pressure as 0 psi or kpa.

Gauge repeatability and reproducibility Gauge repeatability and reproducibility (gr&r) is the evaluation of a gauging instrument's accuracy by determining whether the measurements taken with it are repeatable and reproducible. Repeatability is the variation in measurement obtained with one measurement instrument when used several times by an appraiser while measuring the identical characteristic on the same part. Reproducibility is the variation in the averages of the measurements made by the different appraisers using the same measuring instrument when measuring the identical characteristic on the same part.

Gauge set 1. Two or more instruments attached to a manifold and used for measuring or testing pressure.

2. Two or more instruments used to indicate engine and system conditions, such as oil pressure, coolant temperature, and charging-system indicators.

Gauging point The starting point for measuring a part.

Gauss Measure of flux density in Maxwells per square centimeter of cross- sectional area. One gauss is ic-4 Tesla.

Gauss's Law Karl Friedrich Gauss (1777 - 1855) A German mathematician known for his work in magnetism. Gauss's Law stats that the total of the electric flux out of a closed surface is equal to the charge enclosed divided by the permittivity.

Gear 1. A cogged device that mates or meshes with another.

2. One's personal belongings and/or equipment.

Gear pump A positive-displacement pump that uses an inner drive gear and an outer driven gear, separated on one side by a crescent, to produce oil flow. Gear pumps may use either helical or spur gears and are sometimes called gear-and-crescent pumps.

General availability The point in the product life cycle when production has been ramped-up to sufficient volumes and when product issues have been resolved so that the product is made available to all interested customers.

General engine specifications Specifications that are used to identify a particular style and type of engine.

General over-the-road use A fifth wheel truck designed for pulling multiple standard duty highway equipment such as trailers, flat beds, tankers, and so on.

General property taxes Any tax on real estate or personal property.

Generating Capacity 1. The maximum amount of electric power produced by a generator.

Generator 1. A machine that converts mechanical energy into electrical energy.

2. A source of signals that are transmitted, amplified, of manipulated by electronic circuits.

GEO Geostationary Orbit. Also abbreviated GO.

Geocentric apparent place Geocentric apparent right ascension and declination (α, δ) will be available to assist in recording observations but will not in fact be the interface to the telescope control system. The pole is the precessing-nutating celestial intermediate pole (CIP); the zero point of right ascension will be the true equinox of date and/or the celestial intermediate origin (CIO).

Geocentric Earth centered.

Geodesy The science of the Earth's shape.

Geographic data format (GDF) A transfer file specification for digital roadway and topological map databases. The format includes specifications for database encoding.

Geomagnetic storm A worldwide disturbance of the Earth's magnetic field, distinct from regular diurnal variations.

Geometric centerline A static dimension represented by a line through the center of the vehicle from front to rear.

Geometric dimensioning and tolerancing Geometric Dimensioning and Tolerancing (GD&T) - ANSI-Y14.5 standard for showing the dimensioning and tolerancing on a drawing considering the functions or relationships of part features. GD&T depicts the geometric relationship of part features (instead of the Cartesian relationship), allowing the maximum tolerance which permits full function of the product.

Geometrical axis A line perpendicular to the center of the back surface of an off-axis mirror. It corresponds in usage roughly to the optical axis of an axially symmetric conic mirror. Since the back surface of an off-axis mirror is typically a best-fit sphere or a flat surface, this definition offers a simpler interpretation than one involving the more complex front surface.

Geospace Also called the solar-terrestrial environment, geospace is the domain of Sun-Earth interactions. It consists of the particles, fields, and radiation environment from the Sun to Earth's space plasma environment and upper atmosphere. Geospace is considered to be the fourth physical geosphere (after solid earth, oceans, and atmosphere).

Gerotor A term used for gear rotor pump.

Get off it To release the accelerator and slow down. Also known as get out of it.

Get out of it To release the accelerator and slow down. Also known as get off it.

Get with the program To perform a task properly.

G-force the force that is exerted on a vehicle during acceleration, deceleration, or corning, expressed in units of gravity (g's).

Glass 1. A term used for fibreglass-reinforced plastic.

2. A hard, brittle, transparent substance composed of silicates mixed with potash or suds and lime; used for windows in a vehicle.

Glass beadier An apparatus for using air-propelled glass beads to clean parts.

Glass wrapped A term used to refer to a vehicle with a fibreglass body.

Glass-bead cleaning the cleaning of parts and panels using glass beads propelled with the use of compressed air.

Glass-bead test A method of determining the absolute rating of a filter and its efficiency by introducing a measured quantity of glass beads of varying, but known, diameter, as a contaminant into the fluid, which is then filtered through the element.

Glaze A smooth, glossy surface.

Glazed Polished by heat and friction to a hard, smooth, shiny surface, unlike the original surface finish.

Glitch An intermittent, high-speed error in a circuit.

Global circulation model Calculations used to understand the relative importance of anthropogenic and natural sources of climate change.

Global coordinate system The ATST project has adopted the global coordinate system with the origin of coordinates at the "crossover" point of the mechanical altitude and azimuth axes. The x and y axes are in the horizontal plane, and the z-axis is vertical. "Up" is +z, and the mirror is displaced in the +y direction. The positive x direction is determined by the right-hand rule.

Global interlock system The GIS provides a redundant, stand-alone safety mechanism for personnel and equipment.

Global positioning system / satellites (GPS) Originally developed for the military, this is a set of satellites that constantly broadcast signals down to earth which when received, are used to work out the exact geographical location of the receiver. This has recently appeared as an option for automobiles, and can also be referred to by the abbreviation satnav (satellite navigation) in many countries.

Global warming The gradual warming of the earth's atmosphere due to the greenhouse effect.

Globalization The process whereby competition has expanded geographically through the expansion in freer trade arrangements and improvements in transportation and communications.

Gloss An appearance characteristic that gives the perception of the brightness and luster of a smooth, polished surface.

Glove box A small, enclosed storage area in the front passenger compartment of a vehicle.

Glove compartment A term used for glove box.

Glow plug An electrical plug used to preheat the combustion chamber to aid in starting a cold diesel engine.

Goodies 1. Prizes and awards for race-car winners or show-car participants.

2. High performance equipment on an engine.

Goosed moose A car having a severe forward rake.

Graceful degradation Graceful degradation is the quality of a product, system or design such that when something goes wrong, it happens a little at a time and with plenty of opportunity to take action to correct the problem or at least protect against its worst consequences.

Graceful degradation Graceful Degradation is the quality of a product, system or design such that when something goes wrong, it happens a little at a time and with plenty of opportunity to take action to correct the problem or at least protect against its worst consequences.

Grade 1. A hill; generally a step hill.

2. Sort by size or quality.

Grade labelling Tire ratings that may be used as a guide in buying, enabling the buyer to compare, at a glance, the apparent value of one tire with that of another.

Grashof number Parameter used in convection study to express the ratio of buoyant force to viscous force.

Graticule The grid lines on a screen for measuring oscilloscope traces patterns or waveforms.

Gravimetric energy density Potential energy in a given weight of fuel.

Gravitational waves Einsteinian distortions of the space-time medium predicted by general relativity theory (not yet detected as of November 1995).

Gravity assist trajectory A trajectory in which angular momentum is transferred from an orbiting planet to a spacecraft approaching from behind. The result is an increase in the spacecraft's velocity.

Gravity bleeding A method of purging air from a system by allowing the fluid to force air out of an opened bleeder valve by its own weight.

Gravity casting A process that relies on gravity to pull the material to the bottom of the mold during casting.

Gravity energized A one-way roller clutch that has no mechanical means for holding the roller in contact with the cam and race.

Gravity field survey The mapping of a planet's mass distribution by studying variations in the in the planet's gravity field strength made evident by minute Doppler shifts in an orbiting spacecraft's radio signal.

Gravity The force responsible for the mutual attraction of separate masses.

Gravity waves Certain atmospheric waves within a planet's atmosphere.

Gray market The individual importing of a foreign car into the United States.

Gray market vehicle Cars that have been imported into the country but not through authorized retailers. They may not meet american safety and emission standards. These cars may have significantly lower values than cars imported through normal channels.

Gray water Used wash water from the kitchen and bathroom of a mobile home or recreational vehicle.

Grease fitting The orifice, having a check valve, that a grease gun fits on during the greasing process.

Grease Lubricant consisting of a stable mixture of oil, soap thickeners (usually lithium, sodium, or calcium), and other ingredients for the desired physical or operating characteristics.

Grease plug A small plug in a lubrication hole where there is a danger of over lubrication by high-pressure equipment. It is removed for lubrication and then reinstalled.

Grease seal A spring loaded elastomer seal commonly used to seal bearings. Sometimes called a "lip seal". Not a good choice for sealing the bearing casing of a pump. A labyrinth or face seal would be a better choice.

Greasy friction The friction between two surfaces coated with a thin layer of oil.

Great circle An imaginary circle on the surface of a sphere whose center is at the center of the sphere.

Green 1. Aq colour.

2. A term used for green flag.

Green flag The flag used to signal the start of a race.

Green goo A term used for green Loctite.

Green Loctite A high-temperature anaerobic compound used for securing parts together.

Greenhouse effect A greenhouse is warmed because glass allows the sun's radiant heat to enter but prevents radiant heat from leaving. Global warming is caused by some gases in the atmosphere that like greenhouse glass, hence, the term greenhouse effect.

Greenhouse gases Group of gases which individually act to trap solar energy near the earth. Ghgs for which emission levels have been estimated are carbon dioxide, methane, nitrous oxide, sulphur hexafluoride, carbon tetra fluoride , carbon hexafluoride and hydro fluorocarbons.

Greenhouse Used in automotive circles to describe all of the windows enclosing the passenger compartment.

GREGOR German 1.5-m solar telescope in the Canaries

Gregorian optics station The Upper GOS contains primarily calibration optics (retarders, waveplates, polarizers, etc.). The Lower GOS contains the avefront sensor, alignment targets, the adjustable iris, and other engineering optics.

Gregorian telescope The Gregorian design consists of a concave elliptical secondary mirror, located outside the prime focus and forming an image on the axis of the primary.

Grenade motor An engine that is expected to deliver a very high horsepower for a short time, before it blows up.

Grid 1. A wire mesh, such as that used for a grille.

2. The starting position of race cars based on their qualifying order.

Grid growth A condition where the battery grid grows little metallic fingers, extending through the separators, shorting out the plates.

Grille An ornamental opening in the front of a vehicle through which air is delivered to the radiator.

Grind 1. To use an abrasive wheel to remove metal.

2. The specific contour of a camshaft lobe, such as a quarter-race grind.

3. A long race with emphasis on endurance rather than performance.

Grip A term used for traction.

Grit Fine dust or dirt.

Grocery getter A car used for everyday transportation.

Grommet 1. A reinforced-metal eyelet through which a fastener is attached.

2. A rubber or plastic eyelet inserted in a hole to protect wires that pass through it.

Groove 1. The recess in a part to hold a ring or snap ring.

2. Space between two adjacent tread ribs or lugs; the lowered section of the tire tread.

3. The specific path through a turn in an oval track or in road racing.

4. A rut.

Groove weld A weld made in a groove between the work pieces.

Gross axle-weight rating (GAWR) The maximum allowable fully loaded weight of a given axle.

Gross combination weight (GCW) The total weight of a fully equipped vehicle including payload, fuel, and driver.

Gross horsepower The maximum engine output as measured on a dynamometer. Also known as gross torque.

Gross torque A term used for gross horsepower.

Gross trailer weight (GTW) The sum of the weight of the trailer(s) and the payload.

Gross valve lift The valve lift, including the running-valve clearance.

Gross vehicle weight (GVW) Maximum legal weight at which a vehicle can be operated.

Gross vehicle weight rating The curb weight of the vehicle plus the maximum load it is designed to carry.

Ground A conducting connection, whether intentional or accidental, by which an electric circuit or equipment is connected to earth, or to some conducting body that serves in place of earth. (National Electric Code)

Ground Controlled Circuit A circuit which energizes by applying a ground usually through a control module

Ground effect The reduced air pressure under a vehicle that allows the normal pressure above the vehicle to push the vehicle downward, providing a better grip on the road surface.

Ground electrode A part of the lower end of the spark-plug shell making up the electrode gap, providing a path for current flow while jumping the gap.

Ground Trace The part of a waveform that defines the circuit ground

Ground wave An electromagnetic wave that travels along the Earth's surface; usually emitted by a radio transmitter.

Grounded circuit A condition that allows current to return to ground before it has reached the intended load component.

Grounded Connected to or in contact with earth or connected to some extended conductive body which serves instead of the earth.

Ground-end spring A spring that has flattened ends, usually by grinding, to better fit the perches or retainers.

Grounding A permanent and continuous conductive path to earth with sufficient capacity to carry any fault current liable to be imposed on it, and of a sufficiently low impedance to limit the voltage rise above ground.

Ground-return system A common system of electrical wiring whereby the chassis or frame serves as a part of the electrical circuit.

Ground-truthed When data is collected by remote sensing techniques, ground-truthing confirms that the information is accurate. Ground-truthing is collecting data by non-remote sensing means.

Group injection A method of injecting fuel into the manifold areas of several cylinders of an electronic fuel-injection system while at the same time, actually entering each cylinder as its intake valve is opened.

Group technology An engineering and manufacturing philosophy which identifies "sameness" of parts, equipment, or processes. it provides for the rapid retrieval of existing designs and anticipates a cellular type production equipment layout.

Groupthink An undesirable condition in which all members of a group (e.g. a project team) begin to think alike or pretend to think alike. No members are then willing to raise objections or concerns about a project even though they are legitimate and based on hard data.

Groupware Groupware is any type of software designed for groups and for communication, including email, video conferencing, workflow, chat, and collaborative editing systems. This technology may be used to communicate, cooperate, coordinate, solve problems, compete, or negotiate. While traditional technologies such as the telephone qualify as groupware, the term is ordinarily used to refer to a specific class of technologies relying on modern computer networks, such as email, newsgroups, videophones, or chat. Groupware technologies are typically categorized along two primary dimensions: a)whether users of the groupware are working together at the same time ("realtime" or "synchronous" groupware) or different times ("asynchronous" groupware), and b) whether users are working together in the same place ("collocated" or "face-to-face") or in different places ("non-collocated" or "distance").

Guide 1. A part used to align an assembly or another part.

2. The support for a valve stem in the head.

Guillotine A device equipped with explosive blades used to cut cables, water lines, wires, etc. during separation of spacecraft modules.

Gullwing A coupe-body style with the doors hinged so they open upward.

Gulp valve A valve that opens in an air-injection system to admit extra air into the intake manifold upon deceleration, thus "leaning out" the mixture to prevent backfiring.

Gum Residue that remains when gasoline is allowed to sit for a period of time.

Gumballs 1. Drag racing tires made of especially soft, sticky compound.

2. The rotating light in a hemispherical housing on top of a police car or emergency vehicle.

Gun it To rev up an engine suddenly.

Gun-drilled oil holes Holes that are drilled to allow oil to be fed to the piston pin.

Gurgle method A method of adding refrigerant from small cans to a system without running the engine.

Gusset A triangular piece of metal used to add strength to a corner.

Gut ray That ray in an off-axis optical system that passes through the geometric center of the optical elements. The term corresponds roughly in usage to the chief ray of an axially symmetric optical system.

Gut To remove non-functional parts of a vehicle in order to reduce its weight.

Guts 1. The internal structure of a vehicle.

2. The essential working parts of a device.

Gymkhana An individually clocked maneuverability contest for sports cars over an extremely tight course.

Gyration The circular motion of ions and electrons around magnetic field lines.

Gyroscope A spinning, wheel-like device that resists any force that tries to tilt its axis. Gyroscopes are used for stabilizing the attitude of rockets and spacecraft in motion.

H

Half-Bridge Converter A switching power supply design in which two power switching devices are used to drive the transformer primary.

Half-cycle When the voltage signal form an oxygen sensor crosses over a predetermined threshold. The predetermined thresholds are the rich to lean switch points.

Half-shafts on either side of the differential that transmit power to the drive wheels.

Half-wave rectifier A rectifier that conducts current to the load only during one alternation of a cycle of ac voltage i.e.. An automotive generators out put is half-wave rectified.

Halide Any compound of halogen with another element such as refrigerant

Halide leak detector A device consisting of a tank of acetylene gas, a stove, a chimney, and a search hose used to detect CFC refrigerant leaks by visual means.

Hall- effect sensor A type of electromagnetic sensor used in electronic ignition systems and other sensing circuits. Named for Edwin H. Hall, who discovered the Hall effect

Hall effect The development of a voltage across a semiconductor carrying a current when the semiconductor is in a magnetic field that is at right angles to the current flow.

Hall-effect switch A device that produces a voltage pulse dependent on the presence of a magnetic field. Hall-effect voltage varies as magnetic reactance varies around a current-carrying semiconductor.

Halogen headlamp A sealed-beam headlamp with a small inner bulb filled with halogen which surrounds a tungsten filament. Halogen headlamps may increase luminous intensity at the road surface by 50 to 80 percent, as compared to the long-conventional sealed-beam headlamp systems. Many halogen headlamp systems incorporate high-beam and low-beam in one element, enhancing their serviceability.

Halogen lamp A high-output, white-light lamp used as a headlight. Also known as halogen light.

Halogen leak detector An electronic device used to detect leaks in a refrigerant system.

Halogen light A high-output, white-light lamp used as a headlight. Also known as halogen lamp.

Halogens Refers to any of five nonmetallic and electronegative chemical elements: astatine (At), bromine (Br), chlorine (Cl), fluorine (F), and iodine (I), such as those that may be found in refrigerants.

HALT Highly accelerated life test

Hammer forming To form or shape metal by hitting it with a hammer over, on, or into a base form.

Hand shield A hand-held protective device that has a filter plate and is used in arc welding, arc cutting, and thermal spraying for shielding the eyes, face and neck.

Hand tight To assemble plain or threaded parts as much as can be done by hand, without the aid of a wrench or tool.

Hand Tools Institute (HTI) A trade association that represents automotive hand-tool and tool-box manufacturers.

Hand valve A valve mounted on the steering column or dash, used by the driver to apply trailer brakes independently of the tractor brakes.

Handicap A system that allows two cars from different classes in a drag race to compete with each other.

Handler 1. A driver who is able to get the most out of his or her car.

2. A car that the driver can get the most out of.

Handling The ease of vehicle steering and maneuverability around turns, up hills, etc.

Hang a left To make a left turn.

Hang a right To make a right turn.

Hang it out To deliberately throw the rear end into a slide during a turn.

Hang on unit An under-dash, aftermarket air conditioner. May also refer to other under-dash devices, such as a CB radio.

Hangover Undesired speaker-cone oscillation caused by a transient signal.

Hanle effect Represents the totality of ways in which the scattering polarization can be modified by magnetic fields. The Zeeman and Hanle effects are highly complementary to each other because they respond to magnetic fields in very different parameter regimes.

HAO High-Altitude Observatory

HAP Hazardous Air Pollutant

Hard chrome A heavy-chrome plating applied to metal to increase its durability and resistance to wear.

Hard copy A printed copy of machine output in readable form, such as reports, listings, or graphic images.

Hard disk A magnetic disk capable of storing large quantities of data. Most hard disks reside permanently in their housings; removable hard disk cartridges are less common. Hard disks are the persistent storage memory found in computers, digital cameras and other electronic products. Data access is via a read/write head that moves mechanically across the hard disk surface.

Hard edge A sharp corner where two surfaces meet.

Hard face A seal face either rotating or stationary. The most common materials are silicone carbide, ceramic, tungsten carbide, Stellite, Ni-resist. The hard face must be the wider seal face.

Hard parts the internal engine parts required to rebuild an engine.

Hard pedal 1. A loss in braking efficiency so that an excessive amount of pressure is needed to actuate brakes.

2. A condition whereby the load literally overrides the brakes.

Hard solder Any solder requiring more than 1000 degrees F (538 degrees C) to melt.

Hard spots Areas in the friction surface of a brake drum or rotor that have become harder than the surrounding metal.

Hard throttle A term used when the accelerator is pressed to the floor.

Hardener A chemical component of an epoxy resin that starts a catalytic reaction and causes the mixture to harden.

Hardness Measure of a material's resistance to localised plastic deformation. Most hardness tests involve indentation, but hardness may be reported as resistance to scratching (file test), or rebound of a projectile bounced off the material (scleroscope hardness). Some common measures of indentation hardness are Brinell Hardness Number (HBN) and Rockwell Hardness Number (HRC).

Hardtop A two- or four-door hardtop convertible body type with front and rear seats, generally characterized by the lack of door or "B" pillars.

Hardware 1. The components that make up a computer system, such as a keyboard, a floppy disk drive, and a visual-display terminal.

2. The nuts, bolts, and accessories required to assemble an add-on component.

Hardware configuration item A hardware component of a system, which is designated for configuration management to insure the integrity of the delivered product. It may exist at any level in the system hierarchy, since configuration management must be imposed down to the lowest level where item interchangeability is required. Each is to have (as appropriate) individual design reviews, individual qualification/certification, individual acceptance reviews, and separate operator and maintenance manuals.

Hardware description language A language that describes the physical design, electronic behaviour, logical structure, and system annotation information for circuits. An HDL allows a design to be described in a higher level of abstraction while supporting a logical synthesis path to gate-level implementation.

Hardware software co-design The process and software tools that perform or support hardware/software partitioning, performance evaluation, and design entry for system-level designs that are comprised of both hardware and software elements, as in embedded systems. This includes tools and interfaces that link the design and evaluation steps with code compilation models.

Hardware trigger A trigger is something that causes a data acquisition system to start collecting data. When not generated by software it is known as a hardware or external trigger.

Hardwired The physical connection of two pieces of electronic equipment by means of a cable.

Harmonic balancer A balance shaft or wheel that reduces or eliminates harmonics.

Harmonic damper A term used for harmonic balancer.

Harmonic The rhythmic vibration of a moving part or assembly.

Harmonic vibration Vibtrating in harmony with something near by. This can be a big problem for bearings in stationary or non running equipment.

Harness A group of electrical conductors laced or bundled in a given configuration, usually with several breakouts.

Harness clip A metal or plastic clip used to secure a wiring harness to the car frame or body at various points, providing safe and neat routing.

Harness connector An electrical connector at the end of a wire or harness used to connect the conductor to a device or system.

Hazard analysis Hazard analysis is the detailed examination of a product from the user perspective to detect potential design flaws (possibilities of failure that could cause harm) and to enable manufacturers to correct them before a product is released for use.

Hazard system A driver-controlled system of flashing front-and rear-marker lights to warn approaching motorists of a potential hazard.

Hazardous Material Any material that can cause illness, injury, or death or pollute the air , ground, or water

Head 1. That part of an engine that covers the top of the cylinders and pistons and usually contains the combustion chambers and valve train.

2. That part of a compressor that covers the valve plates and separates the high side from the low side of an air-conditioning system.

Head bolts The bolts that fasten the cylinder heads to the engine block.

Head gasket a gasket used to seal the cylinder head to the engine block.

Head pressure Pressure of the refrigerant from the discharge valve through the lines and condenser to the metering-device inlet.

Head rest / restraint A small padded and adjustable cushion which when present on a car seat can protect against whiplash injuries to the neck.

Head room The distance from the top of an occupant's head to the headliner.

Head The equivalent height of the liquid. 20° C. water is used as the standard where 10 meters (33.9 ft.) of water equals one atmosphere (14.7 psi. or 1 bar). The term head is used instead of pressure in the centrifugal pump business.

Headache rack A rack that is provided at the rear of a truck-cab interior to prevent a load from crashing through the rear window.

Header 1. A vehicle crash.

2. A free-flowing exhaust manifold used on high-performance cars.

Header tank The top and bottom tanks or side tanks of a radiator in which coolant accumulates or is received.

Head-gasket notch A series of one to four notches used to measure the thickness of a gasket to control the piston-to-head clearance on a diesel engine.

Headlamps The lights at the front of a vehicle used to illuminate the road ahead. Also known as headlights.

Head-land ring a top compression ring with an L-shaped cross section.

Headlight delay a system that turns off the headlamps automatically after the operator leaves the vehicle, though the lights were left on.

Headlights a term used for headlamps.

Headliner The interior covering of the roof. Headliners often contain consoles with slots for garage-door openers and other devices, as well as dome lights and wiring for electrical and electronic components attached to the headliner. The covering usually includes a sound-absorbing material.

Heads up display The digital projection of pertinent instrument data onto the lower portion of the windshield on the driver's side. The driver does not have to take his eyes off the road to read his instrument panel.

Head-saver shims Shims that are used to compensate for material removed during head resurfacing to maintain correct overhead-cam drive-chain tension.

Heads-up display (HUD) The images of a vehicle's instrument panel, displayed on the inside surface of the windshield in front of the driver where they can seen without taking the eyes off the road.

Heads-up racer A race driver that is very good.

Heads-up racing Direct, no handicap, competition.

Heat 1. A form of energy released by the burning of fuel.

2. Any temperature above absolute zero.

Heat dam The narrow groove cut into the top of the piston. It restricts the flow of heat down into the piston.

Heat exchanger Device (e.g., a radiator) that is designed to transfer heat from the hot coolant that flows through it to the air blown through it by the fan.

Heat intensity The measurement of heat concentration with a thermometer.

Heat load The "load" imposed on an air conditioner due to ambient temperature, humidity, and all other factors that produce unwanted heat.

Heat of compression An increase of temperature brought by the compression of gas, air, or an air/fuel mixture.

Heat of fusion The amount of latent heat, in Btu, required to cause a change of state from a solid to a liquid.

Heat of respiration The heat given off by ripening vegetables or fruits in the conversion of starches and sugars.

Heat quantity 1. The amount of heat as measured on a thermometer.

2. The amount of heat contained in an object measured in Btu's or calories.

Heat radiation the transmission of heat from one substance to another while passing through, but not heating, intervening substances.

Heat range the measure of a spark plug's ability to transfer heat from the tip of the insulator into the cylinder head.

Heat riser A channel in an intake manifold through which exhaust gas flows in order to heat the manifold, thus aiding in fuel vaporization.

Heat shield A device which protects people or equipment from heat, such as a shield in front of a reentry capsule.

Heat sink 1. A finned, aluminium housing that dissipates heat and serves as a base on which electrical components, such as semiconductor devices, are mounted.

2. A heat dissipating section of the alternator that holds the insulated diodes or all the diodes, and carries heat away from the devices.

Heat tab A small aluminium disk attached to the cylinder head or block having a lead center designed to melt at 260 degrees F (127 degrees C) to indicate if the engine has been overheated.

Heat transfer The movement of heat from a hotter to a colder area by conduction, convection, or radiation.

Heat treatable A metal alloy that can be hardened by heat treatment after being shaped.

Heat treatment A sequence of controlled heating and cooling operations applied to a solid metal to provide the desired properties.

Heat-control valve A thermostatically controlled valve in the exhaust manifold to divert heat to the intake manifold before the engine reaches normal operating temperature.

Heated-air system A term used for heated air-intake system.

Heated-air-intake system A vacuum operated power door in the air-cleaner snorkel that maintains intake air at a near constant temperature by blending ambient air with heated air picked up from a shroud over the exhaust manifold. Also known as heated air system.

Heated-exhaust gas-oxygen sensor (HEGOS) An oxygen sensor that is heated electronically and by engine exhaust so it warms up to normal operating temperature more quickly, allowing the engine to enter closed-loop operation sooner than with a non-heated sensor.

Heater A device used to warm the vehicle interior.

Heater core A small water-to-air radiator-like heat exchanger that provides heat for the passenger compartment.

Heater valve A manual or automatic valve in the heater hose used for opening or closing, providing coolant flow control to the heater core.

Heating The use of an apparatus that produces a relatively high degree of warmth under controlled conditions, by natural or mechanical means, as an aid to ensure personal safety and comfort.

Heating value (TOTAL) The number of British Thermal Units (BTUs) produced by the combustion of one cubic foot of gas at constant pressure, when the products of combustion are cooled to the initial temperature of the gas and air, when the water vapour formed during combustion is condensed, and when all the necessary corrections have been applied.

Heat-riser valve A control valve between the exhaust manifold and exhaust pipe that restricts the flow of exhaust, causing it to flow back through the heat-riser channel under the intake manifold. It acts as an aid to fuel evaporation and to speed engine warm-up.

Heavy-duty coil spring Coil sprints having larger wire diameter, and 3 to 5 percent greater load-carrying capability than a conventional spring.

Heavy-duty shock absorber Shock absorbers having improved seals, a single tube to reduce heat, and a rising rate valve for precise spring control.

Heavy-duty truck Vehicle weighing from 26,001 to 33,001 lbs. Also included off-highway trucks.

Heavy-duty vehicle Any motor vehicle having a gross vehicle-weight rating over 8,500 pounds (3,856 kilograms), that has a vehicle curb weight over 6,000 pounds (2,722 kilograms), or has a frontal area exceeding 45 square feet (4.2 square meters).

Height-sensing proportioning valve A proportioning valve in which hydraulic pressure is adjusted automatically according to the vertical movement of the chassis in relation to the rear axle during braking; sometimes also called a weight-sensing proportioning valve.

Heijunka Keeping total manufacturing volume as constant as possible.

Heim joint A spherical-rod end joint that allows precise adjustment of wheel position.

Heliarc The trade name of a tungsten inert-gas form of welding developed by Linde Welding and Cutting Systems.

Heliarc welding The act of electrically joining two pieces of aluminium or stainless steel using a high-frequency electric weld and an inert gas, such as argon, that is fed around the weld to prevent oxidation.

Helical A device, such as a spring, shaped like a helix or spiral.

Helical compression spring An open-pitch spring, in a variety of configurations, used to resist applied compression forces.

Helical extension spring A spring, usually made of round wire, that exerts a pulling force with various types of end hooks or loops by which they are attached to the load.

Helical gear A gear having teeth cut at an angle across its face, or diagonally to the rotational axis.

Helical spring Springs made of bar stock or wire, coiled into a helical form, including compression, extension, and torsion springs.

Helper spring An additional spring device that permits a greater load on an axle.

Hemi cammer An engine having both hemispherical combustion chambers and an overhead camshaft.

Hemi Half of a sphere.

Hemi head A cylinder head with hemispherical combustion chambers providing room for extra large valves or multi-valve arrangements and providing an ideal shape for smooth, even combustion.

Hemispherical combustion chamber A type of combustion chamber that is shaped like half of a circle. This combustion chamber has the valves on either side with the spark plug in the center.

Henry (H) Unit of measurement of inductance. A coil has one Henry of inductance if an EMF of one volt is induced when current through an inductor is changing at rate of one ampere per second.

Hermaphrodite calliper A combination tool having a calliper leg and a leg from a divider. Used in testing centered work and laying off distances from the edge of a piece or work.

Herringbone gear A gear cut in the V-shaped pattern of a small bony extension of a herring's spine.

Hertz (Hz) The SI unit of measurement for frequency, named in honour of Heinrich Hertz who discovered radio waves. One hertz equals one cycle per second.

Hertz A unit of frequency of any regularly occurring event. Named for Heinrich Rudolf Hertz (1859 - 1894) a German physicist.

Hertz Cycle Named for Heinrich Rudolf Hertz (1857 - 1894) The number of electrical events per second.

Hesitation A momentary lack of throttle response or pause in the rate of acceleration.

Hidden failure A failure which, on its own, does not become evident to the operator or user under normal circumstances.

Hierarchical design Hierarchical design is a design methodology where portions of large designs are divided into manageable sections or sub-blocks that may be created, represented symbolically, designed, and then connected together when completed. This methodology allows different parts of the design to be worked on in parallel.

High definition DVD High definition DVD: next generation of the DVD system commonly in use today. High definition DVD has a storage capacity of approx. 15 GB, whereas the usual DVD today has a storage capacity of "only" 4.7 GB.

High gear The condition achieved from a planetary gear set when any two members, sun gear and planet gears or planet gears and internal gears, are locked together.

High gearing The drive ratio that applies maximum engine output at a relatively high road speed.

High head A term used when the high-side pressure of a system is excessive.

High heat load Refers to the maximum amount of heat that an be absorbed by a refrigerant as it passes through the evaporator.

High Impedance Meter A meter that has a high opposition to the flow of electrical current on the voltage scale usually more than 10 Meg Ohms.

High line Highest specified input operating voltage.

High modulus belt Those belts designed with special reinforcing members with a high degree of elasticity such as steel, fibreglass, or aramid fibre.

High output (hp) A replacement device having a higher output than the original equipment device.

High Pass Filter When a low-level transducer signal is superimposed on a large dc output voltage, a high-pass filter might be useful. This attenuates (removes) low frequencies. Using a cut-off frequency of, say, 4 Hz, will eliminate the dc voltage which has a frequency of zero. A high-pass filter will remove "drift". This can be a particular problem with biological and chemical signals, but not usually with modern electronic signals.

High pedal 1. The position of a brake pedal when it is adjusted near the top of its upward gravel.

2. The position of the clutch pedal when engaged I high gear in the transmission of a Model T.

High riser 1. The raised cylinder heads of the 427 cid Ford engine of the 1960s.

2. The term given to a manifold that places the carburettor above its original design location.

High side A term used for discharge side.

High side drivers Control of the output device by varying the positive side voltage

High spot The outward curvature of a normally flat or smooth surface.

High suction A term used when the low-side pressure of a system is higher than normal due to a malfunction of the system.

High tech A term for developments in technology resulting from new, or new uses of materials and machinery that produce products fully comprehensible only to experts in that particular technology.

High vacuum A vacuum below 500 microns (0.009 psia or 0.66 kpa).

High voltage power supply A power supply with an output voltage of 100V or more. High Voltage power supplies often use air as the primary dielectric medium, avoiding the weight and serviceability problems associated with other dielectrics. Most high voltage supplies are of two types. The first type is simply a step-up power transformer with rectified and filtered (but otherwise unregulated) high voltage secondary. This is the most straightforward, efficient, and lowest cost power supply because of its simplicity. The second basic type is essentially a switcher with a string of voltage doublers that provide the high voltage output. As a switcher, the second type enjoys heat, weight and size advantages over the first type. By employing pulse width modulation, the switcher's output can be regulated. On the negative side, the switcher includes an input energy storage cap that sustains the high voltage output even if the AC power is cut off. This can result in a safety hazard. Product selection requires extra attention to the design features and construction techniques that contribute to safety and protection against failures. Pulse-width modulation and full-wave/ symmetrical cascade high voltage multipliers, with surge limiting resistors, contribute to high efficiency and reliability.

Highly accelerated life test Highly Accelerated Life Test (HALT) is a process developed to uncover design defects and weaknesses in electronic and mechanical assemblies using a vibration system combined with rapid high and low temperature changes. The purpose of HALT is to optimize product reliability by identifying the functional and destructive limits of a product. HALT addresses reliability issues at an early stage in product development.

Hitch receiver A heavy square tube into which the removable shank supporting the hitch ball is inserted.

Hitch weight The load imposed to the tow vehicle's hitch by the trailer tongue, not to exceed 10-15% of the gross trailer weight.

Hohmann transfer orbit An interplanetary trajectory in which a spacecraft is launched into an elliptical solar orbit whose perihelion (inner planet) or aphelion (outer planet) reaches the orbit of the target planet on the opposite side of the Sun. Uses least propellant.

Hold harmless agreement An agreement where one party assumes responsibility.

Hold in Windings A winding that holds the plunger of a solenoid in place after it moves to engage the starter drive

Hold the road To steer and handle well at high speeds.

Hold-down spring A spring tension mechanism that holds a brake show against a backing plate.

Hold-in winding One of two electromagnetic windings inside a solenoid, used to hold the movable core into the solenoid.

Hole theory A theory that states as an electron flows from a negative potential to a positive potential, it leaves behind a hole. According to the hole theory, the hole would move from a positive potential to the negative potential.

Holonic maufacturing systems An HMS is a way of organizing a manufacturing system. The entire range of manufacturing activities from order booking through design, production, and marketing are integrated to realize the agile manufacturing enterprise. In an HMS, key elements such as machines, work centers, plants, parts, products, persons, departments, or divisions have autonomous and cooperative properties. These elements are called "holons". In the system, each holon's activities are determined through the cooperation with other holons, as opposed to being determined by a centralized mechanism.

Honeycomb A converter internal-substrate structure, usually made of ceramic, used to support the catalyst material, such as platinum, that allows maximum catalyst surface exposure to exhaust gases.

Hood scoop An air duct in the hood to allow air to flow directly across the engine.

Hook up 1. To enter a draft in oval-track racing.

2. To set up a car for the best possible traction off the starting line off a drag race.

Hooks Open loops or ends of extension springs.

Hop up To modify an engine for increased performance.

Horizon scanner A scanner which automatically seeks the horizon for purposes of a spacecraft's orientation and control, e.g. one that detects the sharp discontinuity in infrared intensity at the outer edge of the Earth's tropopause.

Horizon The line marking the apparent junction of Earth and sky.

Horizontal Accuracy An indication of how accurately the oscilloscope can measure the time base

Horizontal position 1. Level or parallel to the horizon.

2. The position in which a fillet weld is performed on the upper side of an approximately horizontal surface and against an approximately vertical surface.

3. The position of a groove weld in which the weld axis lies in an approximately horizontal plane and the weld face lies in an approximately vertical plane.

Horizontal Sweep The action of the horizontal system that causes a waveform to be drawn

Horn an electrical noise making device.

Horn brush/slip-ring An electrical contact ring used in the horn circuit, located in the steering wheel, and used to maintain electrical contact when the steering wheel is turned.

Horn relay A relay used to carry the heavy-horn current requiring light current to actuate.

Horse power 33,000 foot pounds per minute. A common method of measuring work.

Hose clamp A device used to attach hoses to the engine, heater core, radiator, and water pump. A popular type of replacement clamp is the high-torque, worm-gear clamp with a carbon steel screw and a stainless-steel band.

Hose-and-line tape A term used for insulation tape.

Hoshin kanri The selection of goals, projects to achieve the goals, designation of people and resources for project completion, and establishment of project metrics.

Host computer A main computer that monitors and controls other computers.

Hot dog A top performer in almost all forms of racing.

Hot drawn A process of forming metal by heating it and pulling it through rollers or dies.

Hot gas The condition of refrigerant as it leaves the compressor until it gives up its heat and condenses.

Hot iron An early term for a hot rod.

Hot patch A patch applied to a tube or tire with use of heat to vulcanize it onto the damaged surface.

Hot rod A passenger car that has been modified and rebuilt for high performance and a distinctive appearance used primarily for straightaway speed or acceleration racing.

Hot setup A combination of modifications and components that enhances maximum performance.

Hot shoe A top race driver.

Hot soak 1. A condition that may occur when an engine is stopped for a period of time after a hot hard run. Causing the gasoline to evaporate out of the carburettor.

2. A period of time where localized combustion heat dissipates through the engine.

Hot spot 1. A localized area where excessive heat can build up, such as around a spark plug, that could cause preignition.

2. An area on an auto body where corrosion is most likely to occur.

Hot tank A large vat containing a hot, caustic chemical solution that is used to clean the engine block and parts.

Hot working The mechanical working of metal at a temperature above it recrystallization point and high enough to prevent strain hardening.

Hotchkiss drive A popular type of drive-line system that features an open or visible drive shaft.

Hot-gas bypass line The line that connects the hot-gas bypass valve outlet to the evaporator outlet.

Hot-gas bypass valve A device used to meter hot gas back to the evaporator, through the bypass line, to prevent condensate from freezing on the core.

Hot-gas defrosting The use of high-pressure gas in the evaporator to remove frost.

Hot-idle compensator A temperature-sensitive carburettor valve that opens when the inlet air temperature exceeds a certain level to allow additional air to enter the intake manifold, preventing an overly rich air/fuel ratio.

Hot-rolled steel (HRS) Steel that is formed while hot.

Hot-tank gloves Long -sleeved, thick rubber gloves for use in the hot tank.

Hot-wire 1. Wire that carries current from the ungrounded terminal of the battery to an electric load.

2. To bypass the ignition switch to start the vehicle.

Hot-wire sensor An electrical device inside the mass-airflow meter that measures air flow and density.

Hour angle The angular distance of a celestial object measured westward along the celestial equator from the zenith crossing.

Hourglass frame A term used for X-frame.

Housekeeping The type of safety in the shop that keeps floors, walls, and windows clean, lighting proper, and containers and tool storage correct.

Housing bore The inside diameter of a bearing housing.

Housing breather A venting device that allows air to enter or leave the axle housing.

Human factors Human factors refers to the characteristics of human beings that are applicable to the design of systems and devices of all kinds. It furthers serious consideration of knowledge about the assignment of appropriate functions for humans and machines, whether people serve as operators, maintainers, or users in the system. And, it advocates systematic use of such knowledge to achieve compatibility in the design of interactive systems of people, machines, and environments to ensure their effectiveness, safety, and ease of performance.

Hunting shift Tendency of an automatic transmission to upshift, downshift, and upshift rapidly.

Hunting The tendency of a system to oscillate around its normal position.

Hurdle rate The minimum return on investment or internal rate of return percentage a new product must meet or exceed for it to be approved for investment with development.

Hybrid 1. A vehicle having two separate propulsion systems, such as a gasoline engine and an electric motor.

2. A vehicle having its major components from more than one source, such as a Ford engine in a Chevrolet chassis.

Hybrid battery A battery that combines the advantages of both the low maintenance and the maintenance-free battery.

Hybrid electric vehicle (HEV) A vehicle combining a battery-powered electric motor with a traditional internal combustion engine. The vehicle can run on either the battery or the engine or both simultaneously, depending on the performance objectives for the vehicle.

Hydraulic balance A method of reducing mechanical seal face loading by reducing the seal face closing area.

Hydraulic brakes A braking system using hydraulic pressure to press the brake shoes against the brake drums.

Hydraulic clutch A clutch that is actuated by hydraulic pressure.

Hydraulic force Occurs any time pressure acts on a seal face area. Force times distance divided by time is a measurement of work done.

Hydraulic lever A means to multiply force using additional stages of mechanical leverage

Hydraulic lifter Valve lifter located between the camshaft and pushrod that uses internal oil pressure to sauce the lifter to expand lengthwise.

Hydraulic pressure A pressure exerted through the medium of a liquid.

Hydraulic principles The use of a liquid under pressure to transfer force or motion, or to increase an applied force.

Hydraulic valve lifter A device using oil pressure to adjust its length and maintain zero valve clearance so that the valve noise is at a minimum. Also referred to as hydro.

Hydraulics 1. The study of liquids and their use to transmit force and motion.

2. Hydraulic ram units, installed in a low rider at each suspension point, to lower or raise the ride height. Also referred to as hydro.

Hydrazine A rocket fuel which burns spontaneously with nitric acid or nitrogen tetroxide. Can also be used as a mono-propellant: when passed through an iridium-bearing catalyst, it decomposes at high temperature into constituent gases of ammonia, nitrogen and hydrogen. Used in small thrusters for orbit modification and attitude control of spacecraft.

Hydrides Chemical compounds formed when hydrogen gas reacts with metals. Used for storing hydrogen gas.

Hydro 1. A term used for hydraulics.

2. A term used for hydraulic-valve lifter.

3. A short term for hydromatic, a transmission built by General Motors.

Hydro cam A camshaft especially designed to operate with hydraulic valve lifters.

Hydro-boost A type of brake power booster that uses hydraulic pressure provided by the power-steering pump to reduce pedal effort.

Hydrochlorofluorocarbon A group of refrigerants that contain the chlorine (CI) atom and the hydrogen (H) atom, which causes the chlorine atom to dissipate more rapidly in the atmosphere.

Hydrodynamic force Generated at the seal faces because, for all practical purposes, liquid is not compressible.

Hydrodynamic seal Special geometric features on the seal face that provide lift by taking advantage of the rotation of one seal face upon the other.

Hydroforming Hydroforming is a metal forming process in which the metal tube blank is formed into a desired shape with the use of internal water hydraulic pressure. Hydroforming is applicable in any industry where complex shapes must be formed with a high degree of precision.

Hydrolyzing action The corrosive action within the air-conditioning system induced by a weak solution of hydrochloric acid formed by excessive moisture chemically reacting with the refrigerant.

Hydrometer An instrument used to measure the specific gravity of a liquid. A battery hydrometer is calibrated to read the expected specific gravity of the batteries electrolyte.

Hydrostatic seal Maintains a controlled gap between the seal faces by balancing the open and closing forces. There is a small amount of leakage across the faces when the shaft is rotating. Used in some compressor applications, but not very practical for the chemicals found in the process industry.

Hydyne A rocket fuel comprised of 60% UDMH and 40% diethylene-triamine.

Hypoid gear The combination of a ring gear and a pinion gear in which the pinion meshes with the ring gear below the centerline of the ring gear. Commonly used in automotive final drives and starter motors.

Hysteresis 1. The property of a magnetic substance that causes magnetization to lag behind the force that produces it.

2. A variable input voltage threshold determined by the logic state of the output of the circuit.

Hysteresis bands The area between two curves on a graph that indicates when a torque converter locks and unlocks. Hysteresis bands exist between graph curves that indicate when a transmission will upshift or downshift.

Hysteresis loop A closed curve that shows, for each value of magnetizing force, two values of the magnetic flux density in a cyclically magnetized material: one when the magnetizing force is increasing, the other when it is decreasing.

Hysteresis loss Energy dissipated due to molecular friction as domains move through cycles of magnetization.

Hz Abbreviation for Hertz.

I

Ideation Ideation is the idea generation phase or stage of new product development. These ideas came come from internal sources such as marketing or engineering as well as external sources such as customers, suppliers, retailers or consultants.

Idle air bleed valve A device used to control the amount of airflow into the idle fuel mixture prior entering the carburettor idle system (M/C) solenoid on

Idle circuit The system within a carburettor that meters fuel when the engine is running at low rpm.

Idle compensator A term used for engine-idle compensator or hot-idle compensator.

Idle limiter A device that controls the maximum richness of the idle air/fuel mixture in the carburettor.

Idle load compensator (ILC) This device is used to control throttle angle during long deceleration such as coasting down a long grade and extends the WOT position to prevent engine stalls.

Idle mixture The air/fuel mixture supplied to the engine while idling.

Idle port An opening into the carburettor throttle body through which the idle system discharges fuel.

Idle relearn The routine of the computer that resets idle requirements when a disconnected battery has been reconnected.

Idle Speed Control Motor Regulates the throttle valve position and is controlled by the computer

Idle speed The speed at which an engine idles in revolutions per minute (rpm), usually between 600 and 850 rpm. The idle speed is specified on the under-hood emissions decal.

Idle system The passages of a carburettor though which fuel is fed when the engine is idling.

Idle The engine speed with no load and the accelerator pedal fully released.

Idle vent An opening in an enclosed chamber through which air can pass under idle conditions.

Idler arm A pivoting component that supports the right side of the steering relay rod in much the same manner as the pitman arm supports the left side.

Idler eccentric A device used in a belt-drive system using the idler pulley as a means of tightening the belt.

Idler gear 1. A gear connecting two other gears in a manner so that they will turn in the same direction.

2. A gear that transmits movement between the drive and the driven gears, but does not affect the speed relationship of those gears.

Idler pulley A pulley that is used to adjust the belt in a belt-driven system.

Ignition advance The moving forward, in time, of the ignition spark relative to the piston position.

Ignition coil A transformer containing a primary and secondary winding that acts to boost the battery voltage of 12 volts to as much as 30,000 volts to fire the spark plugs.

Ignition distributor A term used for distributor.

Ignition map A chart showing the precise advance and retard of the ignition on an electronic control-equipped engine.

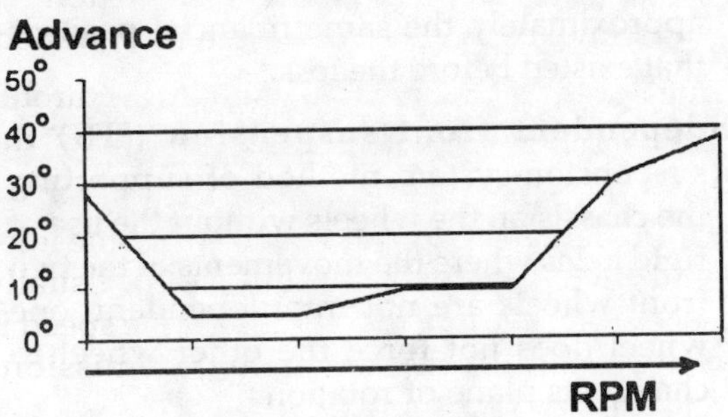

Fig. Ignition map

Ignition points A term used for points.

Ignition Reference Signal Signal used by the ECM to determine engine RPM and crankshaft position.

Ignition reserve The difference between the minimum available and maximum required voltage for proper operation.

Ignition resistor A resistance element in series with the primary circuit to reduce the voltage supplied to the coil during engine operation.

Ignition retard The moving back in time of the ignition spark relative to the position of the piston.

Ignition switch A five-position switch that is the power distribution point for most of the vehicle's primary electrical systems. The spring loaded START position provides momentary contact and automatically moves to the RUN position when the key is released. The other switch detent positions are ACCESSORIES, LOCK, and OFF.

Ignition System Electrical system devised to produce timed sparks from engine spark plugs. Consisting of a battery, induction coil, capacitor, distributor, spark plugs and relevant switches and wiring.

Ignition temperature The lowest temperature at which a combustible material will ignite and continue to burn independent of the heat source.

Ignition timing The timing of the spark, expressed in crankshaft degrees, in relation to top dead center.

Impact strength Energy required to fracture a specimen subjected to shock loading, as in an impact test. Alternate terms are impact energy, impact value, impact resistance and energy absorption. It is an indication of the toughness of the material.

Impact test A method for determining behaviour of material subjected to shock loading in bending, tension, or torsion. The quantity usually measured is the energy absorbed in breaking the specimen in a single blow, as in the Charpy Impact Test, Izod Impact Test and Tension Impact Test. Impact tests also are performed by subjecting specimens to multiple blows of increasing intensity, as in the drop ball impact test, and repeated blow impact test. Impact resilience and scleroscope hardness are determined in non destructive impact tests.

Impedance (Z) Total resistance to flow of an alternating current as a result of resistance and reactance. Also the ratio of voltage to current in AC circuits, containing both resistance and reactance terms, usually expressed as ohms.

Implode The opposite of explode. Bubbles implode in the higher pressure areas of the pump making noise and causing damage to the metal parts. This is normally called cavitation.

In-circuit test In-circuit test is a combination of hardware and software that identifies manufacturing induced faults of printed circuit board assemblies by isolating and individually testing devices using a bed-of-nails fixture. Potential faults include shorts, opens, wrong components, missing components, etc.

Inclination The angular distance between a satellite's orbital plane and the equator of its primary.

Inclined surface 1. A slope or slanted surface 2. A valve rotator.

Included angle 1. Sum of the kingpin or ball-joint inclination angle and the camber angle of a vehicle's front wheel.

2. The sum of the camber angle added to the steering axis-inclination angle.

3. The angle included between the outer surfaces of the link-plate contours of a silent chain that meshes with the sprocket tooth form.

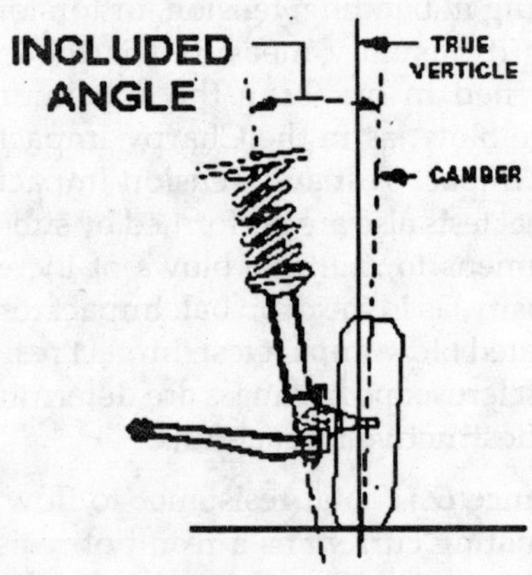

Fig. Included angle

Inclusion A non metallic slug of material which has become entangled in the metal during its manufacture. A severe problem in thin cross section metal bellows manufacture.

Incomplete joint penetration Joint penetration that is unintentionally less than the thickness of the weld joint.

Incompressible A volume of liquid that remains the same under pressure

Incremental development A hardware/software development process that produces a partial implementation and then gradually adds preplanned functionality or performance in subsequent increments. This contrasts with the waterfall model where all functionality is delivered at one time at the conclusion of the project. Similar to the spiral development model.

Indemnity A legal principle specifying that the insured not collect more than the actual cash value of a loss but be restored to approximately the same financial position that existed before the loss.

Independent front suspension (IFS) A suspension-system method of supporting the chassis on the wheels without the use of rigid axles where the movements of the two front wheels are not interdependent; one wheel does not force the other wheel to change its plane of rotation.

Independent publishers Publishers that provide service and repair information on a variety of subjects, such as automobiles.

Independent rear suspension (IRS) 1. A suspension system in which both rear wheels are free to move independently of the other, providing a reduction of unsprung weight as well as overall vehicle-weight reduction.

2. An independent suspension system at the rear of a vehicle.

Independent repair shop Small service outlet offering specialized repair services. They usually do not sell gasoline.

Independent service General or specialty service provided by independent garages on all types and makes of vehicles.

Independent suspension A suspension design that lets each wheel move up and down independently of the others. A vehicle can have two-wheel or four-wheel independent suspension; sportier models have four-wheel independent suspension.

Index 1. To examine and adjust rod journal spacing and alignment.

2. The elapsed time assigned to various classes in drag racing used to calculate starting-time differences for handicap or bracket racing.

3. Installing a camshaft relative to the crankshaft so the valves open and close at precisely the right time.

Index position The index position is a specific location within the range of travel of a mechanism that has a unique sensor or marker to identify it.

Indicated horsepower (IHP) A measurement of power developed by the combustion processes in the cylinders, not considering friction.

Indicated mean effective pressure (IMEP) The average of the pressures developed in the cylinders during the four-stroke cycle, intake, combustion, exhaust, and compression.

Indicated speed The vehicle as indicated on a speedometer.

Indicated torque Torque as calculated from the indicated mean effective pressure, which is an actual indication of output developed within the cylinders.

Indirect costs Costs that are incurred for common or joint objectives that can not be readily identified to a single cost objective and readily treated as a direct cost. A cost that is allocated as opposed to being traced.

Indirect injection (IDI) A system in which fuel is injected into a prechamber, where it is ignited before entering the main combustion chamber.

Indium One of the chemical elements. Atomic number 49. Symbol In. With three electrons in its valance shell it is used as a doping agent to create a "P" type semiconductor.

Induced current Current that flows as a result of an Induced EMF (Electromotive Force).

Inducer A small axial flow vane that attaches to the impeller of a centrifugal pump to increase the N.P.S.H. available.

In-duct sensor A thermistor used in automatic temperature control units for sensing the in-duct returns air temperature.

Inductance (L) Inherent reactive property, measured in henrys, of an electric circuit or circuit element that opposes a change in current flow. Hence, inductance causes current changes to lag behind voltage changes.

Induction 1. A synonym or intake.

2. A process whereby an electromagnet or electrically charged object transmits magnetism or an electric current to a nearby object without making physical contact.

3. The process by which a voltage is induced into a conductor by a changing magnetic field.

Induction coil An electrical device that receives low battery voltage in its primary winding and delivers pulsating high voltage in its secondary winding as the primary field is energized and collapsed by use of a set of points.

Induction hardening A process of hardening the surface of a part by placing electric coils next to it.

Induction motor The most common type used in industry. Has a slippage of 2 to 5 percent compared to synchronous motors.

Inductive circuit Circuit in which an EMF is produced by a changing current.

Inductive reactance An opposing current or voltage created in a conductor whenever there is a charging current flow in that conductor.

Inductive Reluctance A statement of a material's ability to strengthen the magnetic field around it

Inductor A coil or component with the properties of inductance.

Industrial design Industrial Design is the design that is done in companies and consultancies by people trained in industrial design, or in art and design schools in general. Industrial design focuses on the physical form and interactive properties as opposed to the functioning of the product or system.

Industrial-hardened Computer equipment installed to withstand the effects of the manufacturing environment, including dirt, heat, shock, and vibration, as well as insulating the equipment from poor-quality electrical power.

Inert gas A gas that normally does not combine chemically with materials.

Inertia 1. The property of a moving object that causes it to resist any change in speed or direction.

2. The tendency of an object in motion to keep moving and the tendency of an object at rest to remain at rest.

Inertia lock retractor A passive seat-belt system that used a pendulum mechanism to lock the belt tightly during a sudden change in movement.

Inertia weight A small, steel flywheel used to dampen drive-line vibrations on certain applications; generally located on the slip yoke, drive pinion yoke, or near an end of the drive shaft.

Inertial guidance An on-board system for launch vehicles and spacecraft where gyroscopes, accelerometers and other devices satisfy guidance requirements.

Inertial measurement unit An on-board instrument system that measures the attitude of a spacecraft. It includes accelerometers and gyroscopes.

Inertial scavenging An exhaust system that uses the momentum of flowing gases to pull more gasses through.

Infant mortality The high conditional probability of failure due to manufacturing defects, design, installation, or startup procedures during the period immediately after an item enters service.

Inferior conjunction Alignment of Earth, Sun, and an inferior planet on the same side of the Sun.

Inferior planets Planets whose orbits are closer to the Sun than Earth's, i.e. Mercury and Venus. Also called inner planets.

Infield an area enclosed by an oval track or a road course.

Infinite Having no limits whatsoever. Boundless, thus, an infinite number is one more than the largest possible number.

Inflatable tubular restraint This tube of woven material is stiffer and stays inflated longer than a traditional airbag cushion. The tube protects the occupant's head and torso in a side impact, in part by keeping them away from the point of intrusion. The uninflated tube is tucked into the edge of the roof headliner. The tube is attached at the base of the a-pillar in front of the occupant, and at the roofline behind the occupant. When it inflates, the tube angles across the window to keep the occupants head from hitting the window glass or metal side pillar. Because of the longer time the tubular restraint stays inflated, it is expected to also offer protection in a rollover.

Inflation pressure The amount of air pressure required to inflate a tire, measured in psi or kpa.

Information center The electronic display in some luxury vehicles that shows such data as the date, inside and outside temperature, trip distance traveled and distance remaining, time of departure and estimated time of arrival, average and current speed, average and instantaneous fuel economy, fuel used and remaining, and often, an appointment reminder.

Information management task The task of taking a specific product from order-taking through detailed scheduling to delivery.

Information sheet A form that is used to record pertinent information about a vehicle.

Information superhighway A popular buzzword to describe the internet.

Informative inspection A form of inspection used to determine non-conforming product.

Infotainment The telematics market is slowly evolving into "infotainment" (information and entertainment) to combine everything from safety and security to routing and entertainment including Internet access and video games. Vehicle manufacturers increasingly offer infotainment systems either as original equipment or optional packages.

Infrared Electromagnetic radiation of wavelengths between 7500 A, the limit of the visible light spectrum at the red end, and centimetric radio waves.

Infrared exhaust analyzer An instrument able to detect and measure hydrocarbons, carbon monoxide, carbon dioxide, and oxygen levels.

Infrared radiometer A telescope based instrument that measures the intensity of infrared energy radiated by the targets.

Inhibitor A substance used to reduce the rate of a chemical or electrochemical reaction.

Initial graphics exchange specification Initial graphics exchange specification (iges) is a neutral file format used to exchange vector and text data among cad systems. It was the first widely-used neutral file format used for mechanical cad data interchange.

Injection angle The angle at which a spacecraft's return trajectory intersects the Earth's atmosphere.

Injection cylinder A term used for oil-injection cylinder.

Injection valve A term used for fuel injector.

Injector 1. The tube or nozzle through which fuel is injected into the intake airstream of the combustion chamber.

2. A term used for fuel injector.

Injector Bobbin 1. A small plastic device the injector windings are mounted on.

Injector body The housing for a fuel-injection system.

Injector fuel The fuel that is stored inside the injector prior to use.

Injector tube The copper or brass tube used in diesel cylinder heads to col fuel injectors and seal combustion pressures.

Insolation The incoming solar radiation that reaches a planet and its atmosphere or, by extension, any object exposed to solar rays.

Inspect To examine a component or system for any audible or visual signs of a defect.

Inspection maintenance (I/M) The periodic and systematic inspection and maintenance of a vehicle's ignition, fuel, and emissions-control systems, commonly used to refer to state / county emissions inspection and testing programs

Inspection The process of measuring, examining, testing, gauging, or otherwise comparing the unit with the applicable requirements or specifications.

Install Contractor shall construct, erect, or set in place for the intended use and includes to finish and to clean, as applicable in each instance.

Instance 1. As used in product design, an instance is a reference to a geometric object that allows the same geometry to be located at several places in a geometric model assembly without actually copying the geometry. When the original geometry is modified the modifications automatically appear at every instance location

2. With product structures, an instance is a reference to a part. It allows the same part to be used in several assemblies without copying all part information into the assembly.

Instant center An imaginary center point around which a wheel appears to pivot.

Instrument Any item of electrical or electronic equipment which is designed to carry out a specific function or set of functions. For example an electronic balance, a gas analyser or a chromatograph.

Instrument Panel A grouping of indicator lights and gauges designed to inform the driver of curent vehicle operations.

Instrumentation systems Systems of instruments used to observe and control a physical object or process.

Insulated return system The ground wire of a two-wire electrical system.

Insulation Any material that blocks the flow of electricity. Any material containing more than 4 elecrrons in its valance shell.

Insulation resistance The resistance of an insulating material to the flow of current resulting from an impressed dc voltage. Usually measured in megohms.

Insulation tape Tape (usually cork) used to wrap refrigeration hoses and lines to prevent condensate drip.

Insulator 1.A material with high electrical resistance that supports or separates conductors to prevent undesired flow of current to other objects.

2. A material that contains less than four electrons in its valance shell.

Insulator ribs In a spark plug, unevenly spaced circular ridges on the upper section of the insulator to reduce or prevent flashover of the high voltage from the terminal to the lower shell.

Insurance Institute for Highway Safety A highway and vehicle safety research organization.

Insurance institute for highway safety Sister organization to the highway loss data institute. Funded by insurance companies.

Insurance verification Proof that the borrower has auto insurance.

Intake air temperature sensor (IAT) A thermistor device that senses intake air temperature by changing it's resistance as a negative temperature coefficient resister.

Intake manifold A metal component used to duct 1. The air/fuel mixture from the carburettor to the intake parts.

2. Air in an injected engine to the intake ports.

Intake over exhaust (IOE) An engine design having intake valves in the cylinder head and exhaust valves in the block

Intake ports 1. Passages in the cylinder head that direct air/fuel mixture from the intake manifold to the intake valves.

2. Passages in an L-head engine block that direct the air/fuel mixture from the intake manifold to the intake valves.

3. Passages in the head or block of an injected engine that direct air to the intake valves from the intake manifold.

Intake stroke A downward stroke of a piston that draws the air/fuel mixture into the cylinder.

Intake The point at which a fluid or gas enters a pipe or channel. The act of taking in.

Intake valve A valve that opens to admit the air/fuel mixture.

Integer A positive or negative whole number, or 0.

Integral architecture An integral architecture (as opposed to a modular architecture) is a product architecture where 1. the functions of a product are performed using more than one physical building block (e.g., subsystems or subassemblies),

2. a single building block performs multiple functions,

3. the functions are closely coupled and are tightly synchronized,

4. the functions are in close physical proximity, and

5. the interactions between building blocks or interfaces are greater in number and are less well-defined or standardized.

Integral Internal or built-in as a part of the whole.

Integral valve guide A valve guide machined into the cylinder head.

Integral valve seat A valve seat machined into the cylinder head.

Integral-type power steering A steering system with a built-in power assist.

Integrated ABS An antilock brake system in which the ABS hydraulic components, the standard brake hydraulic components, and a hydraulic power booster are joined in a single, integrated hydraulic system.

Integrated child seats May also be called integrated child-safety seats or integrated child-restraint seats. Built-in child seats that fold out of the seatback of a rear seat. Sedans with this option usually have one in the center of the rear seat; minivans may have one or two in the middle seating positions. While nhtsa and every other safety organization stress that any child-restraint seat is better than none, built-in child-restraint seats are considered the safest alternative because they are more securely anchored than a seat attached to seat belts.

Integrated Direct Ignition The type of ignition system used on GM engines that produces the ignition spark without the aid of a distributor

Integrated magnetics A magnetic component in which separate circuit elements share a common core segment.

Integrated product team A cross-functional team consisting of representatives from marketing, engineering, manufacturing, finance. Purchasing, test, quality, finance and any other required disciplines with responsibility for developing a product or product subsystem. This team is empowered to represent the functional disciplines and develop a product by addressing its life cycle requirements including its product and support.

Integrated safety belts Also called integrated seat belts. Lap-and-shoulder belts that are built into the seat rather than anchored to the side pillar and the floor pan. A recent innovation, this arrangement ensures better belt fit and ensures that the belt stays with the seat occupant when he or she moves the seat forward or rearward. Some manufacturers also call this belt-in-seat.

Integration testing Integration testing is conducted to validate that two or more subsystems or components are properly working together. Integration testing usually follows or is conducted in parallel with subsystem or unit testing.

Integration Time The time over which an integrating A-D converter averages the input signal. If chosen appropriately will average over a complete mains cycle thereby helping to reduce mains frequency interference.

Intel The worlds largest manufacturer of computer chips.

Intelligent switching unit (ISU) Provides integration and functionality in the cockpit module with reduced cost, thanks to a new human-machine interface, impulse-coded steering wheel command, integrated low current switches and smart key.

Intelligent transportation systems (ITS) A broad range of diverse technologies, including information processing, communications, control and electronics, which, when applied to a transportation system, can save time, money and lives.

Intelsat Organization of 105 countries (July 1980) owning or operating systems of satellites used by 144 countries and territories around the world for international communications, and by 16 countries for domestic communications.

Inter-axle differential A third differential between the front and rear differentials that permits differences in speed between the front and rear driving axles.

Inter-changeability The basis of mass production; the production of parts that are consistent in size, tolerance, and specifications.

Interchannel competition Rivalry between different channels of the distribution system. For example, independent retailers competing with manufacturers' outlets.

Intercooler Device that cools air as it leaves a turbocharger or supercharger before the air is blown into the engine air intake. Cooling makes the air denser and richer in oxygen, which lets the engine produce more power.

Interface A shared boundary. It might be a piece of hardware used between two pieces of equipment, or a software display communicating between the computer system and the people who use it.

Interface control document This details the specific ways in which two systems interact with each other, including mechanics, electronics, software, and safety interfaces. These documents should be in at least draft form by PDR.

Interface control drawing/document/specification A drawing, document or specification that is used to define and control the physical and functional interface between two or more subsystems within an overall system. These subsystems are typically designed and developed by different parties.

Interface control Interface control comprises the delineation of the procedures and documentation, both administrative and technical, contractually necessary for identification of functional and physical characteristics between two or more configuration items which are provided by different contractors/acquiring agencies, and the resolution of the problems thereto.

Interference angle The difference between the valve face angle and the valve seat angle.

Interference checking The process of identifying if and where two or more geometric objects intersect in either a static or dynamic state.

Interference valve angle The difference between the valve face angle and the valve seat angle, usually one or two degrees.

Interferometer Any of several optical, acoustic, or radio frequency instruments that use interference phenomena between a reference wave and an experimental wave or between two parts of an experimental wave to determine wavelengths and wave velocities, measure very small distances and thicknesses, and measure indices of refraction.

Interior lights The inside vehicle-lighting system, including courtesy, dome, map and instrument- panel lights.

Interior payload The amount of space or material that can be carried inside the vehicle.

Interleaved transformer windings The segmentation of portions of the same winding by another to improve or control inductive coupling.

Interlock An interlock is a hardwired connection between two systems or mechanisms that provides time-critical safety information.

Interlock condition An interlock condition exists if a system or mechanism raises an interlock connection because it has detected a possible safety conflict.

Interlock override An interlock override is a manually set condition to inform a system to ignore a particular interlock condition.

Intermediary Firm or organization that operates between the producer of the goods and the end purchaser. Thus, the members of the distribution channel noted above are intermediaries or "middlemen".

Intermediate gear A small torque increase obtained from a planetary gear set when the sun gear is held and power is applied to the internal gear.

Intermediate lever part of the parking brake linkage under the vehicle that increases application force and work with the equalizer to apply it equally to each wheel.

Intermittent A fault that occurs now and then in an electrical circuit, referring to occasional opens, shorts, or grounds

Intermittent duty A service requirement that demands electrical or mechanical operation for alternate intervals of load and no load; load and rest; or load, no load, and rest.

Intermittent problem A fault that occurs now and then in an electrical circuit, referring to occasional opens, shorts, or grounds

Internal combustion engine (ICE) An engine that converts the energy contained in a fuel inside the engine into motion by combusting the fuel. Combustion engines use the pressure created by the expansion of combustion product gases to do mechanical work.

Internal gear A gear with teeth pointing inward toward the center of the orifice.

Internal impedance The impedance exhibited by a circuit element or component.

Internal rate of return Internal rate of return (irr) is the rate of return at which the present value of benefits equals the present value of costs/investments.

Internal recirculation A loss of efficiency caused by liquid flowing through wear rings or the impeller to volute clearances.

Internal resistance 1. The resistance exhibited by a circuit element or component.

2. The opposition or resistance to the flow of an electric current within a cell or battery; the sum of the ionic and electronic resistances of the cell components.

Internal setup (IED) Die setup procedures that must be performed while machine is in stopped. IED - "inner exchange of die"

Internal-combustion engine (IC) An engine that burns the air/fuel mixture in a chamber inside the engine cylinder, like a conventional reciprocating engine or a rotary engine.

Interoperability The ability of systems, units, or forces to provide services to, and accept services from, other systems, units, or forces and to use the services so exchanged to enable them to operate effectively together.

Interplanetary magnetic field The weak magnetic field filling interplanetary space, with field lines usually connected to the Sun. The IMF is kept out of the Earth's magnetosphere, but the interaction of the two plays a major role in the flow of energy from the solar wind to the Earth's environment.

Interplanetary probe Unmanned instrumented spacecraft capable of reaching the planets.

Interplanetary shock The abrupt boundary formed at the front of a plasma cloud (e.g. from a coronal mass ejection) moving much faster than the rest of the solar wind, as it pushes its way through interplanetary space.

Interpolation A "Connect the dots" processing technique to estimate what a fast waveform looks like based on only a fue sampled points

Interpole A small field pole placed between the main field poles and electrically connected in series with the armature of an electric rotating machine.

Interpreter Software enabling a computer to run programs statement by statement.

Interrupt An Asynchronous external event . When the ECM/MCU detectes the event it causes a predetermined action to occur.

Interrupter A device used to block a signal

Interstallar probe Unmanned instrumented spaceship with artificial intelligence capable of reaching the nearer stars.

Interstellar ark Hypothetical space colony capable of transporting human intelligence to the stars.

Intrachannel competition Rivalry between channel participants at the same level in the distribution channel. For example, department stores competing with one another.

Intranet A corporate wide implementation of the Internet i.e. the network of computers is limited to a particular company or corporation.

Intrinsically safe A system that is incapable of releasing sufficient electrical or thermal energy under any condition to cause ignition of a specific hazardous mixture in the atmosphere, even in its most ignitable concentration.

Intrusive diagnostic test Any on-board test run by the PCM which could have an effect on vehicle performance or emissions.

Intrusive Test This is any on-board test run by the diagnostic management system which will have an effect on the vehicles performance or emission levels

In-vehicle sensor A term used for in-car sensor.

In-vehicle signing On-board display of roadside sign information. The information can be obtained either by short-range transmission from roadside beacons or from on-board data storage. In-vehicle signs are utilized to improve driver effectiveness, especially when driving at night or during inclement weather conditions.

In-vehicle temperature sensor A term used for in-car sensor or in-car temperature sensor.

Inventory The money the system has invested in purchasing things it intends to sell.

Inverse-square law The mathematical description of how the strength of some forces, including gravity, changes in inverse proportion to the square of the distance from the source.

Inversion valve A normally open air valve used primarily in parking and emergency brake systems.

Inverted start When cars are started in the reverse order of their qualifying times, in oval-track racing.

Inverter 1. A device used to convert the direct current produced by a fuel cell to alternating current for use at the motor.

Investigation table A term used for component location table.

Ion engine A rocket engine, the thrust of which is obtained by the electrostatic acceleration of ionized particles.

Ionic bond A primary bond arising from the electrostatic attraction between two oppositely charged ions, such as electrostatic painting.

Ionization Formation of electrically charged particles. Can be produced by high-energy radiation such as light or UV rays, or by collision of particles in thermal agitation.

Iron core A general class of cores that contain iron. Sometimes used to refer to cores made only from steel laminations.

Iron-core coil A coil or transformer which uses an iron core to enhance the magnetic coupling between windings, increasing the magnets flux-density.

Irontite A tradename for the process of filling cracks in cast-iron cylinder heads and blocks using tapered plugs.

Isoblock A hard-rubber insulator block used for motor mounts.

Isobutylene A refinery process petrochemical that is reacted with methanol to form MTBE or with ethanol to form ETBE.

Isolate A technique where a component or system may be separated from the rest of the component or system without the los of fluids or pressures.

Isolate, hold and dump Actions commanded by the ABS microprocessor to regulate hydraulic pressure to an individual wheel brake.

Isolation Between Inputs A transient at an input can also propagate to other equipment connected to that input. This is prevented by providing isolation between inputs.

Isolation The electrical separation between input and output of a power supply by means of the power transformer. The isolation resistance (normally in megaohms) and the isolation capacitance (normally in picofarads) are generally specified and are a function of materials and spacings employed throughout the power supply.

Isolation to Earth or System A high transient voltage at one input may damage not only the input circuit, but the rest of the data acquisition hardware, and, by propagating through the signal conditioning and A-D circuits, eventually damage the computer system as well. You can prevent this type of damage by isolating the input from the earth of the data acquisition and computer hardware.

Isolation transformer A transformer with a one-to-one turns ratio.

Isolation voltage The maximum ac or dc specified voltage that may be continuously applied between isolated circuits.

Isolator A rubber or synthetic device used to separate two parts to reduce noise or vibration.

Isolator pad A rubber or synthetic device used to separate two parts to reduce noise or vibration.

Isomers Chemical compounds having the same molecular weight and atomic structure, but a different molecular structure.

Item A non-specific term used to denote any product, including systems, subsystems, assemblies, subassemblies, units, sets, accessories, computer programs, computer software or parts.

Iteration A technique used in impeller design. You repeat the action until you get to the final result. Very difficult to do until the new computer programs became available.

J

Jacket Usually refers to the heating/ cooling jacket surrounding the stuffing box on some pumps.

Jacketed gasket A gasket having metal grommets around bolt and water holes.

Jacking Modifying the suspension to raise or lower one corner of an oval-track race car in order to provide better handling characteristics.

Jackshaft A shaft found in most overhead cam engines that is used to drive the distributor, fuel pump, and oil pump.

Jam nut A second nut tightened against a primary nut to prevent it from working loose; used on inner and outer tie-rod adjustment nuts and on may pinion-bearing adjustment nuts.

Jet A calibrated passage in the carburettor used to meter fuel flow.

Jet propulsion Reaction propulsion in which the propulsion unit obtains oxygen from the air as distinguished from rocket propulsion in which the unit carries its own oxygen-producing material.

Johnson rod A mythical automotive component that is blamed for any problem that cannot otherwise be explained.

Joint A junction of members that are to be joined or have been joined.

Joint development model A model of partnership with an external manufacturer to jointly design a product that will be produced by that manufacturer. The responsibilities for development and the ownership of the intellectual property are negotiable.

Joint penetration The depth a filler metal extends from its face into a joint, exclusive of weld reinforcement.

Joint tenancy Ownership that is shared by two or more persons.

Joint venture An international business collaboration between foreign interests and private parties from a host country in which two or more parties establish a new business enterprise to which each contributes and where ownership and control are shared.

Joule (J) Unit of energy equal to one watt-second.

Jounce The inward reaction of the spring and shock absorber when a wheel hits an obstruction.

Journal size The finished diameter of a crankshaft journal.

Journal The part of a rotating shaft that turns in a bearing.

Judgment inspection A form of inspection used to determine non-conforming product.

Junction 1. A contact point between two or more conductors in a circuit, also called a node.

2. The region separating two layers in a semiconductor material, e.g. A p-n junction.

Junction block A device on which two or more junctions may be found.

Junction field effect transistor (JFET) A low power semiconductor having a conductive channel whose resistance is controlled by the reverse voltage on the gate channel junction.

Junction transistor A PNP or NPN transistor formed from three alternate regions of p or n type material. Also called a bi-polar transistor.

Junior fueler An old drag-racing term for a lower-class fuel dragster.

Junior stocker An old drag-racing term for a lower-class fuel stocker.

Junk Anything that is so badly worn or deteriorated that it is of no further use.

Junker A vehicle that is ready for the junk box.

Junky Anyone in the vehicle salvage business.

Junkyard A vehicle salvage business that sells used parts.

K

k A symbol for a thousand, from kilo.

Kaikaku Radical Improvement, usually applied only once within a *value stream.*

Kaizen activities Activity through which continuous improvement is sought.

Kanban The Japanese methodology for achieving JIT, often involving the use of kanbans, or cards, to indicate parts status.

Kano model The Kano model, developed by Dr. Noriaki Kano, further refined the notion of quality along two dimensions in contrast to the linear "good-bad" "ok-not ok" dimension in existence all along. The two dimensions were: 1. the degree to which a product or service performs, and

2. the degree to which the user is satisfied. The correlation of quality on two axes further led to three unique definitions of quality, namely: basic quality, performance quality and excitement quality.

Kart A small, open, four-wheeled vehicle with a single cylinder, two- or four-cycle gasoline engine.

Keeper grooves The grooved area on a valve stem to accommodate the keepers.

Keepers Key-like, tapered-metal locking devices used to hold valve retainers in place on the valve stem.

Kemp A custom car or lead sled.

Kerosene A mixture of hydrocarbons distilled from crude petroleum.

Key Performance Indicators (KPIs) KPI refers to the short list of measurable parameters that will indicate how well the business is doing at attaining its goals. In a manufacturing quality scenario, this may be the amount of scrap or rework that gets metered. In a service quality scenario, such as an insurance company, this may be the open inventory of unprocessed claims. In brand management, market share in itself and in comparison with competing brands is sure to be relevant. In logistics, on-time deliveries, empty return loads, or missing items are candidate indicators.

Keyboard An input device used to key programs and data into the computer's storage.

Keyless entry A system for locking and unlocking doors of a vehicle with a central locking system without using the key. Usually, the user controls the locks by pressing a button on a remote key-fob transmitter. Some vehicles have electronic combination locks on the doors near the handle.

Key-off loads A term used for parasitic loads.

Keystone ring Compression piston ring, double tapered and shaped like a keystone.

Keyway A groove milled or machined onto a shaft or into a bore to accept a square, half moon, or a round piece of metal.

Kingpin 1. A pin or shaft on which the steering spindle assembly rotates

2. The pin, mounted through the center of the trailer upper-bolster plate, that mates with the fifth wheel and locks to secure the trailer to the fifth wheel.

Kingpin angle A term used for kingpin inclination.

Kingpin axis The inward tilt of the steering axis from the vertical.

Kingpin inclination A major factor in a vehicle's directional control and stability; the angle of a line through the center of the kingpin in relation to the true vertical centerline of the tire viewed from the front of the vehicle.

Knock A noise within an engine generally caused by detonation or preignition.

Knock off 1. A counterfeit.

2. A cheap auto part packaged to represent a popular brand.

Knock sensor A sensor that signals the engine-control computer when detonation is detected, momentarily retarding ignition timing until detonation ceases.

Knocked-down set (KD set) A package of automotive parts, assemblies, and subassemblies packaged at one location to be assembled at another location.

Knock-off hub A large, single, two- or three-eared wing nut used to retain a wheel; illegal for street use.

Knowledge management The overall management process to capture, organize, manage and disseminate knowledge in an organization to improve enterprise effectiveness by avoiding mistakes and avoiding the time to relearn needed knowledge. Since product development is very knowledge intensive, knowledge management offers tremendous leverage and opportunity for improvement.

Knowledge-based engineering Knowledge-Based Engineering is a set of design automation tools that capture design knowledge and rules to automate the design process.

Knowledge-based system Software that uses artificial intelligence techniques and a base of information about a specialized activity to control systems or operations.

Knuckle A term used for steering knuckle, the part around which each front wheel pivots as it is steered.

Knuckle arm The arm that extends backward from the steering knuckle to provide attachment for the tie rods.

Knurl 1. A series of ridges formed on the outer surface of a piston or the inner surface of a valve guide to help reduce clearance and hold oil for added lubrication.

2. Machined ridges on a part or tool used to provide a friction surface for drive purposes

Kyoto protocol International agreement among industrialized countries to reduce greenhouse gas emissions negotiated in Kyoto, Japan in December 1997.

L

L 1. An acronym for Low, one of the forward gear positions of a transmission.

2. Liter, a unit of metric liquid measure.

Labour per product (LPP/LPU) Also known as Labour Per Unit, is calculated using vague estimates of material, Direct Labour, and overhead cost. The cost per unit is normally used for external reporting only, not management decision. Example: Direct labour (operator) X 30% fringe benefits = LPP = LPU

Labour turnover rate A measure of a plant's ability to retain workers, expressed as a percentage of the production workforce that departs annually-or an annualized rate of employee departures. High turnover rates often indicate employee dissatisfaction with either working conditions or compensation.

Laboratory 1. A laboratory is a test facility that may include chemical, metallurgical, dimensional, physical, electrical, reliability testing or test validation.

2. A laboratory is a research facility that supports development and testing under controlled conditions.

Labyrinth seal A non contacting seal utilizing a tortured path for the escape of the fluid. Utilizes a series of pressure drops to reduce the leakage.

Lacing A method used to mend cracks in cylinder heads and blocks using threaded repair plugs side-by-side and overlapping each other.

Lacquer A type of paint that dries by solvent evaporation that must be rubbed out to produce a gloss.

Ladder bars A device attached to the chassis or frame at one point and to the rear axle at two points on a drag-race car to reduce rear-axle windup.

Ladder chassis A conventional frame design that consists of two side rails, not necessarily parallel, connected to each other by a series of cross members like a ladder.

Ladder frame A term used for ladder chassis.

Lag 1. The incorrect operation of a shock absorber because of aeration due to the mixing of air with oils, causing the shock absorber to produce a poor ride.

2. To fail to keep up; to fall behind.

Lake pipes Straight exhaust pipes with no muffler.

Laminar airflow 1. The smooth, continuous movement of one layer of gas or liquid over another.

2. The movement of a body of air/fuel mixture through the intake manifold and ports as it flows over the boundary layer.

Laminar Flow State of fluid flow, where the particles move smoothly without vortices or other turbulence.

Laminate 1. A structure consisting of two or more layers of material, such as fibreglass.

2. To fabricate a structure of two or more layers of material, such as fibreglass.

Laminated case A container, such as a safety fuel cell, with walls of two or more layers of material.

Laminated Construction The Construction of the armiture from individual stampings

Laminated core An assembly of steel sheets for use as an element of magnetic circuit having the property of reducing eddy-current losses.

Lamp A device to convert electrical energy to radiant energy, which is normally visible light.

Land 1. That part of a piston between the ring grooves that separates and supports the rings.

2. The raised surfaces on a valve spool.

Land speed record (LSR) The maximum speed obtained by 1. A wheel-driven, internal-combustion engine vehicle.

2. A thrust-driven jet or rocket engine vehicle.

Landau A semi-convertible car with a folding top over the rear passenger compartment.

Landau bar Ornamental ogee or S-shaped trim sometimes applied to C-pillars or sail panels.

Landau iron A term used for landau bar.

Landau top A passenger-car top partially covered with vinyl to give it a convertible-like appearance.

Lander spacecraft A spacecraft designed to reach the surface of a planet or moon and survive long enough to telemeter data back to Earth.

Landing gear Retractable supports for a semi-trailer to keep the trailer level when the tractor/truck is detached from it.

Langrangian point In a system dominated by two attracting bodies (such as Sun and Earth), a point at which a third, much smaller body (such as a satellite) keeps the same position relative to the other two. Theoretically, the Sun-Earth system has 5 Lagrangian points, but only two are important: L1, on the sunward side of Earth, about 4 times the distance of the Moon, and L2 at approximately the same distance on the midnight side. The only two lagrangian stable points, L4 and L5, lie in the orbit of the primary body, leading and trailing it by a 60-degree arc. Jupiter's trojan asteroids can be found orbiting around the Jupiter-Sun L4 and L5 points.

Lantern ring A device used to supply lubricant to packing. Usually located in the middle of the packing ring set.

Lanyard Small rope or cord.

Lap 1. A complete trip around an oval track or road course.

2. To gain a full lap over the second place competitor.

Lap joint A joint between two overlapping members in parallel planes.

Lap weld A welding process between two overlapping members in parallel planes.

Lap-and-shoulder belt A safety belt that secures the driver and/or passenger in the seat with a continuous web of material which fits across the lap and crosses the upper body. It keeps the occupant from jerking forward in the event of a crash. Also called three-way belt, three-point belt, or three-point safety harness.

Laser 1. An acronym for light amplification by stimulated emissions of radiation. 2 .A device that produces a concentrated, coherent light beam by stimulated electronic or molecular transitions to lower energy levels.

Lash The clearance between two parts.

Lash-pad adjusters Small round metal pieces of various thickness used to adjust valve clearance.

Late apex Getting to the inside of a turn later than ususal due to a late entry into the corner.

Lateral support system Consists of 6 support links arranged at equal intervals around the outside diameter of M1.

Lateral weight transfer The momentary shift of a vehicle weight from the inside tires to the outside tires, or outside to inside, due to cornering forces.

Lathe-cut seal A fixed seal for a calliper piston that has a square or irregular cross section, not round like an O-ring.

Latitude Circles in parallel planes to that of the equator defining north-south measurements, also called parallels.

Latitudinal form side-to-side

Lattice Structure An orderly arrangment of atoms in a crystalline material.

Launch A good start off the starting line in drag racing.

Launch complex The complex of site, facilities and equipment used to launch a missile or space rocket.

Lay rubber To leave streaks of rubber on the pavement during rapid acceleration. Also lay a patch and lay a scratch.

Lay the crank A reconditioned crankshaft supplied with the appropriate bearings and an installation kit.

Layer winding The method of winding a transformer or choke whereby conductors are layered one top of another, commonly separated by an insulating layer.

Layout 1. For ics, the process of floorplanning, implementing, and verifying the location of transistors and their connections within a chip design.

2. For pcbs, the process of entering, placing, routing, and verifying the location of physical components and their connections within a board design.

Le Mans start A method of starting a race by having the drivers stand across the track and, at the start signal, race to their vehicle, jump in, belt up, start their engines and take off.

Lead 1. To be out in front.

2. An element Pb often used as a body filler and used as a plate meteral

3. a short term for tetraethyl lead a compound formerly used to increase the octane rating of gasoline

Lead customers / users Lead customers or users are those customers or users who are the most advanced users of the product, customers who are pushing the product to its limits, or customers who are adapting an existing product(s) to new uses.

Lead resistance DC resistance in the leads of a circuit element or device.

Lead sulfate A hard, insoluble layer that slowly forms on the plates of a discharging battery that may be reduced only by slow charging.

Lead-acid cell Secondary cell which uses lead peroxide and sponge lead for plates, and sulfuric acid water for electrolyte.

Leaded gasoline Gasoline to which a small amount of tetraethyl lead is added to improve engine performance and reduce detonation, a practice no longer allowed due to EPA regulations.

Leadfoot A person who drives faster than necessary.

Leading link A component of the suspension system that is attached to the chassis behind the wheel and positioned to resist fore-and-aft movement of the wheel.

Leading shoe 1. A brake shoe in a nonservo brake that pivots around a fulcrum in the direction of normal drum rotation. Also known as forward shoe.

2. The first shoe in the direction of drum rotation in a leading trailing brake.

Leading-trailing brake A drum brake that develops self-energizing action only on the leading shoe. Brake application force is separate for the leading and the trailing shoes. Also called a partial-servo or a nonservo brake.

Leaf spring A rear, vehicle-suspension spring featuring one or more flat leaves of spring steel with graduated lengths. It has an "eye" at one end to connect to the vehicle frame and it is connected to the axle with a U-bolt.

promote slippage between the leaves during flexing.

Leak detector 1. A device using visual or audible signals to indicate a leak.

2. A dye-type fluid that may be injected into a system that will indicate the presence of a leak.

Leak Find A trade name for a dye solution that can be injected into a system to find difficult leaks.

Leakage current 1. The ac or dc current

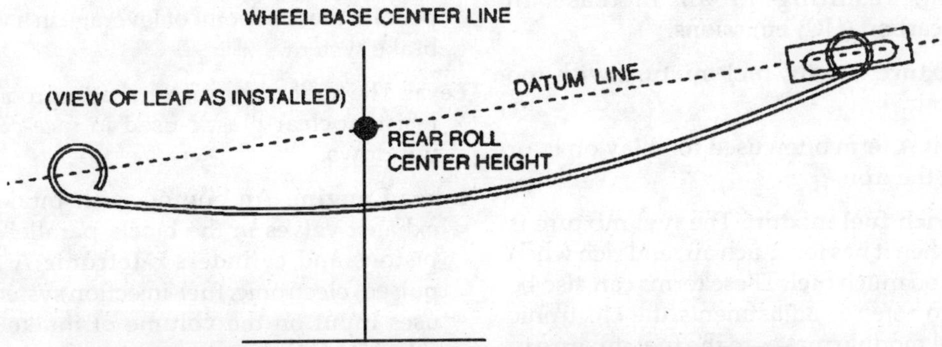

*WHEN THIS DIMENSION DIFFERS BETWEEN THE LEFT & RIGHT SIDES THE HEIGHTS CAN BE AVERAGED TO APPROXIMATE THE HEIGHT OF THE REAR ROLL CENTER

Leaf-spring bushing Bushings that are used to damped noise and vibration from the road to the frame of the car.

Leaf-spring centre bolt A bolt passed through a hole in the centre of each spring leaf for holding the leaves together. The bolt head is used to locate the spring position on the axle-housing.

Leaf-spring hanger The vehicle frame-attachment bracket for the front eye of the rear leaf spring that allows the spring to pivot.

Leaf-spring shackle The small arm, or swing attachment for the rear of the leaf spring located between the frame and spring eye. It allows the spring to shorten and lengthen during normal driving conditions.

Leaf-tip insert Small, replaceable pads of plastic, rubber, or composition placed between spring leaves near their ends to

flowing from input to output and/or chassis of an isolated device at a specified voltage.

2. The reverse current in semiconductor junctions.

Leakage flux Magnetic lines of force that go beyond their intended path and do not serve their intended purpose.

Leakage inductance Self inductance in a transformer caused by leakage flux.

Leaker 1. A vehicle not well prepared.

2. A vehicle leaking oil or coolant.

Lean A term often used for lean mixture or lean out.

Lean Burn A Chrysler electronic engine control that appeared in the mid 1970's. It maintains precise control of the spark timing to allow a very lean mixture to burn properly, reducing emissions using an analogueue computer.

Lean manufacturing Lean manufacturing is a operations philosophy that aims to synchronize production with demand, thereby minimizing inventory and cycle time. Lean manufacturing is supported during product development with approaches such as robust design, mistake-proofing and standardization.

Lean misfire A condition caused by a vacuum leak or open EGR valve that results in combustion, causing one or more cylinders to pass unburned fuel into the exhaust system, resulting in an increase in hydrocarbon (HC) emissions.

Lean mixture An air/fuel mixture with too much air.

Lean on it A term often used for a lay on it, or lay on the iron.

Lean or rich fuel mixture The fuel mixture is lean when it has too much air, and rich when it has too much fuel. These terms can also be used to refer to adjustments the electronic control module makes to the fuel mixture in response to various driving conditions, particularly on engines with variable-valve technology.

Lean out 1. To increase the portion of air or decrease the portion of fuel in an air/fuel mixture.

2. To decrease the portion of nitro in a fuel mixture.

Lean Producing the maximum sellable products or services at the lowest operational cost, while optimizing inventory levels.

Lean Production The system of organization production developed in japan that stresses quick tool changeover times, minimum parts and work-in-progress inventories, high levels of quality and continuous improvement.

Leg room With the front seat adjusted all the way back, the distance from the accelerator pedal's heel point to the back of the front seat cushion.

Lemon A vehicle, generally a new one, with several defects or a defect that can't be resolved.

Level of detail Level of detail - the ability to vary the amount of details displayed in a graphics image to improve performance. For instance, at a distance, models can appear as simple 3d figures, but as users zoom in, a more detailed representation is presented.

Leverage The use of a lever and fulcrum to create a mechanical advantage, usually to increase force applied to an object. The brake pedal is the first point of leverage in a vehicle brake system.

Lexan The trade name for the shatterproof, heat-resistant clear plastic used in race-car rear windows.

L-head engine An engine with intake and exhaust valves in the block, parallel to the pistons and cylinders.**L-Jetronic** A Bosch pulsed, electronic, fuel-injection system that uses input on the volume of intake air to calculate fuel delivery.

Life cycle analysis Life cycle analysis is a method to assist with the quantification and evaluation of environmental burdens and impacts associated with product systems and activities, from the extraction of raw materials in the earth to end-of-life disposal. Lca is increasingly used by industries, governments and environmental groups to assist with decision making for environment-related strategies and materials selection.

Life cycle cost Life cycle cost is the total cost to the customer of acquiring, operating, and disposing of a product/ system over its full life. These costs include development, acquisition, installation, training, operation, support, and disposal.

Life cycle costing Life Cycle Costing is a framework for evaluating design options on the basis of all of the costs that the asset will incur over its lifetime. Present Value Discounting is applied to all of the costs to account for the fact that a dollar today is worth more than a dollar tomorrow. Life cycle costs include the initial purchase cost, ongoing maintenance and periodic updating. Some studies include operating costs.

Life expectancy The anticipated longevity of a part, assembly or component.

Light emitting diode (LED) A semiconductor diode that radiates light when the junction is forward biased and a specified current flows through the diode.

Light end hydrocarbons Term used to denote hydrocarbons from crude distillation that are low density (lighter weight than gasoline) and have low boiling temperatures. Butanes are the most common light end hydrocarbons used in gasoline.

Lighting A market sector served by the Optics division, which manufactures coated optical components for medical and technical illumination, and entertainment applications.

Light-off, mini-oxidation catalytic converter A small catalyst mounted just behind the exhaust manifold that gets hot very quickly after the engine is started, so that it begins working in time to neutralize much of the extra pollution that is produced during cold running.

Lightweight structures Extremely light yet rigid structures, capable of withstanding the heavy launch stresses and temperature excursions encountered in space travel. The Space Technology division (Contraves Space) develops composite materials for satellite platforms and other load-bearing parts, made of carbon or glass-fibre layers around a light alloy core.

Lightweight-skin spare tire A bias-ply, compact spare tire with a reduced tread depth to provide about 2,000 miles (3,218 kilometers) of trad life; designed for emergency use only, and driving speed limited to 50 mph.

Lime deposits A condition that exists in a cooling system when the lime, present in water-based coolants, comes out of solution and coats the engine's water passages.

Limited slip differential A differential in a rear-drive vehicle fitted with a mechanism that limits the speed and torque differences between its two outputs. Limited slip ensures that some torque is always distributed to both wheels, even when one is on a very slipper surface.

Limited-production option (LPO) An item of new car equipment available for a limited market, such as a high-performance police package.

Limited-slip differential A device that helps prevent the drive wheels from skidding or losing traction by diverting power from the slipping wheel to the opposite wheel on the same axle.

Limited-slip differential gear oil A specially formulated gear oil required in limited-slip differentials because of the extreme pressures on the clutch cones or clutch plates and discs.

Line 1. Medium for transmission of electricity between circuits or devices.

2. The voltage across a power transmission line.

Line bearings Position the rotor or shaft radially. Normally of the sleeve type.

Line bored When the drilling or boring is done on a jig that insures every component is in a straight line.

Line conditioner A circuit or device designed to improve the quality of an ac line.

Line contact 1. The contact made between the cylinder and the torsional rings, usually on one side of the ring.

2. The contact made between the valve and the valve seat.

Line mechanic 1. A mechanic who works on the repair line at a dealership.

2. A mechanic who is skilled in a particular automotive system.

Line power supply 1. A power supply in which the ac line voltage is rectified and filtered without using a line frequency isolation transformer.

2. A power supply switched into service upon line loss to provide power to the load without significant interruption.

Line pressure 1. The base pressure established in a transmission by the pump and pressure regulator valve.

2. The pressure present in a line or hose.

Line ream To ream bearings or bushings to size after they have been installed.

Line regulation The change in value of DC output voltage resulting from a change in AC input voltage over a specified range, or from low line to high line or from high line to low line. Normally specified as the + or

Line regulator Power conversion equipment that regulates and/or changes the voltage of incoming power.

Line static pressure A 10 to 15 psi (69 to 103 kpa) hydraulic pressure maintained in a drum-brake system when the brakes are not applied to keep pressure on wheel-cylinder cup lips in order to seal fluid in, as well as air and dirt out.

Line transient A perturbation outside the specified operating range of an input or supply voltage.

Linear power supply An electronic power supply employing linear regulation techniques. "Linears" can provide fast transient response, very precise output regulation, very low output PARD (noise), and excellent isolation from the source voltage. They are also larger (about 10X) heavier (about 10X) and generate more heat (about 5X) compared to an equivalent switcher. Linear supplies typically utilize a 50Hz - 60Hz transformer to provide an isolated lower unregulated secondary voltage that is then regulated with a "linear pass device". The pass device is usually a transistor controlled with a local feedback loop.

Linear pressure drop A straight line pressure drop across the lapped seal faces. Seldom happens.

Linear regulation A regulation technique wherein the control device, such as a transistor, is placed in series or parallel with the load. The output voltage is regulated by varying the effective resistance of the control device to dissipate the power resulting from the difference between input and output voltage time the current.

Linear regulator A power transistor or device connected in series with the load of a constant voltage power supply in such a way that the feedback to the series regulator changes its voltage drop as required to maintain a constant dc output.

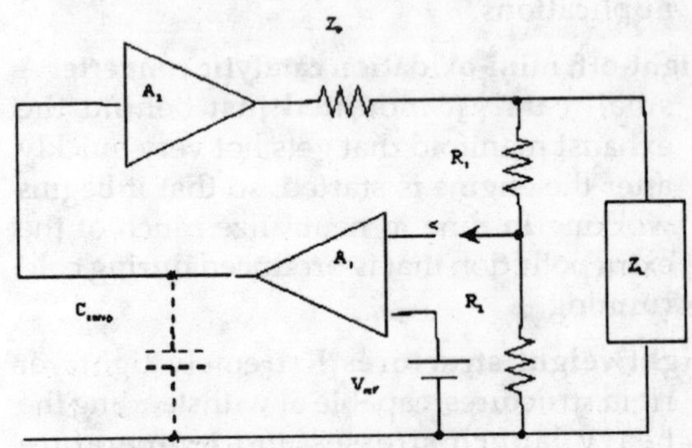

Fig. Linear regulator

Linearity 1. The ideal property wherein the change in the value of one quantity is directly proportional to the change in the value of another quantity, the result of which, when plotted on graph, forms a straight line.

2. Commonly used in reference to Linearity Error

Linearity Ideally an A-D or D-A converter converts the input or output range into equal steps. In practice the steps are not exactly equal. Linearity, or non-linearity, is a measure of how close the steps approach equality.

Load 1. The demand for power placed on an engine.

2. The cargo of a truck or other vehicle.

3. The amount of weight placed on a tire.

4. The heat quantity imposed on an air conditioner.

5. The fuel for a drag racer, such as nitro.

6. Any component, circuit, subsystem or system that consumes power.

7. Any intentional resistance in the circuit.

Load Cell A transducer which converts a force into an electrical signal. It normally comprises four strain gauges in a wheatstone bridge arrangement.

Load decoupling The practice of placing filter components at the load to attentuate noise.

Load range 1. A term used for load rating.

2. The weight a tire will support at a given inflation pressure.

Load rating A term designating the maximum weight that a particular tire is designed to support, usually related to a specific air pressure.

Load regulation 1. Static: The change in output voltage as the load is changed from specified minimum to maximum and maximum to minimum, with all other factors held constant.

2. Dynamic: The change in output voltage expressed as a percent for a given step change in load current. Initial and final current values and the rates of change must be specified. The rate of change shall be expressed as current/unit of time, e.g., 20 amperes A/u second. The dynamic regulation is expressed as a +/- percent for a worst case peak-to-peak deviation for dc supplies, and worst case rms deviation for ac supplies.

Loaded voltage The voltage of the charging system in operation at 2500 RPM with all accessories on

Loading An enrichment of the air/fuel mixture to the point of rough idling, usually resulting in black smoke from the exhaust.

Loading time The available time is derived by subtracting the planned downtime from the available time per day or week, etc.

Load-leveling shock absorber Inflatable shock absorbers, often used with an electronic height-control system, that are pressurized with air to increase their load-carrying capability.

Load-load A method of conducting single-piece flow, where the operator proceeds form machine to machine, taking the part form one machine and loading it into the next.

Load-proportioning valve (LPV) A device used to distribute hydraulic pressure to the front and rear brakes based on the vehicle load.

Lock up torque converter A torque converter that contains a special clutch that forms a solid connection between the engine output shaft and the transmission input shaft when a certain, pre-set speed is attained. This reduces transmission friction losses and increases efficiency.

Lock washer A split-type of washer that helps prevent a bolt or nut from working loose.

Locked rear end A rear axle without a differential where both rear wheels turn at the same speed on curves as well as on a straightaway.

Lockplate A metal washer-like device with tabs that may be bent around a nut or bolt head to prevent them from turning.

Lockstrap A manual adjustment mechanism that allows adjustment of free travel of a lever or rod.

Lock-to-lock The number of turns required to turn the from wheels from one extreme to the other.

Lockup clutch A type of clutch in which the torque converter turbine is locked up with the engine, eliminating slippage.

Lockup The point at which braking power overcomes the traction of the vehicle's tires and skidding occurs, causing loss of control, lost stopping distances, and flat-spotting of the tires.

Lockup torque converter A fluid clutch designed with a clutching assembly in an automatic transmission to improve coupling efficiency at a predetermined vehicle speed.

Loctite A trade name for a line of sealants and adhesives that are popular in the automotive industry.

Logic Analyzer An instrument used to make the logic state of many digital signals visable over time It analyzes the digital data and can represent the data as real-time software execution, data flow values, state sequences, ect

Logic circuits An interconnection of individual combinational and sequential logic circuits to perform decision making, control and computation using digital techniques.

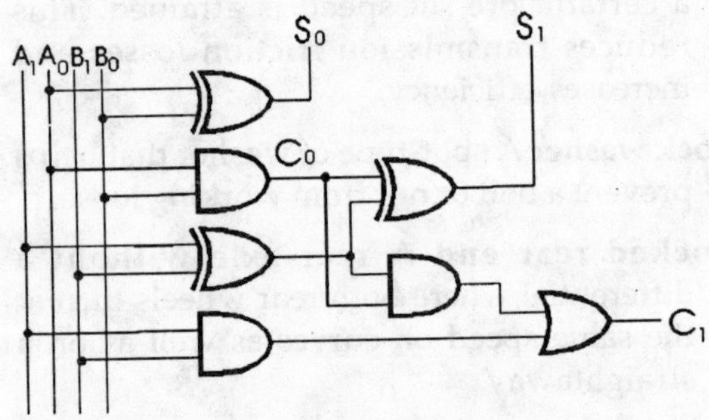

Fig. Logic circuits

Long term adaptive Longft: (adaptive) menory injector pulsewidth compensation stored by the PCM to maintain minimum emissions output

Long term fuel trim The long term fuel injector pulse width compensation the engine controller has stored to maintain minimum emissions output. Long term fuel trim drives Short Term Fuel Trim to maximum operating efficiency.

Long, short arm suspension A front suspension system where the upper control arm is shorter than the lower control arm, allowing each wheel to compensate for changes in the road surface while greatly affecting the opposite wheel in a manner that allows the wheel and tire assembly to rise and fall vertically as it goes over bumps.

Low end power The engine horsepower output during the first 25-30% of engine rpm range.

Low flow A condition that can cause excessive heat inside the pump volute. A temperature rise of 10°C (18°F) across the operating pump is considered excessive. Usually caused by throttling a pump discharge valve.

Low gear A speed obtained from a planetary gear set when the internal gear is held and power is applied to the sun gear, producing an increase in torque.

Low gearing A drive ratio that provides maximum output at a low road speed.

Lubricant Any fluid that will maintain a film thickness of one micron or more at its operating temperature and load.

Lubrication guide book A specially prepared publication, detailing required lubrication services with related information for each make and model of automobile.

Lubrication interval Manufacturer's recommended mileage and/or time limit when periodic lubrication services should be performed as a part of a preventative maintenance program.

Lubrication system The oil pump, filter, hoses, lines and passages in an engine that facilitate the oiling of all moving parts.

Lug bolt A hex-headed, threaded bolt used to hold the wheel on a vehicle.

M

Mach number The relationship between a moving body and the speed of sound in that locality which can vary with temperature, altitude and there fore pressure.

Machinability The relative ease with which materials can be shaped by cutting, drilling, or other chip-forming processes.

Machine 1. A car, usually a late-model car.

2. A device capable of doing work.

Machine vision 1. A computer perception of a visually-based sensory output used to produce a concise description of an image.

2. Devices used for optical non-contact sensing to automatically receive and interpret an image of a real scene in order to obtain information or to control a process.

Machining An operation which shapes metal parts by carving away excess material as chips produced in a sequential process of turning, milling and grinding operations.

Macpherson strut A suspension system that consists of a combination coil spring and shock absorber in one compact unit at each wheel. With this "independent" suspension design, road shocks at one wheel are not transferred to the opposite wheel. Macpherson struts use fewer parts, meaning a reduction on weight and fewer elements that could wear out.

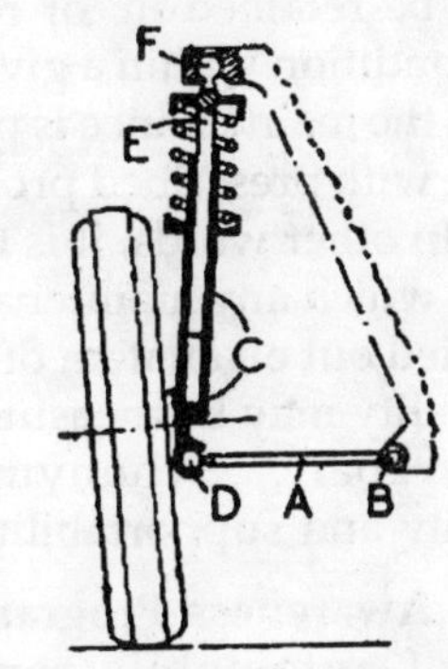

Fig. Macpherson strut

Magnetic amplifier Transformer type device employing a control circuit to vary the magnetic core saturation, thus varying the output voltage of the amplifier.

Magnetic clutch A coupling device used to turn the compressor on and off electrically.

Magnetic drive A type of seal less pump that is pretty much limited to pumping clean lubricating liquids. Similar in concept to a canned pump.

Magnetic field line Lines everywhere pointing in the direction of the magnetic force, used as a device to help visualize magnetic fields. In a plasma, magnetic field lines also guide the motion of ions and electrons, and direct the flow of some electric currents.

Magnetic field The area of influence of a permanent- or electro-magnet that exists between its north and south poles.

Magnetic filter A magnet or magnet assembly located in a fluid system to attract and retain ferrous metal, nickel, and cobalt particles which may be present, composite particles in which a ferro-magnetic material is entrained may also be present.

Magnetic flux density The number of lines of flux per cross-sectional area of a magnetic circuit expressed in Gauss or Tesla. Sometimes referred to as Magnetic Induction.

Maintainability A characteristic of design and installation which inherently provides for an item to be retained in, or restored to a specified condition within a given period of time, when the maintenance is performed in accordance with prescribed procedures and resources. In other words, it is the ease and speed with which any maintenance activity can be carried out on an item of equipment. Maintainability may be measured by mean time to repair. Synonymous with serviceability and supportability.

Maintenance Awareness Program (MAP) A coalition of automobile repair shops, suppliers, car companies, and associations dedicated to building trust between the aftermarket and its customers.

Maintenance management systems Automated software systems for handling maintenance work orders, as well as associated inventory, purchasing, accounting, and human-resources functions.

Maintenance manual A manufacturer's or independent publication containing comprehensive maintenance information for a certain vehicle.

Maintenance specifications A listing of interval oil and lube grades, service points, and capacities for periodic routine preventative maintenance service.

Maintenance-free battery A sealed battery having no practical provision for the addition of water to the cells, since periodic maintenance is unnecessary.

Major diameter 1. The largest in diameter of a cylinder.

2. The largest in the diameter of a bolt, its threads.

Major overhaul An engine overhaul just short of a rebuild, where all worn or damaged parts are rebuilt or replaced.

Major review A formal demonstration and confirmation across the system that supports or constitutes a program milestone event.

Major tune-up A conventional, ignition-engine maintenance procedure that may include points, capacitor cap, rotor, plugs, plug wires, valve adjustment, and a carburettor overhaul.

Make 1. To close a switch.

2. A distinctive name given to a group of cars by a manufacturer, such as Riviera, by Buick.

Make and break 1. The term used when a switch is closed or opened.

2. A pair of contact points, one stationary, and the other operated by a cam that makes the brake, such as in a conventional distributor.

3. To construct something then tear it down.

Manifold heat control A flapper-type valve in the exhaust manifold that diverts heat to the intake manifold.

Manifold pressure A positive pressure in the intake manifold measured in pounds per square inch or bars.

Manifold vacuum The negative pressure in an engine's intake manifold produce by the engine's pumping action and measured in inches of mercury.

Manned maneuvering unit A portable jet-pack device used by astronauts to propel themselves through space independent of a spacecraft.

Manomometer (u-tube) A u-shaped tube with a graduated scale used for measuring the pressure of gases or vapours.

Mantle Middle layer of the Earth; between the crust and the core.

Manual bleeding A two-technician technique for bleeding brakes; one pumps the brakes as the other opens and closes the bleeder screws when required.

Manual control valve A spool valve, located in the valve body, that determines fluid flow from the valve body to various hydraulic circuits controlled by the vehicle driver through the selector lever.

Manual low The position of the units in an automatic transmission when the driver selects the Low or first-gear position of the quadrant.

Manual operation 1. The act of operating the power steering system in the event of an engine system failure; increased effort is required.

2. The ability to override or cancel any automatic or semi-automatic function.

Manual selection The vehicle operator's ability to 1. Move the gear selector lever by hand, such as with a manual transmission.

2. Select any condition or function using a keypad, switch or lever.

Manual slide release The release mechanism for a sliding fifth wheel, which is operated by hand.

Manual transmission A transmission that varies the power and torque through a foot pedal controlled clutch and a floor-mounted or steering-shaft-mounted gear selection lever.

Manufacturability The characteristic of a product's design that facilitates the fabrication of the product's components and their assembly into the overall product.

Manufacturer A person, or corporation engaged in the mass production or assembly of vehicles or any other product.

Manufacturer's service manual A book containing service information and technical data that is provided by the vehicle manufacturer.

Manufacturing automation protocol MAP is a specification for a suite of communication standards for use in manufacturing automation developed under the auspices of General Motors corporation.

Manufacturing cost Includes quality-related costs, direct and indirect labor, equipment repair and maintenance, other manufacturing support and overhead, and other costs directly associated with manufacturing operations. It typically does not include purchased-materials costs or costs related to sales and other non-production functions.

Manufacturing cycle time The length of time from the start of production and assembly operations for a particular (finished) product to the completion of all manufacturing, assembly, and testing for that product or specific customer order. (Does not include front-end order-entry time or engineering time spent on customized configuration of non-standard items.)

Manufacturing execution systems (MES) Software-based systems that provide a link between planning & administrative systems and the shop floor. It can link MRP II-generated production schedules to direct process-control software. An element of computer-integrated manufacturing, MES encompasses such functions as planning and scheduling, production tracking and monitoring, equipment control, maintaining product histories (verifying and recording activities at each stage of production), and quality management.

Manufacturing Process technology (ie. Metal forming, machining, injection moulding, blow moulding, die casting, forgings, electronics/electrical, assembly, coating/plating); capacity utilization (in percent); production volumes; strategic alliances (eg. Joint ventures, technology agreements).

Manufacturing resources planning (MRP) MRP II - Software-based Manufacturing Resources Planning systems that translate forecasts into master production schedules, maintain bills of material (lists of product components), create work orders for each step in the production routing, track inventory levels, coordinate materials purchases with production requirements, generate "exception" reports identifying expected material shortages or other potential production problems, record shop-floor data, collect data for financial reporting purposes, and other tasks depending on the configuration of the MRP II package.

Manufacturing The science of planning, designing, managing, and scheduling people, goods, material, equipment, and machinery to produce a useable and sellable product.

Maquila (maquiladora) Mexican assembly plant located near the U.S.-Mexican border where most production is exported to the United States.

Marbles Small pieces of dirt, rubber, or other debris on the outer edges of the race track, outside the racing line.

Marcel The spring assembly in the clutch-drive plate that absorbs the energy when the clutch lever is released.

Margin or gross margin The return an intermediary achieves on the selling price of the article. That is, if the intermediary buys a product for $1 and sells it for $1.50, the margin is calculated. For example, .50 divided by $1.50, or 33%.

Margin The material between the face and head of an intake or exhaust valve.

Margining The ability to temporarily shift output voltage by a specified amount (often +/- 5%) for system testing.

Maria Dark areas on the Moon, actually lava plains, once believed to be seas.

Masking 1. To cover up, as in prior to painting.

2. The modifying of intake ports with dams to redirect the air flow.

Masking area A preparation stage in a paint shop in which the parts that are not to be painted are covered with paper and held in place with masking tape.

Mass Air Flow Measurement of the volume and density of the air entering the engine.

Mass airflow sensor (MAS) A device found at the inlet of the intake manifold of an electronic fuel-injection system which supplies the computer with input as to the volume of air entering the manifold, using the temperature differential between a heated platinum wire or a plastic film and the passing air, to generate a signal of varying voltage.

Mass airflow sensor Device that measures the flow of air entering the throttle housing.

Mass customization Mass customization is the evolutionary step beyond mass production in manufacturing whereby products can be customized to the needs of individual customers while achieving most of the economies of mass production. It is based on principles such as product line rationalization, standardization of components, modular design, postponement of customization to late in the production cycle, multi-function assemblies, use of software to customize product operation, etc.

Mass production The manufacture of a product in large quantities.

Mass ratio Ratio of the total mass of a rocket vehicle to the mass remaining when all the propellant is consumed.

Mass The quantity of matter in a body. It can be determined by measuring the force of gravity (weight) acting on it and dividing this by the gravitational acceleration at that point. Thus, the mass of a given body remains the same everywhere, while its weight changes with the gravitational attraction.

Master control A primary or main control.

Master production schedule An expression of the operational plan of production by period, indicating the timing and size of end item quantities (used in MRP).

Master/slave operation Interconnection of two or more regulated supplies in which one (the master) controls the others (the slaves).

Mastering The process of making a metal template original from which optical storage media (CD, DVD) can be replicated.

Material expanders Fillers that are used in the place of active material in a battery; the primary difference between a 3- and 5-year battery.

Material Safety Data Sheet (MSDS) A sheet required of the product manufacturer by OSHA, containing specific information about hazardous materials to be displayed in any workplace where such materials are used.

Materiel A generic term covering systems, equipment, stores, supplies and spares, including related documentation, manuals, computer hardware and software.

Mating ring Another name for the hard face in a mechanical seal. It can be either rotating or stationary.

Matter 1. Anything that occupies space and possesses mass or weight

2. All things in nature are composed of matter.

Max Q Maximum dynamic pressure; the point during launch when the vehicle is subjected to its greatest aerodynamic stress.

Maxi-fuse A large, cartridge-type fuse having a higher current-capacity use because they are less likely to cause an underhood fire when overloaded.

Maximum load 1. The highest allowable output rating specified for any or all outputs of a power supply under specified conditions including duty cycle, period and amplitude.

2. The highest specified output power rating of a supply specified under worst case conditions.

Maximum power transfer Condition that exists when resistance of load equals internal resistance of source.

Maxwell's Law James Clerk Maxwell (1831 - 1879) did revolutionary work in electromagnetism and kinetic theror of gasses. Maxwell showed that a few relatively simple mathematical equations could express the behaviour of electric amd magnetic fields and their interrelated nature.

Mean cycles before failure (MCBF) Mean Cycles Between Failures; the average number of equipment cycles between failures; total equipment cycles divided by the total number of failures during those cycles (includes both product and non-product cycles).

Mean Diameter The middle diameter. Usually refers to the center of the seal face. The term is commonly used with metal belows seals to describe the middle diameter of the bellows plate.

Mean effective pressure (MEP) The average pressure developed within an engine cylinder during a four-stroke cycle.

Mean motor scooter A top-performing car in drag racing.

Mean solar time Time based on an average of the variations caused by Earth's non-circular orbit.

Mean stress Algebraic difference between maximum and minimum stress in one cycle of fluctuating loading, as in a fatigue test. Tensile stress is considered positive and compressive stress negative.

Mean time between failures Mean time between failures (mtbf) is a measure of the reliability of a product or piece of equipment. It is equal to the number of failures in a given period divided by the total equipment uptime in that period. It represents the average time between failures for a repairable product for a defined unit of measure (e.g., operating hours, cycles, miles, etc.).

Mean time to repair (MTTR) The average time required to repair a product.

Mean time to repair Mean time to repair (mttr) is a measure of maintainability of a product or piece of equipment. It is equal to the total or the estimated downtime of the product or equipment in a given period divided by the number of failures or the number of repairs performed in that period.

Mechanical fade Brake fade due to heat expansion of a brake drum away from the shoes and linings. Mechanical fade does not occur with disc brakes.

Mechanical seal A positive sealing device used to seal all fluids (liquids and gases). The primary seal is a set of lapped seal faces that are installed perpendicular to the shaft.

Mechanical systems The telescope Mechanical Systems include drive motors, encoders, brakes, over-travel stops, cable wraps, and other equipment used to position and control the movement of the Telescope Mount Assembly.

Mechanic's wire Bendable steel wire used to attach or fasten parts or other objects together.

Mechatronics The introduction of electronic controls into mechanical components.

Media oriented system transport (MOST) A versatile, high-performance and low-cost multimedia network technology based on synchronous data communication. It is ideal for multimedia applications such as audio, video, navigation, and communication in the automotive sector.

Medical payments coverage Optional insurance that pays for medical and/or funeral expenses for the driver and passenger of a vehicle.

Medium-duty truck Vehicle weighing from 14,001 to 26,000 lbs.

Memorandum of agreement The document which specifies agreements between involved parties relative to a specific interface. The MOA delineates responsibilities of each organization and identifies formats and data elements required for a successful interface.

Memory The faculty of an electronic device to record and store data and/or instructions for future action on a command.

MEMS Micro-electrical-mechanical systems allow sensors, electronics, and actuators to be batch fabricated on a single chip. MEMS can perform extremely delicate operations such as drilling holes with bits finer than the human eye can see unaided.

Menu A list of choices for selecting a confifuration or feature

Mercosur The southern cone common market of Argentina, Brazil, Paraguay, and Uruguay. Mercosur enacted a common external tariff on January 1, 1995 for almost 85 percent of tariff items of the four countries; most trade among members will be duty-free.

Meridian Great circle that passes through both the north and south poles, also called line of longitude.

MES Manufacturing Execution Systems use network computing to automate production control and process automation. They do so by downloading recipes and work schedules, and uploading production results, thereby bridging the gap between business and plant-floor or process control systems.

Mesosphere A division of the Earth's atmosphere extending from altitudes ranging 18-30 miles to 48-55 miles.

MET Materials, energy & toxicity. A metric to measure environmental impact of a product. This metric was developed by developed by tno, a research organization in the netherlands.

Metal bellows Used in mechanical seal designs to eliminate the need for a dynamic elastomer.

Metal deactivator Gasoline additive used to neutralize the effect of copper compounds found in gasoline

Metal fatigue A breakage of the metal caused by the bending and flexing of a metal part beyond its endurance limit.

Metal forming Solid metal and molten metal process such as casting, forging, stamping and machining.

Metallic lining Break friction material made from powered metal that is formed into blocks by heat and pressure.

Metallic paint A paint which contains tiny dots of metal in its makeup to give it extra quality.

Metallizing Deposition of single or multiple ultra-thin reflecting layers of metal (e.g. Silver) onto optical storage media. Reflective metallizing bounces the laser beam that reads and writes data on the medium.

Meteor The luminous phenomenon seen when a meteoroid enters the atmosphere, commonly known as a shooting star.

Meteorite A part of a meteoroid that survives through the Earth's atmosphere.

Meteoroid A solid body, moving in space, that is smaller than an asteroid and at least as large as a speck of dust. Nearly all meteoroids originate from asteroids or comets.

Meter (m) Basic metric unit of length equal to 3.28 feet, 1.09 yards or 39.37 inches. Related units are the decimeter (dm) at 10 per meter, the centimeter (cm) at 100 per meter, the millimeter (mm) at 1000 per meter and the kilometer (km) at 1000 meters.

Meter Mode Setting an oscilloscope to display digits as with a multimeter

Metering device A device for metering the proper amount of refrigerant into an evaporator. The two types for automotive air conditioning systems are thermostatic expansion valve (TXV) and orifice tube (OT)

Metering valve A hydriodic control valve used primary with front disc breaks on RWD vehicles. The metering valve delays pressure application to the front brakes until the rear drum brakes have started to operate providing better control and stopping power.

Methanol (CH_3OH) An alcohol containing one carbon atom. It has been used, together with some of the higher alcohols, as a high-octane gasoline component and is a useful automotive fuel in its own right.

Mica A mineral with excellent insulating and heat resistant qualities, hence it is used for dielectrics and insulators.

Michael Faraday Michael Faraday (1791 - 1867) an English Physicist The measurement of the capacity of a condenser of capacitor is masured in units called FARADS for Faraday

Microgravity An environment of very weak gravitational forces, such as those within an orbiting spacecraft. Microgravity conditions in space stations may allow experiments or manufacturing processes that are not possible on Earth.

Micrometeoroid Meteoroid less than 1/250th of an inch in diameter.

Micrometeoroid protection Shielding used to protect spacecraft components from micrometeoroid impacts. Interplanetary spacecraft typically use tough blankets of Kevlar or other strong fabrics to absorb the energy from high-velocity particles.

Microprocessor 1. A data processor with electronic integration capabilities.

2. A digital computer or processor built on a single integrated circuit chip

3. A microprocessor can perform functions of arithmetic logic and control logic itis the basic building block of a microcomputer system such as thoes used in automotive applications.

MIMOSA Machinery Information Management Open Systems Alliance MIMOSA advocates open exchange of equipment condition related information between condition assessment, process control and maintenance information systems through published, consensus, conventions. This to gain greatest value by combining vital condition information from multiple sources for collective evaluation, reaching accurate determinations of current condition and projected lifetime and communicating results in a useful, understandable form.

Minimum flow A condition that can cause excessive heat inside the pump volute. A temperature rise of 10°C (18°F) across the operating pump is considered excessive. Normally caused by throttling a pump discharge valve.

Minimum forward reduction Simple planetary-gear-set combination with the ring gear as input, the planetary carrier output, and the sun gear held.

Minimum ground clearance The distance between the ground and the lowest point of the vehicle chassis (usually the axle). A vehicle can drive over any object shorter than its minimum ground clearance.

Minimum idle air Air that passes the throttle plate to the throttle bore opening not subject to computer control

Minimum load 1. The lowest specified current to be drawn on a constant voltage power supply for the voltage to be in a specified range.

2. For a constant current supply, the maximum value of load resistance.

Minimum operating temperature The lowest ambient temperature at which the power supply will continuously operate safely and within specifications.

Minimum peak The lowest voltage recorded over a specific time period

Minimum premium The smallest premium that an insurance company will accept for payment on a policy.

Minimum starting temperature The lowest ambient temperature at which a power supply will turn on and operate safely.

Minimum thickness The lower limit if the thickness required after metal has been removed from an item such as a brake disc, to retain its integrity.

Min-Max A feature that increases its sampling rate to the maximum level and displays the maximum, minimum and adverage voltage levels recorded

Minor diameter 1. The smallest diameter of a cylinder.

2. The smallest diameter of a bold thread.

Minor overhaul An engine-repair procedure that includes a valve job, replacing the piston rings, the rod bearings, and the gasket set.

Minor tune-up An engine-maintenance procedure performed on distributor ignition vehicles that includes only the replacement of parts found to be defective.

Missile An object or a weapon that is fired, thrown, dropped, or otherwise projected at a target; a projectile.

Missing teeth 1. A condition found in a synchronous belt drive that may be caused by under-tensioning, misalignment, or excessive shock loads.

2. Teeth broken off a gear.

Mistake-proofing Mistake-proofing - improving product designs, tooling designs, or processes to prevent mistakes from being made or to quickly and easily detect or mitigate the effect of a mistake. Mistake proofing involves six principles: elimination, replacement, prevention, facilitation, detection, and mitigation.

Mitsubishi Jet Valve A tiny third valve that admits nothing by air to churn up the air/fuel charge and promote lean running and a complete burn.

Mobile Air Conditioning Society (MACS) A nonprofit organization founded in 1981 for the dissemination of distribution of comprehensive technical information, training, and communications to its members consisting of automotive air-conditioning shops, installers, distributors, supplies, and manufacturers in the United States and Canada.

Mock-up A full-size replica or dummy of a vehicle, e.g. a spacecraft, often made of some substitute material such as wood to assess design features.

Modbus An industrial networking system that uses peer-to-peer communications. More details on Modbus are in Issue 32 of Monitor.

Mode A term generally used to describe a particular set operating characteristics.

Mode door A diverter door within the duct system for directing air through the heater and / or evaporator core.

Model line Group of motor vehicles having the same platform or model name.

Model name Word, group of words, letter, number or similar designation assigned to a motor vehicle by a marketing division of a motor vehicle assembler.

Model The style of the vehicle produced by the manufacturer (ford mustang, chrysler lebaron, honda civic).

Model year The year of vehicle manufacture, as designated by the manufacturer; not consistent with the calendar year.

Modem An acronym for modulator-demodulator; a modem is a device and/or program that enables a computer to transmit data over telephone lines.

Modesty panel The panel below the bumpers that conceals the chassis components. Also known as modesty skirt.

Modified macpherson strut A type of macpherson strut suspension that uses shock struts with coil springs mounted between the lower arms and spring pockets in the cross member, to absorb minor road vibrations by the chassis rather than fed back to the driver through the steering system.

Modified strut A strut suspension where the coil spring is not part of the assembly and is independently located between the lower control arm and the frame.

Modular architecture A modular architecture (as opposed to an integral architecture) is a product architecture where 1. the physical building blocks (e.g., subsystems or subassemblies) perform one or a small number of functions in their entirety,

2. the interactions between the building blocks or interfaces are minimal, well-defined, and generally fundamental to the primary functions of the product, and

3. the building block elements are discrete, interchangeable and individually upgradeable.

Modular design Modular design consists of combining standardized building blocks or "modules" in a variety of ways to create unique finished products. Thus, even though the parts and assemblies may be standardized, the finished product is unique.

Modular Form of construction in which hardware or software units, often with differing functions, are quickly interchangeable.

Modular wheel A wheel made of different sections that are bolted or riveted together.

Modulation The variation of a property of an electromagnetic wave or signal, such as its amplitude, frequency, or phase.

Modulator 1. A device that varies the frequency amplitude and phase of electromagnetic waves.

2. A device that regulates hydraulic line pressure in a transmission to meet varying load conditions.

Module A self-contained part of a hardware item designed as a single replaceable unit, with a specific integral electronic function. It should require no installation other than mechanical mounting and completion of electrical connectors.

Modulus in bending Ratio of maximum fibre stress to maximum strain, within elastic limit of stress-strain diagram obtained in flexure test. Alternate term is flexural modulus of elasticity.

Modulus of elasticity Rate of change of strain as a function of stress. The slope of the straight line portion of a stress-strain diagram. Tangent modulus of elasticity is the slope of the stress-strain diagram at any point. Secant modulus of elasticity is stress divided by strain at any given value of stress or strain. It also is called stress-strain ratio.

Modulus of rigidity Rate of change of strain as a function of stress in a specimen subjected to shear or torsion loading. It is the modulus of elasticity deter-mined in a torsion test. Alternate terms are modulus of elasticity in torsion and modulus of elasticity in shear. Apparent modulus of rigidity is a measure of the stiffness of plastics measured in a torsion test. It is "apparent" because the specimen may be deflected past its proportional limit and the value calculated may not represent the true modulus of elasticity within the elastic limit of the material.

Modulus of rupture Ultimate strength determined in a flexure or torsion test. In a flexure test, modulus of rupture in bending is the maximum fibre stress at failure. In a torsion test, modulus of rupture in torsion is the maximum shear stress in the extreme fibre of a circular member at failure. Alternate terms are flexural strength and torsional strength.

Modulus of toughness The work done on a unit volume of material as a simple tensile force is gradually increased from zero to the value causing rupture is de fined as the Modulus of Toughness. This may be calculated as the en-tire area under the stress-strain curve from the origin to rupture. Toughness of a material is its ability to absorb energy in the plastic range of the material.

Moisture ejector A valve mounted to the bottom or side of the supply and service reservoirs that collects water and expels it every time the air pressure fluctuates.

Mold 1. A hollow form into which molten metal is poured to form a part.

2. A fungus.

3. To form or shape an object.

Mold-bonded linings A pad assembly made by applying adhesive to the pad and then pouring the uncured lining material onto the pad in the mold. The assembly is cured at high temperatures to fuse the lining and adhesive to the pad.

Molded connector A male or female electrical connector usually having one to four wires that are molded into a one-piece component.

Molten Carbonate 1. A type of fuel cell electrolyte that contains carbon, Oxygen and another element. Solid at room temp., it must be melted in order to function.

Molten carbonate fuel cell (MCFC) A type of fuel cell that contains a molten carbonate electrolyte. Carbonate ions (CO3-2) are transported from the cathode to the anode. Operating temperatures are typically near 650°C.

Moly lube A lubricant popular for coating engine parts during rebuilding to avoid an initial dry start up.

Molybdenum An element (Mo) used in some steel alloys to add hardness and strength.

Molybdenum disulfate A combination of molybdenum (Mo) and sulfur (S) sometimes added to oil and grease to improve their lubricating qualities.

Moment of inertia Referring to rotation about an axis. In the pump business it refers to a formula that describes the shape of the shaft. A solid shaft would have a different moment of inertia than a hollow shaft.

Momentum The force of continuing motion. The momentum of a moving object equals its mass times its speed.

Monitor 1. An electronic sensor, such as the lambda sensor.

2. In computer terms a monitor is a display screen.

Monkey motion The excessive movement of a mechanical device, such as an improperly adjusted carburettor linkage.

Monolithic catalytic converter A catalytic converter which has its catalytic materials coating a ceramic honeycomb, as distinguished from the pellet-bed converter.

Monolithic Substrate The ceramic honeycomb structure used as a base to be coated with a metillic catalyst material for use in the converter

Monolithic timing The use of an electronic timing device to make precise adjustments in engine timing when the engine is running.

Mono-propellant A rocket propellant consisting of a single substance, especially a liquid containing both fuel and oxidizer, either combined or mixed together.

Mould Hollow form, matrix or cavity into which materials are placed to produce goods of desired shapes.

Mount and drive Pulleys, mounting plates, belts, and fittings necessary to mount a compressor and clutch or any other engine driven assembly on an engine.

Mount base Also known (informally) as the Yoke. The Mount Base is comprised of the major structural elements that support and hold the OSS and allow it to rotate and track the sun.

Mount base assembly transfer optics This will include the beam path from the Altitude Crossover Plane, M5, M6 and up to the Azimuth Crossover Plane. The beam envelopes and the gut ray move with the Mount Base Assembly.

Mountain motor A big block V-8, usually Chevrolet or Ford, enlarged to 500 cid or more; often to as much as 600 to 800 cid.

Mounting grommet 1. The replaceable rubber bushings used at both ends of the shocks for mounting to reduce the transfer of sounds and jarring motions between the shocks and their mounting points on the vehicle.

2. A reinforced eyelet through which a fastener is attached.

Mounting position The position at which a device is mounted, such as horizontal or vertical.

Muffler The exhaust system device in the tailpipe that reduces engine noise. Some vehicles have more than one muffler along the tailpipe.

Mule 1. A yard tractor not suitable for over-the-road service.

2. A road race car used for practice or testing.

Multec Injector An injector that is designed with a high degree of anti-clogging qualities.

Multi port Fuel injection Individual injectors for each cylinder are mounted in the intake manifold

Multidisciplinary teamwork The timely and cooperative application of all appropriate disciplines in an open-communication, shared-information environment to effect people functioning as a team to achieve optimum solutions (people, product, and process) satisfy customer needs, objectives, and requirements.

Multi-fuel vehicle A vehicle having a modified engine that runs on more than one fuel, such as LPG and gasoline.

Multi-leaved spring A flat suspension spring having more than one leaf.

Multi-link suspension Independent suspension controlled with several link arms that restrict undesired motion of the suspension for a smoother ride and more precise handling.

Multimedia The use of computers to present text, graphics, video, activation, and sound in an integrated way.

Multimeter A multi-range meter used to measure voltage, current, and resistance

Multi-physics analysis Multi-physics analysis involves the use of computer-aided engineering tools to model multiple physical effects of a design such as structural, thermal, fluid/air flow, acoustic, electromagnetic, etc. This type of analysis allows consideration of the interaction of multiple effects with one another such as fluid flow and thermal.

Multiple disc A clutch with several driving and driven discs.

Multiple pass A term that applies to a recovery/recycle unit that removes refrigerant from an air-conditioning system and circulates it through the recovery/recycle unit to remove

Multiple security deposits A lessee may choose to pay additional security deposits (or monthly payments up-front) in exchange for a lower monthly lease payment via a reduced money factor, which reduces the monthly charges.

Multiple-disc clutch 1. A clutch that consists of alternating friction and steel discs that are forced together hydraulically to lock one transmission part to another.

2. A clutch having a large drum-shaped housing that can be a separate casting or a part of the transmission housing.

Multiple-leaf spring Leaf springs having a series of flat, steel leaves of varying lengths clamped together with a center bolt extending through them to maintain their position.

Multiplex 1. The ability to transmit two or more signals over a single channel simultaneously.

2. The ability to select one of a multiple of input signals to appear at an output during a particular time period.

Multiplexer A mechanical or electrical device for sharing a circuit by two or more coincident signals.

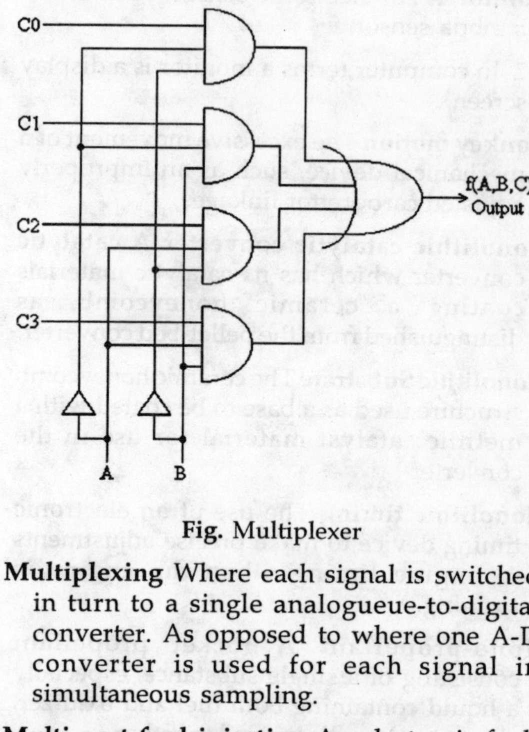

Fig. Multiplexer

Multiplexing Where each signal is switched in turn to a single analogueue-to-digital converter. As opposed to where one A-D converter is used for each signal in simultaneous sampling.

Multi-port fuel injection An electronic fuel-injection method that uses individual injectors to spray fuel directly into each intake port, bypassing the intake manifold. Also called multi-point fuel injection.

Multistage rocket A rocket having two or more stages which operate in succession each being discarded as its job is done.

Multistrike Ignition An idle ignition strategyy where the spark plugs are fired multiple times per TDC event to insure idle ignition.

Muscle Car A muscle car, by the strictest definition, is an intermediate sized, performance oriented domestic model vehicle, powered by a large V8 engine, offered at an affordable price. Usually from 1958 through 1977 models.

Mutual inductance (M) A measure of the amount of inductive coupling between two coils.

Mylar A registered trade mark of El Dupont De Nemours & Co. For polyethylene terephathalate. The polyester film is a common insulating material used in transformers, capacitors, etc.

N

Nafion A sulfuric acid in a solid polymer form, usually used as the electrolyte of a PEM fuel cell.

Nagara system A production system where seemingly unrelated tasks can be produced by the same operator simultaneously.

Named insured The person or company that is designated as the insured.

Nanotechnology Nanotechnology is molecular manufacturing or, building things one atom at a time. A nanometer is one billionth of a meter and approximately ten atoms fit inside one nanometer. Utilizing the physical properties of atoms and molecules, nanotechnology proposes the construction of nanosize devices possessing extraordinary properties.

Natural Gas A naturally occurring gaseous mixture of simple hydrocarbon components (primarily methane) used as a fuel.

Naturally aspirated An engine that uses atmospheric pressure to force the air into the cylinders.

Nautical mile The distance spanned by one minute of arc in latitude, defined internationally as 1,852 meters (6,076.1033 feet).

Necking knob A know attached to the steering wheel that permits rapid one-hand steering.

Necking Localised reduction of cross-sectional area of a specimen under tensile load. It is disregarded in calculating engineering stress but is taken into account in determining true stress.

Negative Temperature Coefficient Type of thermistor in which resistance decresses as temperature increases

Negative terminal The side of a battery or electrical device nearest ground potential, often identified by a minus (-) sign.

Negative wheel slip Wheel lockup that occurs when too much braking force is applied to a wheel and the tire skids on the pavement. ABS controls negative wheel slip by modulating the hydraulic pressure to the wheel or wheels that is/are skidding.

Negligence Failure to use a certain degree of care.

Nematic Describes a fluid that is a liquid crystal with a threadlike form. It has light slots that can be rearranged by applying small amounts of voltage

Neoclassic The modern design of a car to resemble a classic of the late 1920's and early 1930's.

Net horsepower The maximum engine horsepower as measured on a dynamometer.

Net present value Net present value is a financial analysis technique that discounts a series of cash inflows (revenue) and outflows (investments and expenses) to determine the suitability of an investment. Npv is an evaluation technique used to screen new product development projects as part of portfolio management.

Net shape processing Any manufacturing process which creates an object in its finished form without the need for finish machining or other actions.

Net torque The maximum engine torque as measured on a dyno.

Neutral 1. The position of an automatic transmission when the engine is disengaged from the drive train.

2. A condition of a planetary gear set when all members, gears, and planet carriers, are free to rotate.

Neutral flame An oxyfuel gas flame that has neither oxidizing nor reducing characteristics.

Neutral junction The center connection to which the common ends or a Y-type stator winding are connected

Neutral start switch A term used for neutral safety switch.

Neutral The ac return somewhere connected to ground, but which should not be used for ground because it is a current-carrying path.

Neutral wire A grounded conductor in a three-wire electrical system that does not carry current unless the system is unbalanced.

Neutralized A term used for allowing a system to settle in a non-stressed position. Used in reference to exhaust systems, powertrain mounts, and suspension parts and mounts.

Neutron Atomic particles having approximately the same mass as a hydrogen atom; very penetrating.

New product development New product development is the business process for developing new hardware, software and service products for the enterprise. It includes all activities from development of the idea or concept for the product, the development of the product and its processes, and the launch of the product into production and into the market place.

New product introduction 1. New product introduction refers to the set of activities that occur once a product has been developed and is ready to be introduced into the marketplace.

2. New product introduction is used by some organizations synonymously with new product development.

Newton That force which gives a mass of 1 kilogram an acceleration of 1 meter per second per second; equal to 100,000 dynes.

Newtonian fluid A fluid that does not change viscosity as it is agitated.

Newton's laws of gravity 1. Every mass in the universe exerts an attractive influence on every other mass.

2. For two objects, the force of gravitation acting between them is directly proportional to the product of their masses, and inversely proportional to the square of the distance between their centers of mass.

Newton's laws of motion There are three laws: 1. A body in motion remains in motion, and a body at rest remains at rest, unless some outside force acts on it.

2. A body's acceleration is directly proportional to the force applied to it, and the body moves in a straight line away from the force.

3. For every action, there is an equal and opposite reaction.

Next Generation Manufacturing Systems Systems used to link customers, suppliers, and the many functions within a company with other companies in the enterprise. Next Generation Manufacturing Systems assume that next generation manufacturing enterprises are organized into distributed work units, that the work units have a high degree of autonomy, and that they behave intelligently.

Nitrogen (N2) A diatomic colourless, tasteless, odorless gas that constitutes 78% of the atmosphere by volume.

Nitrogen A colourless, odorless, tasteless gas that makes up 78% of the atmosphere and is part of all living tissues.

Nitrogen oxide Any chemical compound of nitrogen and oxygen (nox) as a by-product of combustion that forms smog in the presence of sunlight.

Nitrogen tetroxide A liquid oxidizer that reacts spontaneously with hydrazine.

Nitromethane A mixture of nitric acid and methane which is used to fuel top fuel dragsters and funny cars; is also called nitro or top fuel.

Nitrous oxide A non-flammable, non-explosive gas (N2O) used as an oxidizing agent with gasoline or methanol to increase the rate and efficiency of combustion thereby increasing the horsepower.

Nitrous Oxide Systems (NOS) A popular manufacturer of nitrous oxide injection systems.

NLGI An acronym for the National Lubricating Grease Institute, Usually associated with grease, The higher the NLGI number, the firmer the grease; #000 is very fluid whereas # 5 is very firm. The consistency most recommended is NLGI #2 (soft)

Noble metals A rare and expensive group of metals that resist corrosion such as gold (Au), iridium (Ir), platinum (Pt), palladium (Pd), ruthenium (Ru), and rhodium (Rh), some of which may be found in catalytic converters.

Noctilucent clouds Weakly-luminous clouds, seen at night at heights of about 50 miles (80 km) above the Earth.

Noise The aperiodic random component on the power source output which is unrelated to source and switching frequency. Unless specified otherwise, noise is expressed in peak-to-peak units over a specified bandwidth.

No-load test A starting-motor test to determine the rpm and current draw of a starter motor. The mechanical tachometer is placed on the drive end of the armature shaft and the motor is then connected to a proper power source. An ammeter, attached to the power means, determines current required.

Nomax A popular trade name for a fabric that is used in race drivers' apparel.

Nominal group technique Nominal Group Technique, similar to brainstorming, is used by teams to generate ideas on a particular subject. Team members are asked to silently write down as many ideas as possible. Each member is then asked in turn to share one idea which is recorded. After all ideas are recorded, they are discussed and prioritized by the group.

Nominal size A designation used for general identification, such as the inside diameter of a tube.

Nominal stress Stress calculated on the basis of the net cross section of a specimen without taking into account the effect of geometric discontinuities such as holes, grooves, fillets, etc.

Nominal value The stated or objective value of a quantity or component, which may not be the actual value measured.

Nominal voltage The stated or objective value of a given voltage, which may not be the actual measured.

Non lubricant The fluid that will not maintain a film thickness of at least one micron at its operating temperature and load. A concern with mechanical sealing.

Non repetitive pulse A ramdom electrical signal, with no specific pattern or frequency

Non synchronized Relating to injector control, these are fuel pulses not related to the RPM reference. ECM control of the fuel injectors in not directly related to ignition reference signals.

Non volatile memory Memory that is retained in the computer not affected by applyed voltage

Non-coherent Communications mode wherein a spacecraft generates its downlink frequency independent of any uplink frequency.

Nonconductor 1. A material that does not conduct electricity.

2. A material that is resistant to the conduction of heat.

Nonconformance A nonconformance is product or material which does not conform to the customer requirements or specifications.

Noncorrosive flux A soldering paste, usually composed of resin-based materials, that does not chemically attack the base metal.

Non-destructive inspection / test The inspection or test of the product or part that retains the product's or part's physical and operational integrity.

Non-developmental item Non-developmental item is a broad generic term that covers material available from a wide variety of sources with little or no developmental effort by the purchaser.

NDIs include: Items obtained from a domestic or commercial marketplace; Items already developed and in use by users; Items already developed to other standards and in agreement with the purchaser.

Nonparallel drive shaft A drive shaft with working angles of equal length; the companion flanges and/or yokes are not parallel.

Nonpolarized gladhand A gladhand that can be connected to either the service- or the emergency-gladhand.

Non-renewable The termination of the insurance policy at the expiration date.

Non-scheduled time A period of time when the equipment is not scheduled to be used in production, such as un-worked shifts, weekends, and holidays.

Non-servo A type of drum brake in which the brake shoes work independently of each other.

Non-slip differential A differential that is designed so that when one wheel slips, the major portion of the driving power is transferred to the non-slipping wheel.

Nonsynchronized mode A mode of operation on a throttle body, fuel-injection system in which the injector is pulsed every 1 2 . 6 milliseconds, independent of a distributor reference pulse.

Non-value added Activities or actions taken that add no real value to the product or service, making such activities or action a form of *waste*.

Normal evacuation An evacuation to remove air and moisture from a system with only one pump down.

Normal wear pattern The wear pattern of a part of assembly that is normal or expected.

Normally aspirated An engine that is not fitted with either a turbo or supercharger.

Normally closed (nc) A term referring to a device, such as an electrical switch, that is closed in its normal position.

Normally open A term referring to a device such as an electrical switch, that is open in its normal position.

Nose 1. Frond end of a vehicle.

2. The uppermost part of a camshaft lobe.

3. The front of a semi-trailer.

Nose over An engine or other component that is starting to self-destruct.

Nose plug A threaded, tapered plug used to repair cracks in the diesel engine cylinder head around the injector holes.

Nose shroud A cover on the nose of a rocket or spacecraft which jettisons before insertion into orbit.

NOT gate (inverter) A digital logic circuit in which its signal output is the inverse or its input.

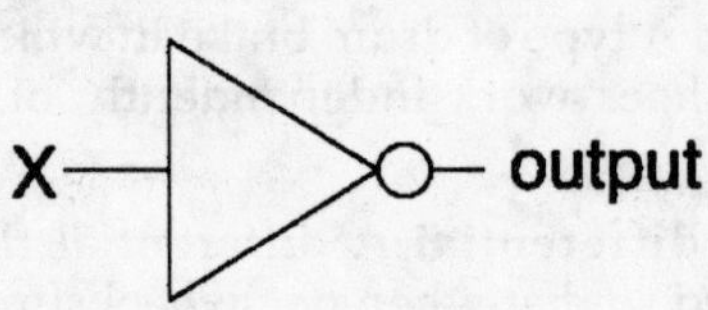

Fig. NOT gate

No-tilt convertible fifth A fifth wheel with fore/aft articulation that can be locked out to produce a rigid top plate for applications that have either rigid and/or articulating upper couplers.

Nucleus The core or central part of the atom, consisting of protons and neutrons and having a positive electrical charge

Number failure code A term used for failure code.

Number of cylinders The total number of cylinders (1, 2, 4, 6, or 8) contained in an engine.

O

Object The term object is used to mean a collection of attributes that represent either a physical or logical artefact. For example an object can represent all the information required for an item or a drawing. The key feature of an object is that it represents data that can be manipulated as a group, so copying an item object copies all the attributes associated with the object in one action. Objects are specialized into classes. An object representing a particular type of bolt, for example, could be in the class of objects called bolts. Objects have 'methods'. Methods are ways in which the object can be accessed, modified, displayed, etc. Another feature of objects is inheritance. One class can be based upon another, the new class is called a subclass. The new class will inherit all the attributes of the other class. So you could have a class called bolt with attributes, length, thread type and pitch. You can then create a subclass called brass. The new class would inherit the attributes length, thread type and pitch from the bolt class and have a new attribute 'brass'.

Object-oriented database A database used to store objects that form the bases of object-oriented computing, in which data as well as references to the procedures used to perform operations on that data are combined.

Object-oriented software Results from a kind of modular programming. Each object is a software package containing a collection of related procedures and data that can be reused, shortening application development time. Objects also make it easier to customize software systems to mirror actual business processes, and to do so without negatively impacting the ability to migrate to later software releases.

Observatory Control System (OCS) The OCS is the highest level software system, responsible for coordinating observations and providing system services.

Observed horsepower The brake output of an engine as observed on a dynamometer.

Occultation The passage of a celestial body across a line between an observer and another celestial object; and the progressive blocking of light, radio waves, or other radiation from a celestial source during such a passage.

Occupational safety glasses Protective eye wear designed with special hit-impact lenses, frames, and side protection.

Occurrence Any event that resulted in a loss or damage to the insured or the insured's property.

Octane A gasoline's ability to resist detonation. The higher the octane number, the greater the fuel's resistance to detonation.

Octane enhancer Common term designating components that are added to gasoline to increase octane and reduce engine knock. Examples are toluene, ethanol, ETBE, and MTBE.

Octane number The number representing the average blend of iso-octane with an index of 100, and other hydrocarbons with an index of 0, usually between 87 and 92.

Octane rating A unit of measurement on a scale intended to indicate the tendency of a fuel to detonate or knock based on the percentage of isooctane in the fuel. The higher the rating, the higher the percentage of isooctane and therefore the greater the resistance to detonation offered by the fuel.

Octane requirement increase (ORI) The increase in octane requirement that results from the build up of combustion chamber deposits.

Octane requirement The minimum octane rating of a fuel required to operate an engine without a spark knock.

Odometer A mechanical or electronic counter in the speedometer that indicates trip or total miles accumulated on the vehicle.

Odometer rollback The illegal practice of rolling a vehicle's odometer back to indicate that it travelled fewer miles than it actually has.

Odometer rollover Occurs when the vehicle's mileage exceeds the mechanical limits of the odometer - usually 99,999 miles. This must be certified by the seller, under the truth in mileage act.

Off time The part of an electrical signal during which an electrical device is de-energized

Off-beam instrument set Any local instrument set operating without access to the light beam.

Offset 1. A term used for wheel offset.

2. A condition when two parts are not directly in line with each other.

Offset current The direct current that appears as an error at either input terminal of a dc amplifier when the input current source is disconnected.

Offset Error If you get a reading other than zero for a zero condition then you have an offset error: every reading will be inaccurate by this amount.

Offset link A type of chain link that utilizes bent, or offset link plates and is assembled with a pin at one end and a bushing and roller at the other end, so as to act as a combination link.

Offset rod A rod on which the bran section is not directly centered over the bearing housing.

Offset section A factory-assembled section of roller chain, made up of a roller link and an offset link, and used to connect strands of chain having an odd number of pitches.

Offset voltage The dc voltage that remains between the input terminals of a dc amplifier when the output voltage is zero.

Offset yield strength Arbitrary approximation of elastic limit. It is the stress that corresponds to the point of intersection of a stress-strain diagram and a line parallel to the straight line portion of the diagram. Offset refers to the distance between the origin of the stress-strain diagram, and the point of intersection of the parallel line and the 0 stress axis. Offset is expressed in terms of strain (often 0.2%).

Onboard fire-extinguishing system A fire-extinguishing system permanently installed in the driver's compartment of a vehicle; required by most race-sanctioning bodies.

On-Demand Test A Ford self-test The Key On/ Engine Off and Engine Running Positions which are technician initiated, are from a program within the processor

Open course That part of a race track that extends beyond the finish line into a shut-off area.

Open end lease A lease in which the lessor is not responsible for the value of a vehicle when the lease is done. Under this arrangement the lessee is responsible for the difference between the current actual market value at the end of the lease and the residual value. Also known as a "finance" lease.

Open Loop A mode of computer operation in which fuel delivery calculations are based primarily on preprogrammed values, not Oxygen sensor input

2. The Oxygen sensor value is not considered when in Open Loop

Open structural member A flat body panel having an open access from the rear.

Operating temperature range The range of ambient baseplate or case temperatures through which a power supply is specified to operate safely and to perform within specified limits.

Operational amplifier (OP-AMP) A high gain differential input device that increases the magnitude of the applied signal to produce an error voltage.

Operational concept definition document The source of all DRD requirements that cannot be derived directly from the SRD. This includes information on the philosophy and concept that underpins a subsystem, trade studies, meeting notes and presentations, etc. It does not include DRD information. Not every subsystem will have a complete OCDD (for example, the enclosure) but, if one is requested, Systems Engineering will pull the relevant pieces together to make an OCDD. The main components of this document shall be completed by CoDR, even if a formal OCDD does not exist.

Operations Work or steps taken to transform material from raw materials to finished product.

Opportunity cost The cost of using money one way as opposed to using it in another, more economically advantageous way. For example, the loss of interest experienced as a result of removing money from a savings account to be used as a down payment on a vehicle.

Opposed engine An engine with cylinder banks at 180 degrees, such as the Volkswagen flat four.

Opposite lock Turning the steering wheel in the opposite direction of a turn to control or correct oversteer.

Optical flat A high quality glass lapped flat on one side and used with a monochromatic light to read seal face flatness.

Optical horn A Chrysler term for a flash to pass dimmer switch feature.

Optics support structure Also known as the OSS, or (informally) as the Telescope Tube. The OSS is comprised of the major structural elements that support and hold M1 and its cell; the secondary mirror and its support equipment; and other mirrors and/or optical devices used to relay the optic beam to the ATST scientific instrumentation. The OSS may also provide support to a variety of other related equipment, such as calibration optics, Lyot stops and filters, and various mechanical equipment (e.g., heat stop, mirror covers, etc).

Organic brake lining Brake linings that are made of a carbon-based compound combined with non-organic magnesium silicate and/or glass and synthetic fibres, replacing asbestos which as been determined to be hazardous to health and the environment.

Organic friction material A friction material having organic binders substantially formulated with nonmetallic fibres. Used with both breaks and clutches.

Organization A unit within a company or other entity (e.g., Government agency or branch of service) within which many projects are managed as a whole. All projects within an organization share a common top-level manager and common policies.

Organization for economic co-operation and development (OECD) An organization whose purpose is to provide its 29 Member countries with a forum in which governments can compare their experiences, discuss the problems they share and seek solutions which can then be applied within their own national contexts. Each member country is committed to the principles of the market economy and pluralistic democracy.

Orthogonal array Orthogonal array - an array to represent a fractional factorial experimental design in design of experiments.

Oscillation damper A shock absorber may be considered a damper that controls energy stored in the springs under load.

Oscillation The rotational movement, either in a fore and aft or side-to-side direction around a pivot point, such as in a fifth wheel design in which such articulation is permitted.

Oscillator This semiconductor device creates a rapid back-and-forth movement of voltage.

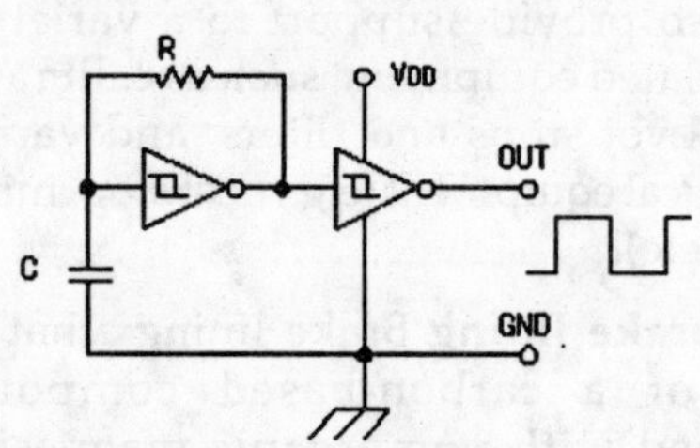

Fig. Oscillator

Oscilloscope An instrument that produces a visible image of one or more rapidly varying electrical quantities with respect to time or with another electrical quality.

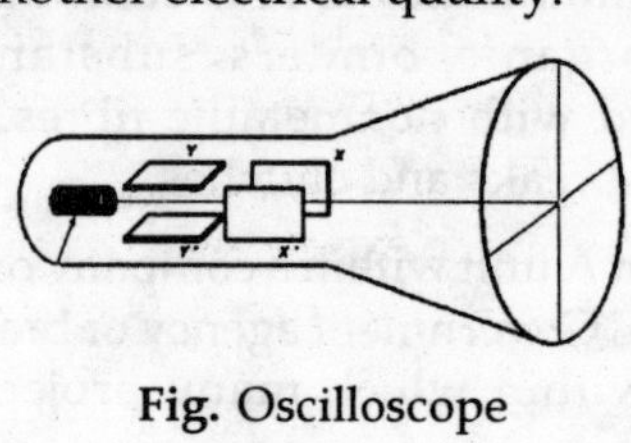

Fig. Oscilloscope

Out of phase A term that relates to a drive shaft having two universal joints that are not in phase with each other, producing a jouncing or jerking motion as the drive shaft turns.

Out of round The condition of a circular part when it is not circular, as in slightly oval shaped.

Out of square The condition of a square or rectangular part when its vertical sides are not at right angles to its horizontal sides.

Out of the box A vehicle, device, or component that is absolutely stock without modification.

Out of the chute Off the starting line in a drag race. Also known as out of the gate and out of the hole.

Outboard brakes A brake assembly that is mounted at the outer or wheel end of an axle half shaft.

Output compare A function provide by the computers timing system, whereby an external event is generated when an internal counter value matches a predefined value.

Output current limiting A protective feature that keeps the output current of a power supply within predetermined limits during overload to prevent damage to the supply or the load.

Output driver An electronic switch in a computer used to turn ON and OFF an actuator in a vehicle.

Output Driver Module A module capable of controlling seven different outputs with differing loads. It contains circuit protection and data bus communications and fault detection

Output ripple and noise Output ripple is the periodic AC component imposed on the output voltage of a converter. Output noise refers to unwanted variations in the converter output that are unrelated to the switching frequency. Ripple and noise are typically specified together as a peak-to-peak value over a specified bandwidth. Ripple and noise is sometimes referred to as

Periodic And Random Deviation (PARD). Care must be taken when measuring output noise and ripple not to induce errors in the test instrumentation. Attach a twisted wire pair (about 1 foot in length) with three twists per/inch between the converter outputs and an appropriate load. Connect a 33 uf electrolytic capacitor across the load. Using an oscilloscope with a minimum bandwidth of 20 mhz and a probe with the ground clip disconnected, measure the ripple at the connection of the load and twisted pair wires. This method eliminates the 'common mode noise" that interferes with measurements made directly at the converter output pins.

Output State Check A Ford performance test in which the Key On/Engine Off mode after the continuous codes have been sent which forces the processor to activate some actuators for additional diagnostics

Output voltage accuracy The maximum allowable deviation of the DC output from its ideal or nominal value. Output voltage accuracy is sometimes called "output voltage tolerance" or "output voltage setpoint" and is commonly given as a percentage of output voltage.

Overall equipment efficiency (OEE) The overall performance of a single piece of equipment, or even an entire factory, governed by the cumulative impact of 3 factors: Availability, Performance Rate (which includes machine speed), and Quality Rate. (Ex: 50% Availability (0.5) x 70% Performance Rate (0.7) x 20% Quality Rate (0.8) = 30%OEE) Use to measure how effective in relationship to the planned production schedule.

Overcharge 1. A term used when too much refrigerant or oil is added to the refrigeration system.

2. A condition where the charging system is supplying too much voltage/current to the battery.

Overcooling A general term used if the engine does not reach design operation temperature in a predetermined time period

Overcord The covering bonded to the tensile member of a synchronous belt that protects it form frictional wear if a backside idler is used or if power is transmitted from the backside of the belt.

Overcurrent device A device capable of automatically opening an electric circuit, both under predetermined overload and short-circuit conditions, either by fusing of metal or by electro-mechanical means.

Overdrive A transmission gear with a ratio below 1:1, which improves fuel economy by reducing engine revolutions per minute at highway speeds. On a five-speed manual transmission, the fourth and fifth gears are overdrive. On a four-speed automatic transmission, the fourth gear is overdrive. When an overdrive gear set is engaged, the output shaft turns at a higher rate than the input shaft, reducing engine revolutions at cruising or highway speeds.

Overdrive band A transmission brake device that is engaged in overdrive.

Overhead camshaft engine An engine in which the camshaft is mounted over the cylinder head.

Overhead position The position in which welding is performed from the underside of the joint.

Overhead valve (OHV) An I-head arrangement where the valves are located over the piston in the cylinder head.

Overhead valve engine (ohv) An engine with both intake and exhaust valves placed directly over the piston. In this design, the camshaft is located in the block, and the valves are actuated by pushrods and rocker arms.

Overheat 1. To become excessively hot.

2. A general term used if the engine exceeds design operating temperature.

Oversteer The opposite of understeer, this is when the vehicle turns much more than was desired by the driver. It is a condition more commonly found when cornering forcefully in RWD than FWD vehicles because sudden power to the rear wheels can cause them to slide sideways.

Overstressing Application of high fluctuating loads at the beginning of a fatigue test and lower loads toward the end. It is a means for speeding up a fatigue test.

Overtime Includes pay received for the number of hours worked in excess of the standard workday or workweek.

Overvoltage 1. The potential difference between the equilibrium of an electrode and that of the electrode under an imposed polarization current.

2. A voltage that exceeds specified limits.

Overvoltage protection (OVP) A power supply feature which shuts down the supply usually by firing a crowbar or clamping the output when an output voltage exceeds a preset level.

Oxygen sensor A device found in the exhaust manifold, which generates a small voltage dependent on the amount of oxygen present in the exhaust stream, used as a signal to the engine-control computer to determine the amount of fuel necessary to maintain a proper air/fuel ratio.

Oxygenate In the petroleum industry a term used to denote octane components containing hydrogen, carbon, and oxygen in their molecular structure. Includes ethers such as MTBE and ETBE as well as alcohols such as ethanol and methanol.

Oxygenated gasoline Gasoline containing oxygenates such as ethanol or MTBE. Provides chemical enleanment of the A/F charge thereby improving combustion and reducing tailpipe emissions of CO.

P

Packaging A collective term that covers a swathe of technologies for encasing semiconductor chips and creating electrical and other links between integrated circuits and the "outside world".

Packing The soft rings that mechanical seal replace to stop leakage. Packing must leak because it works on the theory of a series of pressure drops to reduce the stuffing box pressure to the point where the leakage is acceptable. A minimum of five rings of packing is required to do this.

Pad A brake lining and metal back riveted or molded together. Two pads provide stopping friction by rubbing against both sides of a rotor or inside drum when the brakes are applied.

Pad hardware Miscellaneous small parts such as antirattle clips and springs as well as support clips that hold brake pads in place and keep them from rattling.

Paddock area A designated area behind or near the pit where race cars are tuned and prepared for qualifying and racing.

Pad-wear indicator A mechanical or electrical warning device on disc brakes that indicates need for pad replacement.

Paint A material, generally containing binders, solvents, and pigments, that is applied as a liquid to a surface, and forms a solid film for decoration and/or protection.

Paint shop The area in a repair facility where the vehicle is refinished by refinishing technicians or painters.

Paint, Body and Equipment Association (PBEA) A trade association of manufacturers, wholesalers, and distributors of vehicle paint and body-shop equipment.

Panel cutter Special "snips" used to cut through body sheet metal, designed to leave a clean, straight edge that can be easily welded.

Panel truck An enclosed light truck or van having no windows in the cargo area.

Panhard rod A rod that is attached to the vehicle frame at one end and to the axle at the other end to prevent the chassis from moving side to side relative to the axle; used on a beam axle or de Dion axle type rear suspension.

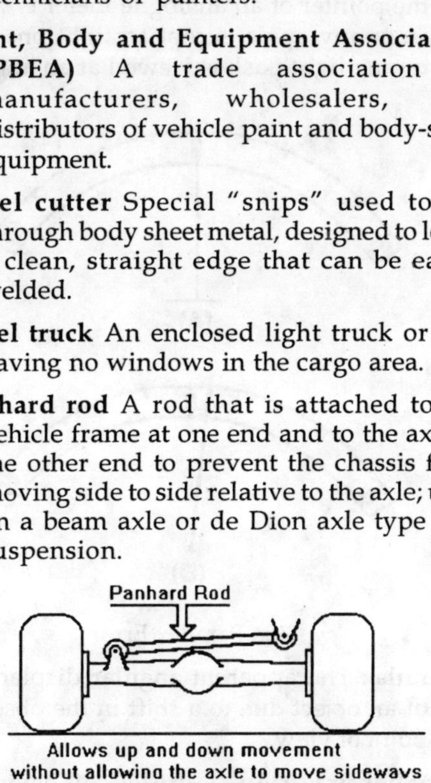

Fig. Panhard rod

Paper Sheets of fibrous absorbent materials made from organic fibres that will "swallow up" liquids and may well disintegrate under such action if the liquid is a solvent for the binder. Manufacturer documents used to obtain vehicle titles.

Parabolic reflector A reflector in the shape of a concave parabola used for radar and other microwave antennas.

Parachute An apparatus used to retard free fall, consisting of a light, usually hemispherical canopy attached by cords and stored folded until deployed in descent.

Parade lap A lap or laps in closed-course racing before the pace lap to give the drivers an opportunity to warm up their tires and engines and to give the fans a chance for a good look at the cars at a slow speed.

Parallax Error This is an error that occurs if the pointer of an analogue meter is viewed from any angle except straight on, or if a timing light flash is viewed at an angle.

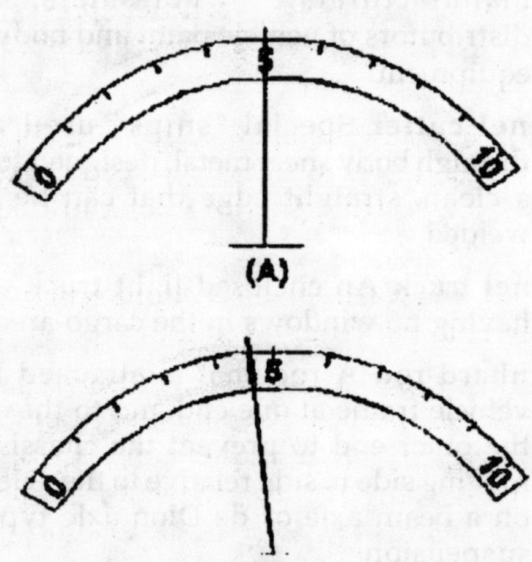

Fig. Parallax Error

Parallax The apparent angular displacement of an object due to a shift in the observer's point of view.

Parallel circuit An electric circuit having two or more paths that the current flows through at the same time. The current is divided with more of it flowing through the path of least resistance.

Parallel development Design and production engineers at a company and at its suppliers work jointly from beginning to end, not in a sequential manner.

Paralleling Connecting the output of two or more supplies together, either for the purpose of delivering additional current oı creating redundancy. The reliability of a power system is critical to overall system performance. Indeed, the requirements for power systems are generally higher than those for the system overall. A cost effective way to provide very high levels of reliability is to connect a number of independent power supply units in parallel, such that if one power supply fails, the remaining power supplies will continue delivering sufficient current to supply the maximum system load without any interruption. This architecture is often referred to as an (n+x) redundant power supply. If any one of several paralleled supplies can fail without causing the entire system to fail, the power system is said to be (n+1) redundant. If any two of several paralleled supplies can fail without causing the entire system to fail, the power system is said to be (n+2) redundan. By allowing the live removal or insertion of power supply boards to and from the system power bus, a "hot pluggable" redundant power supply system provides a practical way to achieve the zero down time required in critical applications such as mainframe computer systems and telecommunications systems. In practice, (n+x) redundancy can be difficult to achieve. Connecting power supplies in parallel and ensuring that the failure of one parallel supply unit does not affect the remainder presents particular difficulties, ruling out paralleling schemes such as master-slave configurations entirely.

Parent metal The original metal in a body panel to which another metal panel has been added.

Pareto analysis/ diagram An analysis/ diagramming technique using frequency of occurrence to identify and display results generated by each identified cause. This analysis is commonly used to decide where to apply initial effort for maximum effect.

Pareto principle The Pareto principle suggests that 20% of a set of independent variables is responsible for 80% of the result. In quantitative terms, for example, 80% of the problems come from 20% of the causes (machines, raw materials, operators etc.). Therefore, effort aimed at the right 20% can solve 80% of the problems.

Parity An error-checking scheme whereby additional bits are added to data to aid in detecting common errors in data transmissions.

Park contacts Electrical contacts, inside a windshield-wiper motor assembly, to provide current to the motor after the control switch has been turned to the OFF or PARK position, allowing continued motor operation until reaching the park position.

Park Neutral Switch Used on vehicles equipped with automatic transmissions to indicate the transmissions shifted position.

Parking orbit Orbit in which a space vehicle awaits the next phase of its planned mission.

Parking-brake cable A cable or cables that transmit brake-actuating force in the parking-brake system.

Parking-brake equalizer A device used to equalize the pull between the parking brake actuator and two rear wheels.

Parking-brake system A brake system, intended to hold a vehicle stationary, in which one or more brakes may be held in the applied position without continued application of force to the control.

Part classification Classification of parts or other elements of a product by their geometry, material, function and/or the processes used to manufacture them. Part classification is used to find components or subassemblies to use in a product design and to aid in standardization efforts.

Part model A data model that contains the complete geometric and functional representation of a part and its characteristics. A comprehensive part model would also contain related analysis, configuration, manufacturing and support data.

Part number (PN) The alphanumeric designating a part as listed in a catalogue or parts list.

Partial loss Any loss that does not completely destroy the property or exhaust the insurance.

Partial Oxidation Fuel reforming reaction where the fuel is partially oxidized to carbon monoxide and hydrogen rather than fully oxidized to carbon dioxide and water. This is accomplished by injecting air with the fuel stream prior to the reformer. The advantage of partial oxidation over steam reforming of the fuel is that it is an exothermic reaction rather than an endothermic reaction and therefore generates its own heat.

Participatory design Participatory design refers a democratic approach to design that encourages participation in the design process by a wide variety of stakeholders, such as: designers, developers, management, users, customers, salespeople, distributors, etc. The approach stresses making users not simply the subjects of user testing, but actually empowering them to be a part of the design and decision-making process. This is accomplished through direct involvement with the product development team on major projects for one or a small number or customers or through frequent customer or user review and feedback during the development process using mechanisms such as focus groups, web-based customer participation, usability studies, etc.

Particle A very small piece of dirt, metal or other debris which can be contained in oil, air, or fuel that can be removed with a filter.

Particulate 1. A solid matter, mainly soot from burned carbon, in an internal combustion engine's exhaust.

2. A form of solid air pollution such as microscopic solid or liquid matter that floats in the air.

Particulate emissions Solid particles, such as carbon and lead, found in vehicle exhaust soot. A problem, especially in diesels.

Particulate trap An emissions control device in the exhaust system of a diesel engine which is used to capture particulate before they can enter the atmosphere.

Parting edge The excess flash material that is found around the edge of a part that has been cast in a two-piece mold.

Partition coefficient The ratio of the solubility of a chemical in water as compared to its solubility in oil.

Part-lib Parts library an international standard that will offer the capability for computer-sensible representation and exchange of part library data.

Parts changer One who, without proper diagnostic skills, randomly changes parts until the remedy to a problem is found.

Parts chaser 1. A vehicle used for errands.

2. A person who runs errands, especially for buying parts.

Parts counterman One very knowledgeable in part numbers, and/or determining requirements who generally works in a parts-supply wholesale ore retail outlet.

Parts distribution A wholesale, discount, or retail establishment that sells parts to the trade or do-it-yourselfer.

Parts manager One responsible for ensuring that the in-house stock of parts is adequate so the customer's parts are readily available to the service technician when needed.

Parts per million (ppm) 1. A unit used to measure the amount of contamination in a substance, such as moisture in refrigerant.

2. A measurement of the emissions of a motor vehicle given as the number of parts of a particular chemical within one million parts of exhaust gas.

Passenger car A four-wheeled, motor-driven vehicle that carries ten passengers or less, intended for used on streets and highways.

Passenger presence detection (PPD) A sensor system that is able to detect the presence of a passenger. Different systems are known as infrared scanning weight sensing, capacitive measurement.

Passenger vehicle Four wheeled motor vehicle that also includes mini-vans and sport utility vehicles.

Passivated The protective oxide layer (ceramic) has formed on the metal surface.

Payload Revenue-producing or useful cargo carried by a spacecraft; also, anything carried in a rocket or spacecraft that is not part of the structure, propellant, or guidance systems.

Peak and hold Full current applied to an injector to raise the pintle, then reduced to hold the pintle up for the remainder of the pulse width

Peak Detection An acquisition mode available with digital oscilloscopes that enables you to observe signal details that may otherwise be missed

Peak inverse voltage (PIV) The maximum reverse voltage which a semiconductor junction can withstand without permanently damaging the junction

Peak Maximum value of a waveform reached during a particular cycle or operating time.

Peak out The engine speed at which maximum horsepower is developed.

Peak output current The maximum current value delivered to a load under specified pulsed conditions.

Peak-to-peak amplitude The signal amplitude measured between the maximum positive and negative peaks excursions of an ac signal.

Peak-to-peak Voltage The voltage measured from the maximum point of a signal to its minimum point

Peal-and-hold injector A peak-and-hold injector is typically used for TBI and some port low-resistance injector systems.

Pearl paint A type of paint that is similar to metallic paint, but instead of minute metal particles it uses mica. Mica is a kind of semi transparent, crystalline mineral that absorbs and reflects light in prismatic fashion. This gives a dramatic, multi-dimensional effect to the paint. Sometimes called "pearl coat."

Peer review The review of work products during their development that is performed by peers to identify defects for removal.

Peer-to-Peer Communication A communication between networked devices in which any device can initiate data transfer.

Pellet thermostat An engine-coolant thermostat having a wax pellet as a power element, which grows when heated and shrinks when cooled, connected through a piston to a valve.

Penetrating oil A special, thin oil that penetrates rust and/or tight threads and helps in the removal of stubborn nuts and bolts.

Pentroof Combustion Chamber A flat or pancake shaped combustion chamber that has the overhead valve heads that are almost parallel to the top of the piston and the spark plug at or near the center of the chamber. This design provides a very short distance for the flame front to travel, excellent air flow characteristics and efficient combustion.

Percent of grade A value that is determined by dividing the height of a hill by its length, often used to determine the power requirements of trucks or for determining maximum pay load.

Perimeter frame A conventional chassis frame design that is similar in construction to a ladder frame, having full-length side rails that support the body at its greatest width, providing optimum protection to passengers in the event of a side impact.

Period of revolution Time of one complete cycle in orbital motion - referred to as a year when applied to Earth.

Period of rotation Time of one complete cycle - referred to as a day when applied to Earth.

Permanent magnet A piece of ferrous material, such as steel, that retains its magnetic properties without the use on an electric current.

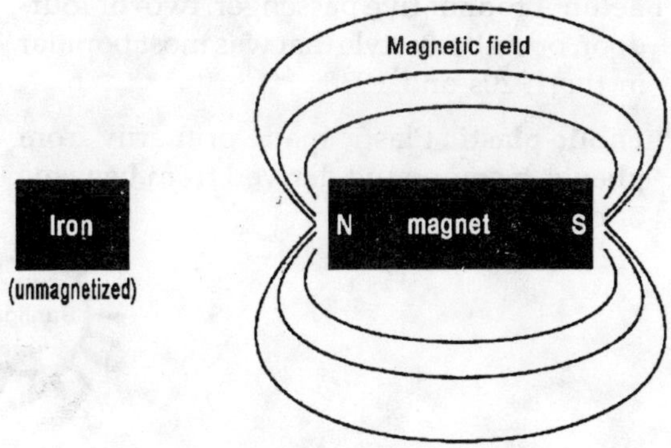

Fig. Permanent magnet

Permanent magnet generator A reluctance sensor. A sensor that generates a voltage signal by moving a conductor through a permanent magnetic field.

Permanent strainer A device generally of a Y-configuration having an accessible cylindrical strainer element used in horizontal and vertical lines. The element, retained by a plug end which may be plain or fitted with a valve, can be opened for "blow-through" cleaning.

Permatex A tradename for a brand of engine and transmission sealants.

Permeability The measure of the ease with which magnetic lines of force can move through a material compared with air.

Permissible exposure limit (PEL) The maximum length of time a person should be exposed to a hazard or hazardous material established by OSHA and expressed as a time-weighted average limit or ceiling exposure limit.

Personal injury protection (pip) A broad type of medical payments insurance. Usually offers protection for expenses incurred up to a dollar amount.

Perturbation Modifications to simple conic section orbits caused by such disturbances as air drag, non-uniformity of the Earth, and gravitational fields of more distant bodies such as the Moon.

Phaeton A four or five passenger, two- or four-door, open-body style that was most popular in the 1920s and 1930s.

Phenolic plastic Plastic made primarily from phenol, a compound derived from benzene or carbolic acid.

Photodiode A semiconductor diode operated as a photo conductor. Normal operation is with its junction reverse biased.

Photoelectric principle The production of electricity created by light striking certain sensitive materials, such as selenium or caesium.

Photolithography The process of projecting a photomask pattern or image with UV light onto a wafer coated with photosensitive emulsion.

Photomasks A market sector served by the Unaxis Wafer Processing division. A term given to the photolithographic masks used in chip manufacturing.

Photometer An optical instrument that measures the intensity of light from a source.

Phototransistor A bipolar transistor that has gain and produces collector current when light shines on its collector-base junction.

Photovoltaic cells Crystalline wafers called solar cells which convert sunlight directly into electricity without moving parts.

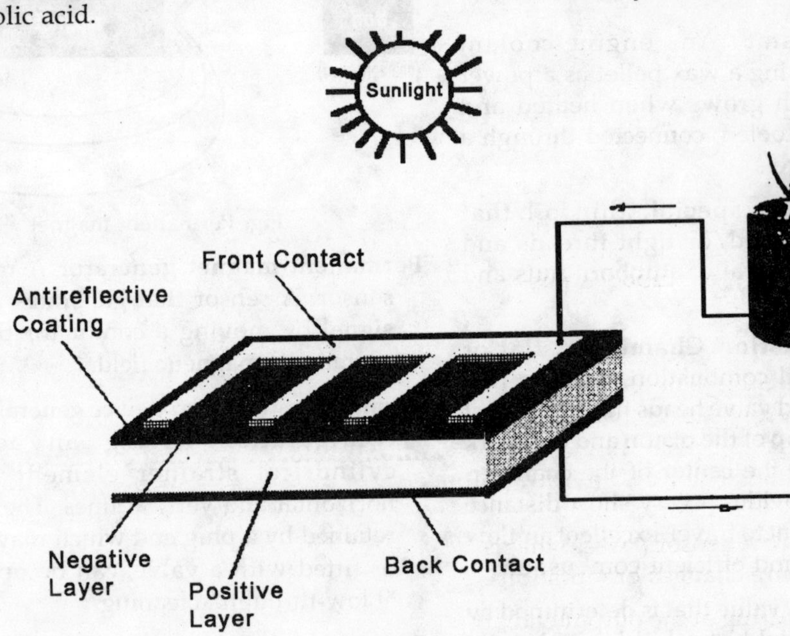

Fig. Photovoltaic cells

Photovoltaic diode A device having a junction of two dissimilar metals that produces an electrical signal proportional to the amount of light that strikes it.

Physical configuration audit An engineering inspection of a configuration item to verify that the item "as-built" conforms to the "as-designed" documentation.

Physical hazards Any personal hazards, such as excessive noise, temperature, pressure, and rotating equipment.

Physical properties Properties that pertain tot he physics of a material, such as melting point, density, electrical and thermal conductivity, specific heat, and coefficient of thermal expansion.

Physics of failure Analysis to determine the physical causes for the failure of electronics components or assemblies.

Pi filter A filter consisting of two line-to-line capacitors and a series inductance in all configuration used to attenuate noise and ripple.

Pi The mathematical constant which is equal to the ratio of the circumference of a circle to its diameter. Approximately 3.1415927

Pickup tube 1. A tube extending from the outlet of the receiver almost to the bottom of the tank to ensure that 100% liquid is supplied to the liquid line or metering device.

2. A tube used to transfer fuel or oil from the storage tank.

Piezoresistive A sensor whose resistance varies in relation to pressure or force applied to it. A piezoresistive sensor receives a constant reference voltage and returns a variable signal in relation to its varying resistance.

Piezoresistive Sensor A sensor (knock) that is sensitive to pressure changes or vibrations.

Pilot bearing The bearing at the output end of the engine's crankshaft that supports the transmissions input shaft.

Pilot hole A small hole drilled into a part to serve as a guide for a larger sized drill bit.

Pilot model The first vehicle of a new design to be constructed on the assembly line to test the assembly procedures and fixtures before full-scale production begins.

Pilot operated A small valve used to energize or regulate a large valve.

Pilot production The initial limited-quantity production of the production-ready version of the product design used to confirm readiness for large quantity production.

Pilot shaft A shaft that is used to align parts during assembly.

Pinion 1. A small gear such as at the end of the steering shaft of a rack-and-pinion steering system.

2. A small ring gear such as the planet gears in a planetary gear set.

3. The small gear on the end of a starter drive which rotates the engine flywheel ring gear for starting.

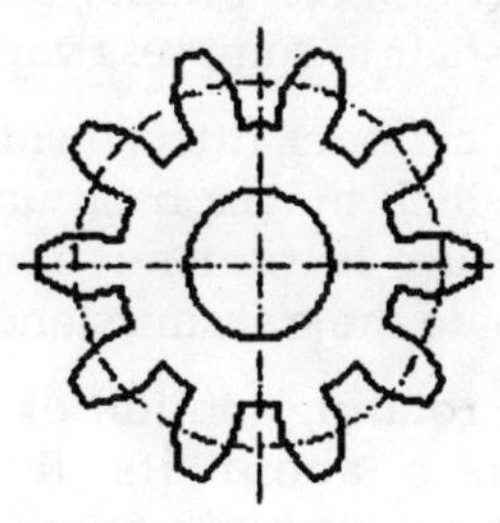

Fig. Pinion

Pinion bearing A bearing that is used to support the pinion gear in the differential housing.

Pinion Factor A calculation using the final drive ratio and the tire circumference to obtain accurate vehicle speed signals

Pinning Using a series of pins to repair and crack in the casting by screwing them into a series of predrilled holes in a manner so that they overlap each other from one end of the crack to the other.

Pipe strain The strain on the pump volute caused by the piping. It will cause excessive mechanical seal movement and can cause contact between rotating and stationary pump and seal components.

Pipeline management Pipeline management is the process of managing new development projects that are currently in the pipeline (both proposed and approved). This addresses the management of capacity and resources to undertake the selected projects and the coordination of cross-functional resources to optimize throughput.

Piston 1. An engine part that reciprocates in the cylinder and transfers the force of the expanding gases via the piston pin and connecting rod to the crankshaft.

2. A round, caliper component in a disc brake that is moved outward by fluid pressure to press the pads against each rotor face.

3. An aluminium or sintered-iron component of a drum brake inside a wheel cylinder that supports the cylinder cup.

4. That part of a compressor that is driven by a crankshaft to compress vapour.

5. A small cylinder fitted inside a hollow cylinder which moves back and forth by fluid pressure to transmit reciprocating motion inside the transmission.

Pitch 1. The rotating motion of a vehicle's spring mass about its lateral axis, compressing the springs at one end of the vehicle and extending the springs at the other end.

2. The uniform spacing of adjacent elements of a series of points, lines, joints, planes, blades, or teeth, as in a broach or gear.

3. The center-to-center dimension between chain joints.

4. For in a sprocket, the dimension between the centers of rollers bedded against the bottoms of adjacent tooth spaces.

5. The distance between corresponding points on adjacent threads.

6. The angle of a setting of some tools.

7. The distance from center-to-center of wire in adjacent coils in an open-wound spring.

Pivot A pin or short shaft upon which another part rests or turns, or about which another part rotates or oscillates.

Pivot bushing A rubber bushing in the front suspension located at the inner end of the control arms where pivoting occurs.

Pixel The smallest unit of colour in a display. Frames are made up of lines and columns of pixels. The number of pixel lines in each frame expresses video resolution.

Pizza cutters The extremely narrow front wheels of a drag-race car used to reduce rolling resistance.

Planning A condition that exists when a tire is unable to stay in contact with the ground or pavement due to water. Also known as hydroplaning or aquaplaning.

Planned downtime The amount of time officially scheduled in the production plan, which includes, no orders, changeovers and planned maintenance.

Planning and scheduling systems A new generation of planning and scheduling tools, which unlike MRP II, includes constraint models that deal with both materials and capacity. These technologies can be applied along a continuum extending from short-term plant-floor scheduling to strategic planning of supply chains.

Plant-level return on net assets (ROA) Where possible, this should be calculated as: plant-level profitability (PLP) divided by average net assets (ANA) employed. PLP can be calculated as annual value of shipments minus materials costs and manufacturing costs. ANA should include fixed assets, inventory, receivables, and cash minus accounts payable and other current liabilities.

Plasma A gaslike association of ionized particles that responds collectively to electric and magnetic fields.

Plasma arc welding (PAW) An arc welding process that uses a constricted arc between a non-consumable electrode and the weld pool (transferred arc) or between the electrode and the constricting nozzle (non-transferred arc) with shielding obtained from the ionized gas issuing from the torch, which may be supplemented by an auxiliary source of shielding gas.

Plasma detector A device for measuring the density, composition, temperature, velocity and three-dimensional distribution of plasmas that exist in interplanetary regions and within planetary magnetospheres.

Plasma display panel (PDP) Flat panel display using plasma electronic technology.

Plasma engine A rocket engine in which thrust is obtained from the acceleration of a plasma with crossed electrical and magnetic fields.

Plasma spray A method of putting a hard coating on a base material. Not recommended for mechanical seal faces.

Plasma wave An oscillation or wave in a plasma that falls in the audio range of frequency.

Plasma wave detector A device for measuring the electrostatic and electromagnetic components of local plasma waves in three dimensions.

Plasmasphere The region of the atmosphere consisting of cold dense plasma originating in the ionosphere and trapped by the Earth's magnetic field.

Plastic deformation Deformation that remains after the load causing it is removed. It is the permanent part of the deformation beyond the elastic limit of a material. It also is called plastic strain and plastic flow.

Plastic filler A two-part,. putty-like material that may be used to fill dents in body panels.

Plastic gasket compound A plastic paste that is squeezed out of a tube to make a gasket in any shape or configuration.

Plastic injection molding A technique for producing plastic parts by forcing molten plastic into a mould.

Plastic moulding A process that converts organic-based materials, by means of a general-purpose press and purpose-built tooling under controlled heat and pressure, and injects the hot material into a die cavity shaped in the final form of the intended part.

Plastic range If you stress metal enough, it stretches through the elastic range, its yield point and ends up in the plastic range where it loses its memory.

Plasticity Tendency of a material to remain deformed, after reduction of the deforming stress, to a value equal to or less than its yield strength.

Plastigauge The tradename for a small, thin, plastic strip that is used to help determine the clearance between main and/or connection rod bearings and the crankshaft journals.

Plate 1. A rectangular sheet of spongy lead in a battery.

2. A platform on which an assembly may be mounted.

Plate group All of the positive plates and all of the negative plates that make up one battery cell.

Plated-thru-hole Plated-thru-hole is a method of obtaining electrical connection between components and substrate (printed circuit board) by soldering component leads (or pins) inserted in plated through-holes.

Platform Primary load-bearing structural assembly of a motor vehicle determining the basic size of the motor vehicle, and is the structural base that supports the driveline and links the suspension components of the motor vehicle.

Platinum (Pt) A rare, valuable metallic element which is highly resistant to corrosion, and is used as a catalytic agent in automotive catalytic converters of the oxidizing type.

Plenum 1. A chamber containing air under pressure.

2. A chamber containing a gas such as air or an air/fuel mixture under higher than atmospheric pressure.

Plenum chamber An area, filled with air at a pressure that is slightly higher than the surrounding air pressure, such as the chamber just before the blower motor.

Plug nozzle A doughnut-shaped combustion chamber which discharges engine gases against the surface of a short central cone (the plug). Adapted in the form of an LH2 cooled heat shield, it can be used as a combination rocket/ aerodynamic braking device.

Plug weld A weld made in a circular hole in one member of a joint, fusing that member to another member.

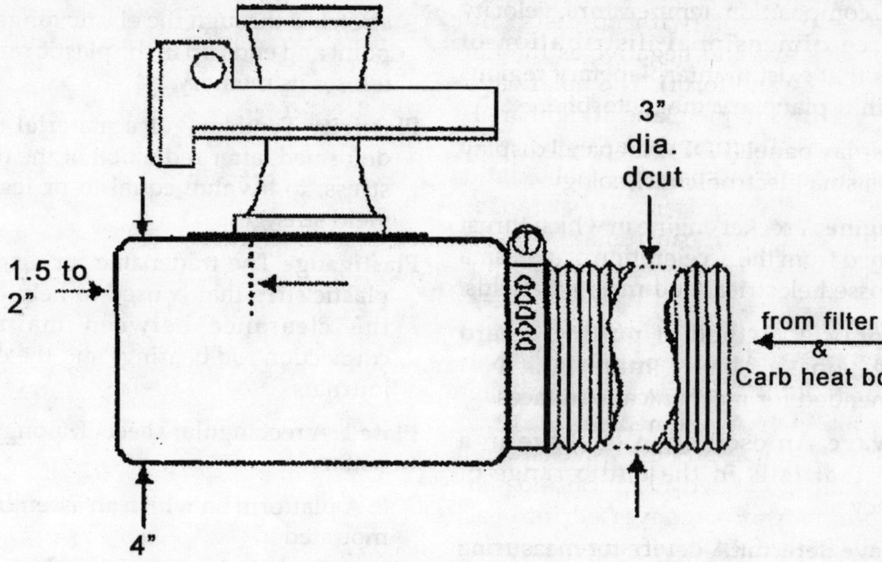

Fig. Plenum chamber

Plenum-blower assembly Located on the engine side of the firewall, this assembly contains air ducts, air valves, and a blower that permits the selection or air from the outside or inside of the car and directs it to the evaporator or to the heater core if desired.

Plexiglas A brand name of a transparent, lightweight plastic material often used for windows in an enclosed race car.

Plies The cord plies that surround both beads and extend around the inner surface of the tire to enable the tire to carry its load.

Plug 1. A spark plug.

2. A threaded, tapered metal pin used to repair cracks in castings, such as an engine block.

Plunger That part of a hydraulic valve lifter that moves up or down depending on oil pressure.

Pluto Ninth planet from the Sun, considered by many a minor planet.

Plutonium-238 A form of the radioactive element plutonium, characterized by high energy emissions.

Pogo stick The air- and electrical-line support rod mounted behind the cap to keep the lines from dragging between the tractor and trailer.

Poincaré sphere A reference sphere used to represent all possible states of polarization. All linear polarizations will lie on the equator and circular polarizations will correspond with the poles. Named for the French mathematician Henri Poincaré.

Point of intersection Apex of the extended centerline of the pivot axis and the extended centerline of a front wheel.

Polarization With respect to light radiation, the restriction of the vibrations of the magnetic or electric field vector to a single plane. In a beam of electromagnetic radiation, the polarization direction is the direction of the electric field vector (with no distinction between positive and negative as the field oscillates back and forth). The polarization vector is always in the plane at right angles to the beam direction. Near some given stationary point in space the polarization direction in the beam can vary at random (unpolarized beam), can remain constant (plane-polarized beam), or can have two coherent plane-polarized elements whose polarization directions make a right angle. In the latter case, depending on the amplitude of the two waves and their relative phase, the combined electric vector traces out an ellipse and the wave is said to be elliptically polarized. Elliptical and plane polarizations can be converted into each other by means of birefringent optical systems.

Policy A guiding principle, typically established by senior management, which is adopted by an organization or project to influence and determine decisions.

Policy deployment The selection of goals, projects to achieve the goals, designation of people and resources for project completion, and establishment of project metrics.

Polished Intake and exhaust passages that have been finished to the smoothest possible surfaces to remove the rough surfaces left by factory casting.

Pollutant 1. Any substance that adds to the pollution of the atmosphere.

2. Any substance in the exhaust gas from an engine or evapourating from the fuel tank or carburettor.

Poly alkaline glycol (PAG) The synthetic lubricant of choice in many automotive air conditioners used with HFC-134a refrigerant.

Poly ester A synthetic lubricant that is used in some HFC-134a air-conditioning systems.

Polyglycol A mixture of several alcohols. Polyalkylene-glycol-ether brake fluids that meet specifications for DOT 3 and 4 brake fluids.

Polymer A compound used as a binder for solid rocket propellant systems; more generally, a compound consisting of repeating structural units.

Portable cable An electrical cable used to transmit electrical power to the mobile equipment.

Portal A web site that provides a comprehensive set of services and information for a particular audience such as product development personnel. The goal of such a site is to be a one-stop web resource for its target audience.

Ported vacuum Engine vacuum, available above the throttle plates of a carburettor, used to advance ignition timing when the throttle is opened above its idle position.

Ported-vacuum switch (PVS) A valve which permits or stops the passage of a ported vacuum to a vacuum-operated component, such as a distributor advance mechanism that may be thermostatically operated, or controlled by electric current or the movement of a mechanical component.

Portfolio management The process of managing new product ideas, proposed projects and current projects under development as a portfolio to 1. Maximize the value of the portfolio,

2. Keep it in balance, and 3.

Align it with company strategy. By characterizing and reviewing the projects in a company's portfolio as a whole, a big picture is presented and used to prioritize and select projects.

Porting To enlarge the intake and exhaust passages.

Positioned weld A weld made in a joint that has been placed in a position to facilitate making the weld.

Positive 1. One of the two poles of a magnet.

2. One of the two terminals of an electrical device, such as a battery.

3 A pressure greater than atmospheric.

Positive absolute-suction throttling valve A suction throttling valve having bronze bellows under a nearly perfect vacuum, which is not affected by atmospheric pressure.

Positive back-pressure EGR valve A common type of exhaust gas recirculation valve which uses exhaust system backpressure to sense engine load, thus more accurately metering the amount of exhaust recycled.

Positive crankcase ventilation (PCV) An engine emissions-control system, operating on engine vacuum, that picks up crankcase gases and meters them into the intake stream to be burned.

Positive crankcase ventilation valve An emission device that routes oil pan vapours to the intake manifold to be burned during combustion. Also known as the pcv valve.

Positive displacement pump Called a PD pump. Gear, sliding vane, progressive cavity, lobe etc. the capacity determined by the pump speed. The maximum head is determined by the horsepower available and the casing strength.

Positive engagement starter A Ford type of starter motor that uses the magnetic field strength of a field winding to engage the starter drive into the flywheel ring-gear

Positive equity The vehicle's market value is greater than the amount the borrower owes on it.

Positive offset A wheel rim placed outward of the centre of the mounting flange.

Positive Temperature Coefficient An increase in resistance due to an increase in temperature.

Positive Temperature Coefficient Device An electrical component in which the resistance increases as the temperature surrounding it increases

Positive terminal The terminal to which electrons flow in a complete electrical circuit often identified by the symbol +.

Positive wheel slip The excessive wheel spin that occurs during acceleration as a wheel loses traction and spins on the pavement.

Positive wiping seal A seal that maintains constant contact with a stem or shaft to wipe off excess oil.

Post ignition Ignition that occurs after the engine-ignition system is shut off, due to carbon buildup in the combustion chamber.

Postcombustion Emissions Device A device such as the catalytic converter that controls emissions after the combustion event

Postponement Postponement, also known as "delayed differentiation," is a supply chain strategy that delays product differentiation at a point closer to the customer. This involves designing and developing standard or generic configurable products that can be customized quickly and inexpensively once actual consumer demand is known. Postponement is a key element of design for leadtime and design for the supply chain.

Potassium Hydroxide 1. A solution of the elements potassium, hydrogen, and oxygen that serves as the electrolyte for a type of fuel cell. Chemically KOH.

Potential difference The voltage difference between two points, calculated algebraically.

Potential energy Energy of a body or system with respect to the position of the body or the arrangement of the particles of the system.

Potentiometer (pot) A three-wire variable resistor that acts as a voltage divider to produce a continuously variable output signal proportional to a mechanical position.

Potting An insulating material for encapsulating one or more circuit elements.

Precision elastic limit The stress that produces a permanent strain of one part in 106 in a material tension test.

Precision engineering Sensing, control, and fabrication at scales comparable to or smaller than optical wavelengths.

Precision insert bearing Bearings that may be installed in an engine without having to be bored, reamed, honed, or ground.

Precision technology A Unaxis core competence. A blend of applied mechanics, IT, physics and chemistry for delicate processes. Interdisciplinary exchange drives development of innovative equipment and products.

Precombustion chamber A second combustion chamber placed directly off the main combustion chamber, used to ignite a rich mixture of air and fuel which then ignites a lean mixture in the main combustion chamber.

Precombustion Emissions Device A device that control emissions before the combustion event as with the O_2 sensor.

Predelivery inspection (PDI) The process of inspecting, adjusting, and fine tuning a new car prior to delivery to the customer.

Predetonation The undesirable "knock" or "ping" that occurs when the ignition of the air-fuel mixture occurs before the ignition spark. Also known as "pre-ignition".

Predictive control A type of automatic control in which the current state of a process is evaluated in terms of a model of the process and controller actions are modified to anticipate and avoid undesired excursions.

Predictive Maintenance (PdM) Practices that seek to prevent unscheduled machine downtime by collecting and analyzing data on equipment conditions. The analysis is then used to predict time-to-failure, plan maintenance, and restore machinery to good operating condition. Predictive maintenance systems typically measure parameters on machine operations, such as vibration, heat, pressure, noise, and lubricant condition. In conjunction with computerized maintenance management systems (CMMS), predictive maintenance enables repair-work orders to be released automatically, repair-parts inventories checked, or routine maintenance scheduled.

Preformed radiator hose A large-diameter hose connecting the radiator to the engine cooling system that is moulded in the proper shape to fit on a certain engine.

Preheater A glow plug that is used to heat the precombustion chamber of a diesel engine before it is started.

Preheating Heating the weld area of a metal before welding, to avoid thermal shock and stress.

Pre-ignition The ignition of the air/fuel mixture in the combustion chamber by means other than the spark, usually caused by hot spots in the combustion chamber due to sharp edges, carbon accumulation, or spark plugs with a heat range that is too hot.

Preliminary design review The second major review in the life of an observatory project. The PDR forms the basis for determining whether the preliminary design approach for the observatory is acceptable to start detailed design.

Preload 1. The transfer of weight from one side of the vehicle to the other to compensate for the lateral weight transfer that occurs when cornering.

2. The pressure applied to a part during assembly or installation.

Pre-regulation The initial regulation circuit in a system containing at least two cascade regulation loops.

Prerun 1. To examine and appraise and off-road race route.

2. To test a rebuilt engine before installing it in the vehicle.

Prerunner 1. A vehicle that is built to be used to examine and evaluate the off-road route of an upcoming race.

2. A vehicle often used as an off-road race-chase vehicle for parts, fuel, and other supplies.

Press fit The fitting of two parts with very close tolerances, one inside the other, generally by using a hydraulic press. Also known as force fit.

Pressure 1. Force applied to a specific area, force per unit area or fore divided by area, usually expressed in psi, psig, or psia (kpa or kpa absolute).

Primary resistance A calibrated resistance wire or ballast resistor in a conventional ignition system inserted between the ignition switch and coil to reduce the primary voltage at low engine speeds.

Primary seal A seal between the compressor shaft seal and the shaft to prevent the leakage of refrigerant and oil.

Primary shoe The brake shoe toward the front of the vehicle in a dual-servo drum brake having a lining that may be shorter, thinner, and have different frictional characteristics from the second shoe.

Primary side control A name for an off-line switching power supply with the pulse-width modulator in the primary.

Primary surfacer A type of paint having a high-solids content which fills small imperfections in body wok and which usually must be sanded.

Primary The low-voltage circuit of an ignition system.

Primer A type of paint applied to a surface to increase its compatibility with the topcoat or to improve the corrosion resistance of the substrate.

Primer sealer A type of paint that provides adhesion for the topcoat and seals old, painted surfaces.

Priming spring A coil spring that holds the movable slide in the maximum position in a vane-type pump.

Primitive Used in a chemical sense, indicating an unmodified material representative of the original composition of the solar nebula.

Prindle The standardized shift pattern for an automatic transmission, PRNDL.

Pro rallies Road rallies which are very competitive and are run at high speeds on roads closed to the public.

Process A natural phenomenon marked by gradual changes that lead toward a particular result. A series of actions or operations leading to an end. A continuous operation or treatment in manufacturing. Continuous and regular production executed in a definite, uninterrupted manner.

Process automation Includes objectives of control and also those of enterprise management. This requires an integrated approach to plant operations and enables a variety of applications such as production scheduling, inventory control, performance monitoring, statistical process control, maintenance management and environmental audit.

Process capability 1. (statistical definition) process capability is the repeatability and consistency of a manufacturing process relative to the customer requirements in terms of specification limits of a product parameter.

Process control Automatic monitoring and control of a process by an instrument or computer programmed to respond appropriately to feedback from the process.

Process development Defining and developing a manufacturing process to accommodate the specific requirements of a given product while meeting process quality and cost objectives.

Process kaizen Continuous improvement through incremental improvements.

Process manufacturing The manufacturer of products such as chemicals, gasoline, beverages and food products that typically are produced in "batch" quantities rather than discrete units. Many process operations require inputs such as heat, pressure, and time (for thermal or chemical conversion).

Process planning Process planning is the analysis and design of the sequence of processes, resources requirements needed to produce products into workable instructions for manufacture. It also includes the specification and selection of tools, fixtures, equipment and inspection/ test requirements.

Process simulation Use of a mathematical model by a computer program to envision process design scenarios with real-time visual and numerical feedback. Process optimization and the ability to forecast potential problems are the results.

Producibility Is a measure of the relative ease and economy of producing a product, using conventional and flexible manufacturing methods. The process must not sacrifice functionality, performance, effectiveness or quality.

Product A product is a given set of items. The set could comprise of system, subsystem, hardware, or software items and their documentation.

Product architecture The scheme by which the functional elements of a product are arranged into physical building blocks (e.g., subsystems or subassemblies) and interact with each other to perform the overall function of the product. Product architectures can be modular or integral.

Product brief / project brief A summary document that communicates essential information about the product to be developed and is used to guide the development effort. Product briefs contain a description of the product including distinguishing characteristics; critical technology and its status; a description of the intended market or customer; a product strategy and its basis of competition; the target cost and target price; essential project information (boundary conditions) such as development cost, development schedule, milestones, and required resources/ personnel; and significant risks.

Product data A representation of information about a product in a formal manner suitable for communication, interpretation, or processing by human beings or by computers.

Product data management / product information management Product data management and product information management are synonymous terms referring to a software tool that maintains a repository of product data and provides functions such as access control, managing parts and assemblies, bill of material structuring, version control/engineering change control, workflow management/ process management, web publishing and reviewing, view and markup, database backup, and user administration.

Product development team Product development team - a team consisting of representatives from marketing, engineering, manufacturing, finance. Purchasing, test, quality, finance and any other required disciplines with responsibility for developing a product or product subsystem. This team is empowered to represent the functional disciplines and develop a product by addressing its life cycle requirements including its product and support.

Product family A product family is a set of individual products that share common technology and address a related set of market applications. Product families include a number of products or product lines targeted at somewhat different markets or usage situations.

Product identification number (PIN) A four-digit code that is used to identify hazardous material during transportation.

Product launch Product launch begins once a product has been developed and a decision is made to proceed with production and marketing. It consists of all of the steps to plan and prepare for production of the product including ramp-up to full-volume production or general availability; the steps to plan, promote, market and sell the product, and the steps to prepare for servicing and support of the product.

Product life cycle 1. The product life cycle from a marketing perspective is typically defined by its sales volume profile and broken down into the following phases: introduction, growth, maturity and decay.

2. The product life cycle from the broader enterprise and user perspective is defined by phases of its overall life: concept, development, production, operation, support, and disposal.

Product lifecycle management Product Lifecycle Management (PLM) generally refers to the process of managing all data related to the product over its life cycle. It broadens the concept of product data management (PDM) to address managing the product configuration and the availability of product data into its later lifecycle stages of production, operation, support and disposal and to address managing the product's related process data. ARC Advisory Group defines six components of PLM: innovation and portfolio management, project and program management, collaborative design, product data management, manufacturing process planning, and service and support management. Michigan's PLM Development Consortium defines it as "an integrated, information driven approach to all aspects of a product's life from its design inception, through its manufacture, deployment and maintenance and culminating in its removal from service and final disposal. PLM is the integration of business systems to manage a product's lifecycle." CIMdata Inc., says that PLM represents "a business approach to solving the problem of managing the complete set of product and plant definition information and the processes through which it passes. The PLM process includes creating and changing that information, managing it through its life and disseminating and using it throughout the lifecycle of the product."

Production engineering Planning and control of the mechanical means of changing the shape, condition of materials toward greater effectiveness and value.

Production launch Production launch begins at the point that a product has been developed and is ready to begin production and proceeds through ramping-up production to full volume or general availability. It is the narrower set of product launch activities focusing on production and excluding marketing, support and service.

Production part approval process Production part approval process (ppap) is a big three automotive process that defines the generic requirements for approval of production parts, including production and bulk materials. Its purpose is to determine during an actual production run at the quoted production rates whether all customer engineering design record and specification requirements are properly understood by the supplier and that the process has the potential to produce product consistently meeting these requirements.

Production readiness review Production readiness review is a design review conducted prior to putting a product into production. This review assesses whether all needed product and process data has been completely generated, that the production process has been validated, and that the company is ready to begin production (either pilot production, low-rate initial production, production ramp-up, or full-rate production).

Productive Time A period of time when the equipment is performing its intended function. This includes regular production (including loading and unloading of product), rework, work for third parties, and engineering runs done in conjunction with production.

Productivity A measurement of output for a given amount of input(s). In this document, labour productivity (value added per employee) is used as the measure of productivity. Increases in productivity are considered critical to raising living standards.

Program 1. A complete structured sequence of program statements directing a computer to implement an algorithm.

2. All software programming necessary to solve a problem.

Program manager The role at a high enough level in an organization that the primary focus is the long-term vitality of the organization, rather than the short-term project and contractual concerns and pressures. In general, a senior manager for engineering would have responsibility for multiple projects. Perceptual Map A Perceptual Map is a visual method for comparing customer perceptions of different products considering two different characteristics of those products. It is used to show relationships between marketplace competitors and the criteria used by buyers in making purchase decisions and recommendations. Perceptual maps may be used for market segmentation, concept development and evaluation, and tracking changes in marketplace perceptions.

Programmed protection system A bypass valve system to protect the catalyst and their containers from being destroyed due to overheating conditions that may be caused by certain modes of operation or from an engine malfunction.

Programmer The part of an automatic temperature control system that controls the blower speed, air mix doors, and vacuum diaphragms.

Programming The control of a power supply parameter, such as output voltage, by means of a control element or signal.

Progressive linkage A carburettor linkage system used on multiple carburettors to progressively open the secondary circuits.

Progressive-rate spring A spring used in a vehicle that stiffens under load.

Projection display An Optics division business unit. Projection Display produces optical components and modules for colour management in data and video projectors, and rear-projection tvs.

Projector-beam headlights A headlight that uses a spherical reflector to tightly control the light beam. The bulb or light source directs the light inward, toward the reflector at the back of the headlight assembly, which then projects it forward from the vehicle. These lights are more powerful, accurate and expensive than standard sealed-beam and halogen headlights, and are generally found on sport and luxury models.

Prometheus An acronym for Program for European Traffic with Highest Efficiency and Unprecedented Safety.

Promotion level In a product data management system, information is assigned to promotion levels based on the approvals it has received. These are defined and named by the system administrator. Examples are review, preliminary release, prototype release, and production. Each promotion level has its own set of authorizations for access and approvals.

Propane enrichment A service procedure used to set idle mixture where a metered amount of propane gas is added to the intake stream and a resulting rpm increase is observed.

Propellant A chemical or chemical mixture burned to create the thrust for a rocket or spacecraft.

Propeller shaft 1. A term used for drive shaft.

2. A hollow, steel shaft that connects the transmission output shaft to the differential drive pinion yoke through universal joints at each shaft end.

Property class A number stamped on the end of a metric bolt to indicate the hardness of the bolt.

Property damage liability The protection of the property when not under control of the insured.

Proportional limit Highest stress at which stress is directly proportional to strain. It is the highest stress at which the curve in a stress-strain diagram is a straight line. Proportional limit is equal to elastic limit for many metals.

Proportioning valve A valve in the brake hydraulic system that reduces pressure to the rear wheels to achieve better brake balance.

Protective atmosphere A gas or vacuum envelope surrounding the work pieces, used to prevent or reduce the formation of oxides and other detrimental surface substances, and to facilitate their removal.

Protective tube A shock absorber component used to keep dirt and road dust away from the seals and piston rods.

Protest 1. To object to a rule or conditions prior to the competition.

2. To accuse a competitor of a rules violation after a race competition.

Protocol An agreed-upon format for transmitting data between two devices.

Prototype A physical model or representation of the new product concept or design. Depending upon the purpose, prototypes may be non-working models or representations, functionally working, or both functionally and geometrically complete and accurate. Prototypes (physical, electronic, digital, analytical, etc.) Can be used for the purpose of, but not limited to: a) assessing the feasibility of a new or unfamiliar technology, b)assessing or mitigating technical risk, c) validating requirements, d) demonstrating critical features, e) qualifying a product, e) qualifying a process, f) characterizing performance or product features, or g) elucidating physical principles.

Prove-out circuit A function of the ignition switch that completes the warning light circuit to ground when it is in the START position, so the warning lights will be on during engine cranking to indicate to the driver that the bulbs are working properly.

Psychological inertia Psychological inertia is the tendency of persons to formulate opinions or attitudes, make decisions or seek known or familiar solutions to problems based on their current frame of reference, experience, and training.

P-type semiconductor material (P) A semiconductor material in which holes are the majority carriers and there is a deficiency of electrons.

Pu Plutonium An element with the atomic number of 94.

Public-domain software Refers to any program that is not copyrighted. Public-domain software is free and can be used without restrictions.

Puck 1. A brake pad.

2. A valve-adjusting shim.

Puffer A supercharger.

Pugh concept matrix The pugh concept matrix is used to a) evaluate multiple design concepts and select the preferred concept alternative and b) synthesize the best elements of other concepts into an improved concept (which may be a hybrid or variant of the best of other concepts. The pugh matrix is useful because it does not require a great amount of quantitative data on the design concepts, which generally is not available at this point in the process.

Pull A condition whereby the vehicle veers to one side or the other when the brakes are applied.

Pull circuit A circuit that brings a tractor cab from a fully tilted position up and over the center.

Pull rod A body-working tool in which the curved end of the rod is inserted into the drilled hole to pull out dents.

Pull up resistor A fixed resistor in a voltage divider circuit for sensor input signals. A variable resistor sensor changes its voltage drop proportionally to the quantity being measured, and the pullup resistor drops the rest or the reference voltage

Pull-down menu A menu of commands or options that appears when you select an item with a mouse.

Pulley A wheel-shaped, belt-driven device used to drive engine accessories.

Pull-in windings One of two electromagnetic windings inside a solenoid used to move the core.

Pull-type clutch A type of clutch whereby the release bearing is pulled toward the transmission.

Pulsar Discovered in 1967. Pulsars emit radio signals the pulsations of which are extremely precise. The evidence suggests that pulsars are fast-spinning neutron stars.

Pulsation A surge that may be felt in the brake pedal during low-pressure braking.

Pulse A common waveform shape that has a fast rising edge, a width, and a fast falling edge.

Pulse Air Injection Reaction System Pulse air directed in to the exhaust system to reduce exhaust emissions

Pulse generators An electromagnetic unit that generates a voltage signal used to trigger the ignition control module that controls (turns on and off) the primary ignition current of the electronic ignition system.

Pulse Modulated A circuit that maintains average voltage levels by puulsing the voltage on and off.

Pulse period A term used for pulse width.

Pulse Rate The number of pulses that take place over a spicific amount of time

Pulse ring A ring on the crankshaft to trigger a sensor used with an onboard computer control ignition timing.

Pulse time 1. The time duration from the point on the leading edge which is 50% of the maximum amplitude to a point on the trailing edge which is 50% of the maximum amplitude of the signal or wave form.

2. A term used for pulse width.

Pulse Trains A repetitive series of pulses

Pulse Width Modulation 1. A method of electronic control where the frequency of the pulses remains the same, but the percentage of "on time" compared to "Off time" varies

2. The characteristics of a continuous on-and-off cycling solenoid for a fixed number of times per second while the frequency of the cycles remains constant the ratio of on-time to off-time varies or is modulated

Pulse Width The amount of time the pulse take to go from low to high and back to low again, conventionally measured at 50% of full voltage

Pulsed air system A shortened term for pulsed air-injection system.

Pulsed air-injection system A device found in the exhaust emission system that injects pulses of ambient air into the exhaust as an aid in the burning of exhaust gases.

Pulse-delay variable resistor A resistor in the electrical circuit of the wiper system used to delay the wiper motion from 0 to 25 seconds.

Pulse-width modulation (PWM) A method of regulating the output voltage of a switching power supply by varying the duration, but not the frequency, of a train of pulses that drives a power switch.

Pump A device that transfers gas or liquid from one place to another and creates an increase in pressure or volume.

Pump rotors Parts of a positive displacement pump that uses rotating rotors to develop flow.

Pump screen A filter to remove debris from a fluid before it enters the pump.

Pumpdown A term used for evacuate, as in an air-conditioning system. Also known as pulldown.

Pumping brakes A term that refers to the practice of depressing the brake pedal several times in quick succession.

Pumping ring Used with a convection system to get circulation between two mechanical seals. Absolutely necessary if oil is used as a barrier fluid because of oil's poor specific heat..

Punch it 1. To apply full throttle.

2. A term used for plant it or punch off.

Punch off A term used for punch it or plant it.

Puncture-sealing tire A specialty tire made to permanently tread punctures up to 3/16 inch (4.76 mm) in diameter with a resultant rubber-compound sealer applied to the inside of the tire by the manufacturer.

Pup converter A term used for light-off, mini-oxidation catalytic converter.

Pur sang A French term for "pure blood" used to describe a particularly fine car.

Purchase option The lessee's right to purchase the vehicle, if he so chooses, at lease end. It's a good idea to negotiate a purchase-option price at the same time you negotiate the capitalized cost and residual value, if possible. Some contracts lock-in a pre-determined value for the vehicle.

Purchasing Total materials purchased for manufacturing/assembly into auto parts (ie. Steel, plastics, rubber, textiles, etc.).

Pure stock A vehicle that conforms to all original factory specifications.

Purge 1. To remove moisture and/or air from a system or a component by flushing with a dry gas, such as nitrogen (N).

2. To remove all refrigerant from an air-conditioning system.

Purge valve A valve in some evaporative emission-control system charcoal canisters to limit the flow of vapour and air to the carburettor during idle.

Purity test A static test that may be performed to compare a suspect refrigerant pressure to an appropriate temperature chart to determine its purity.

Push A condition in cornering when the slip angle of the front tires is greater than the slip angle of the rear tires and the front end of the vehicle tends to break loose and slide toward the outside of the turn.

Push angle 1. The travel angle when the electrode is pointing in the direction of weld progression.

2. Also used to partially define the position of guns, torches, rods, and beams.

Push bar A bar in the front of the push car to prevent damage to either vehicle when push-starting a race car.

Push circuit A circuit that raises a tractor cab from the lowered position to the desired tilt position.

Push start To start a vehicle by pushing it with another.

Pusher A rear-engine vehicle.

Push-pull converter A power switching circuit that uses two or more power switches driven alternately on and off.

Pushrod A general term for any rod that transfers force in compression. In a conventional overhead valve layout, pushrods are used to transfer reciprocating motion from the cam followers to a more distant part of a valve train, typically the rocker arms. Pushrods are eliminated in overhead camshaft designs.

Pushrod tube The passage from the engine block to the cylinder head that contains a pushrod.

Push-type clutch A type of clutch in which the release bearing is not attached to the clutch cover.

PVD (Physical vapour deposition) A coating process in which molecules from a solid body of material are dislodged by positively charged particles in an electromagnetic field and deposited onto a substrate. PVD can be used to make electrical contacts and connections, for example.

Pyroconductivity Electric conductivity that develops with changing temperature, and notably upon fusion in solids that are practically nonconductive at atmospheric temperatures.

Pyrotechnics The use of electrically initiated explosive devices to operate valves, ignite solid rocket motors, and explode bolts to separate from or jettison hardware, or to deploy appendages.

Q

Q The energy-storing characteristic of an electronic circuit, system or device, equal to reactance divided by resistance. Q determines rate of decay of stored energy; the higher the 0, the longer it takes for the energy to be released.

Quad Driver 1. A dipole (DIP) device which contains circuit drivers to control outputs

2. an integrated circuit used to turn certain components on and off

3. A quad driver has four separate out puts (switches) and can turn on and off up to four circuits and or components independently such as four fuel injectors

Quality assurance A planned and systematic pattern of all actions necessary to provide adequate confidence that management and technical planning and controls are adequate to: Establish correct technical requirements for design and manufacturing; Manage and design activity standards, drawings, specifications, or other documents referenced on drawings, lists, or technical documents.

Quality circle A small group of people who normally work as a unit and meet frequently for the purpose of uncovering and solving problems with the quality of the items produced, process capability, or process control.

Quality function deployment A structured planning and decision-making methodology for capturing customer requirements (voice of the customer) and translating those requirements into product characteristics, part characteristics, process plans, and quality/process control requirements using a series of matrices.

Quarter elliptic spring One half of a semi-elliptic spring having one end attached tot he chassis frame and the other attached to the axle.

Quarter mile 1. A distance of 1,320 feet (404) meters).

2. The standard distance for a drag strip.

Quarter mile drag strip The most popular standard distance for a competitive drag-racing strip.

Quarter panel A sheet metal panel that covers the area from the rear-door opening to the taillight area, and from the bottom of the surface to the base of the roof, or from the headlamp area to the front-door opening, and from the bottom of the surface to the base of the hood.

Quarter race cam A camshaft that has been regrown to provide better than standard performance.

Quarter-speed cam A camshaft that rotates once for each four revolutions of the crankshaft.

Quartz A mineral found in nature as a hexagonal cross section. It is pointed at one end and has a fractured base where it broke from the rock formation in which it grew.

Quartz crystal A thin slab cut from quartz ground to the proper thickness so it will vibrate at a desired frequency when supplied with energy.

Quartz halogen headlamps A headlamp bulb having a quartz envelope holding the tungsten filament and filled with an inert gas containing iodine or another of the five halogen gases. The gas serves to remove the tungsten deposits from the bulb wall and redeposit them on the filament, preventing blackening of the bulb surface and reduction of light output. This kind of cycle requires very high filament operation temperatures which necessitates the use of quartz instead of glass. These lamps produce more lighting power per watt of electrical power than standard sealed beam headlamps.

Quartz-halogen A type of high output lamp containing a gaseous halogen; used for headlights, driving lights, or fog lights.

Quasars Quasi-stellar objects. They are believed to be among the most distant objects in the observable Universe, emitting more energy than some of the most powerful galaxies.

Quasi-Static alignment system The QSA maintains diffraction-limited optical performance during observations as changing gravity flexure moves the optical components.

Quench area Any internal portion of a combustion chamber which causes combustion to cease because of the temperature drop in the air/fuel charge where it meets this area.

Quench The introduction of a fluid outside the seal to cool the product, or dilute any leakage across the seal faces.

Quick change Any system or sub system that is designed for ease and speed for changing or replacing.

Quick changeover The ability to change tooling and fixtures rapidly (usually minutes), so multiple products can be run on the same machine.

Quick charger A battery charger that is designed to charge or boost a battery in a short time.

Quick coupler A coupler that allows hoses to be quickly connected and/or disconnected. Most shop air hoses, for example, are equipped with quick couplers.

Quick release valve A device used to exhaust air as close as possible to the service chambers or spring brakes.

Quick take-up master cylinder A dual master cylinder that supplies a large volume of fluid to the front disc brakes on initial brake application, which takes up the clearance of low-drag calipers.

Quick take-up valve The part of the quick take-up master cylinder that controls fluid flow between the reservoir and the primary low pressure chamber.

Quick-changeover techniques A variety of techniques, such as SMED (single-minute exchange of dies), which reduce equipment setup time and permit more frequent setups, thus improving flexibility and reducing lot sizes and manufacturing cycle times.

Quick-turn prototyping Production on a quick turnaround basis of a small quantity of products that are used to prove the design.

Quill shaft A hollow shaft with another shaft inside it.

R

R & D An acronym for research and development.

Race 1. That element of a one-way roller clutch providing the cylindrical surface through which the rollers and cam transmit torque.

2. A grove, edge, or track on which a rolling or sliding part moves.

Race for the pink To race for actual ownership of the competitor's vehicle.

Rack and pinion A type of steering assembly that has a gear or pinion at one end of the steering shaft engaging in a horizontal-toothed bar or rack having tie rods at either end that are attached to the steering arms.

Rack and pinion steering A steering gear in which a pinion on the end of the steering shaft merges with a rack of gear teeth on the major cross member of the steering linkage. When the steering wheel is turned, the pinion gear turns, moving the rack to the left or right, thus steering the wheels.

Rack and pinion steering The steering wheel is connected to a pinion gear that meshes with a toothed bar, also called a rack or linear gear. As the pinion turns, the rack moves side to side, moving the steering linkage and causing the front wheels to turn left or right. The ends of the rack are linked to the steering wheel with tie rods.

Radian Unit of angular measurement equal to the angle at the center of a circle subtended by an arc equal in length to the radius. Equals about 57.296 degrees.

Radiation 1. A natural process by which energy is transmitted.

2. The transfer of heat through a soled, liquid, or gas medium.

Radiation belt The region of high-energy particles trapped in the Earth's magnetic field, also known as the Van Allen belts.

Radiation dose The amount of energy per unit of mass of material deposited at each point of an object undergoing radiation.

Radiation Energy in the form of electromagnetic waves or particles.

Radiator A heat exchanger used to remove heat from the coolant in the cooling system containing a vertical- or horizontal-finned tubing section connected between two tanks.

Radiator cap A term used for radiator pressure cap.

Radiator core The center of the radiator, made of tubes and fins, used to transfer heat from the coolant to the air.

Radiator fan A term used for fan.

Radiator hose An oil- and ozone-resistant synthetic-rubber hose that connects the radiator to the thermostat outlet housing and water pump inlet housing.

Radiator hose clamp A term used for hose clamp.

Radiator pressure cap A cap that seals in pressure from hot expanding coolant until a predetermined limit is reached, then the valve opens, allowing excess pressure to escape, generally to a coolant-recovery tank.

Radiator shutter system An engine temperature-control system that controls the amount of air flowing through the radiator by use of a shutter system.

Radiator The copper or aluminium device in front of the engine through which hot engine coolant is circulated and cooled. The liquid is then recirculated back through the engine block to cool it.

Radio astronomy The science of astronomy using radio waves instead of light waves.

Radio choke A small coil of wire installed in the power lead leading to a pulsing unit such as an IVR to prevent radio noise (reduce RFI).

Radio data systems (RDS) An information transmission system that was defined by the European Broadcasting Union.

Radio frequency choke (RFC) Coil with specified impedance to of currents.

Radio frequency interference (RFI) A form of electromagnetic interference created in the ignition secondary circuit which disrupts radio and television transmission.

Radio guidance A system which is dependent on outside signals for information.

Radio The least energetic form of electromagnetic radiation, having the lowest frequency and the longest wavelength.

Radiofrequency Interference Radio and television interference caused by electromagnetic energy

Radioisotope thermoelectric generator A device that converts the heat produced by the radioactive decay of plutonium-238 into electricity by an array of thermocouples made of silicon-germanium junctions. The Pu-238 is contained within a crash resistant housing.

Radiused 1. A procedure that is used to reduce the radius diameter at the area the valve stem meets the valve head.

2. A valve type that is ground with a radial grinder to aid in air flow around the valve.

3. Wheel wells that have been cut into a circular shape.

Rake A suspension or structural design to lower one end of the vehicle in relation to the other end.

Ram air Air forced through the radiator and condenser coils by the movement of the vehicle or the action of the fan.

Ram air cleaner An air cleaner for high-performance cars that opens an air scoop on the hood to provide a ram effect when the throttle is wide open.

Ram induction An intake manifold designed to cause a resonant effect at a specific predetermined engine speed.

Ram tuning The tuning of an intake manifold to ensure that the passages are of sufficient length to cause a resonant effect at a specific predetermined engine speed.

Ramp angle An angle formed by lines adjacent to the static-loaded radius of the front and rear wheels intersecting at the point of the lowest ground clearance under the middle of the vehicle.

Random function determination A value analysis methodology that list basic and secondary functions performed by a component or product in a verb-noun format.

Random intermittent welds Welds on one or both sides of a joint in which the weld increments are made without regard to spacing.

Range shift cylinder Located in the auxiliary section of the transmission, this component, when directed by air pressure via low and high ports, shifts between high and low range of gears.

Range shift lever Located on the shift knob, this lever allows the driver to select low- or high-gear range.

Range The maximum and minimum allowable full-scale signal (input or output).

Rapid application development Rapid application development) is a way of developing a system by completing an initial working part of the system, and then incrementally adding to it every few months. Instead of waiting to finish the entire system, the system owners can put the system into use earlier.

Rapid manufacturing Rapid manufacturing refers to the use rapid prototyping technologies to directly manufacture low volumes of parts.

Rapid prototyping 1. Rapid prototyping refers to various technologies such as stereolithography and selective laser sintering that can rapidly create parts for visualization, product mock-ups, or functional product prototypes or produce rapid tooling to manufacture small to medium volumes of parts. Rapid prototyping processes involve devices, ranging from office modellers to four-ton machines, that accept 3d cad files, slice the data into cross-sections, and construct layers from the bottom up, bonding one on top of the other, to produce physical prototypes.

2. More generally, it is the process of quickly generating prototypes or mock-ups of what a product system will look like. Rapid prototyping may be done with paper prototypes such as sketches, low-fidelity physical prototypes, cad visualization, rapid application development, or video prototyping.

Rapid tooling Rapid tooling refers to the use rapid prototyping technologies to fabricate tooling in a much shorter period of time than conventional tooling. Rapid tooling technologies include methods such as RTV moulds, high-speed milling, centrifugal casting, etc.

Rated output current The maximum continuous load current a power supply is designed to provide under specified operating conditions.

Ratio 1. The relative amounts of two or more substances in a mixture.

2. The comparison of two numbers as in teeth on gears.

Ratio valve A device used on the front or steering axle of a heavy-duty vehicle to limit the brake application pressure to the actuators during normal service braking.

Rationality A term associated with comprehensive component testing in which signal inputs from components are compared to inputs from other components to verify conditions coincide.

Ravigneaux gear train A planetary gear train with two sun gears, three long and three short planetary pinions, planetary carrier, and ring gear.

Raw-materials turn rate A measure of asset management that typically is calculated by dividing the value of total annual shipments at plant cost (for the most recent full year) by the average raw-material value at plant cost. Plant cost includes material, labour, and plant overhead.

Rayleigh scattering Scattering by particles very small compared with the wavelength of the radiation being considered. A feature of Rayleigh scattering is that the scattered flux is inversely proportional to the fourth power of the wavelength. Thus in the visible region, blue light is scattered more strongly by the molecules of the air than longer wavelengths, accounting for the blue colour of the sky.

Reactance The impedance that a pure inductance or a pure capacitance provides to current flowing in an ac circuit.

Reactant A chemical substance that is present at the start of a chemical reaction.

Reaction bonded A method of manufacturing silicone carbide mechanical seal faces. These materials are sensitive to caustic and some other high pH liquids.

Reaction control system System of thrusters used to control spacecraft attitude.

Reaction disk The components in the vacuum power booster that provide pedal feel or feedback to the driver.

Reaction member The member of a planetary gearset that is held in order to produce an output motion. Other members react against the stationary, held member.

Reaction time 1. The amount of time required to physically apply the brakes of a vehicle after mentally being aware of the necessity to do so.

2. The amount of time required from the moment the Christmas tree light turns green until the vehicle trips the starting-line timing light.

Reaction wheels Electrically-powered wheels mounted in three orthogonal axes aboard a spacecraft. To rotate the vehicle in one direction, you spin up the proper wheel in the opposite direction. To rotate the vehicle back, you slow down the wheel.

Reactive A component that exhibits the property of either capacitance or inductance.

Reactive maintenance (RM) Action after an event (opposite of proactive). A reactive maintenance department acts when breakdown occurs, or work order is requested.

Reactivity The characteristics of a material that cause it to react violently with another material, such as air, heat, or water.

Reactor Device or process vessel in which chemical reactions (e.g., catalysis in fuel cells) take place.

Read vapour pressure (RVP) A method of determining vapour pressure of gasoline and other petroleum products. Used as an indicator of the volatility of gasoline. The Read Vapour Pressure of gasoline is the pressure of the vapour in a sealed gas tank when the fuel is raised to 100o F and varies with altitude.

Read-only memory A solid-state storage chip that is programmed at the time of its manufacture, cannot be programmed by the computer user, and retains its program when the computer's power is turned off.

Readout The action of a radio transmitter sending data either at the same time as data are acquired or by playback from an electronic memory.

Real power The actual power consumed by a circuit irrespective of power factor,

Rear-axle assembly A group of parts that operate to transfer driving torque from the drive shaft to the rear wheels, and includes mounting pads for the rear springs and shocks.

Rear-camshaft plug A plug driven in the rear of the block behind the camshaft to contain the oil supplied to the rear camshaft bearing.

Rear-end torque The twisting reaction of a rear-axle assembly in a direction opposite to that of wheel rotation when power is applied, controlled by the control arms in coil-spring models and by the leaf springs in leaf spring models.

Rear-spring shackle A device bolted to a rubber bushing in the rear main leaf eye, and having an upper shackle bolt that extends through a similar rubber bushing in the rear-spring hanger, providing fore and aft movement with variations in spring length.

Rear-suspension rebound stop A function of the rear shocks as they limit the distance the rear-axle assembly can move downward.

Rear-wheel drive The drive train in which power is applied through the rear wheels only.

Reassembly Putting the disassembled parts of a device back together, replacing those parts as necessary.

Rebound clip Metal clamps placed at three or four internals around multi-leaved springs to prevent the leaves from becoming separated on rebound.

Rebound The outward extension of the springs and shocks in a vehicle suspension system.

Rebound travel The downward movement of a wheel from its normal position when the spring and shock are expanding, as the sudden drop of a wheel into a depression and a weight transfer away from the wheel.

Rebound valve A calibrated piston valve mounted on the shock piston that provides variable resistance to fluid flow during rebound.

Receiver An electronic device that receives incoming radio signals and converts them to perceptible forms.

Receiver dehydrator A combination container for the storage of liquid refrigerant and a desiccant.

Receiver tank The second tank in a radiator is often referred to as a receiver because it receives coolant after passing through the many tubes of the radiator core. The receiver, on automatic transmission equipped cars, contains a transmission oil cooler.

Receptacle A contact device, installed at an electrical outlet, extension cord, or drop light for the connection of an attachment plug and flexible cord to supply power to portable equipment.

Recharging 1. The action of forcing electrical current into a battery, reversing the chemical reaction between the plates and electrolyte.

2. Filling an air-conditioning system with refrigerant and repairs are made.

Reciprocating A back-and-forth, up-and-down, or to-and-fro motion, such as that of a piston in a cylinder.

Reciprocating engine A power plant in which the piston move in a reciprocating motion.

Reciprocating motion The motion of an object between two limited positions in a straight line back and forth or up and down.

Recirculate door A door in the plenum that regulates the amount of reticulated air flow.

Recirculating ball A steering mechanism in which the steering shaft turns a worm gear causing a toothed metal block to move back and forth, turning the front wheels. Ball bearings reduce friction between the worm gear and the metal block.

Recirculating-ball steering gear A low-friction steering gear box having a worm-gear meshing with a cross shaft sector through about 40 recirculating balls that travel in matching grooves inside the ball nut and outside the worm shaft which acts as a rolling thread.

Recirculating-ball-and-nut steering gear A type of steering gear having a nut meshing with a worm gear sector, with balls that circulate between the nut and worm threads.

Reclaim To process used refrigerant to new product specification by means that may include distillation. This process requires that a chemical analysis of the refrigerant be performed to determine that appropriate product specifications are met.

Recombination battery Sometimes called dry-cell batteries because they do not use a liquid electrolyte solution.

Reconditioned A used part that has been repaired, tested, and determined to be in good order. Also known as recon.

Reconditioning reserve Another name for the security deposit when leasing a vehicle.

Record length The number of waveform points used to create a record of a signal

Record Throughout the PMS, the requirements to "record" information are interpreted to mean "set down in a manner that can be retrieved and viewed." The result can take many forms including, but not limited to, hand-written notes, hard-copy, or electronic documents, and data recorded in computer-aided software engineering (CASE) and project management tools.

Recovery cylinder A cylinder for the storage of used R-12 and/or R-134a must meet DOT specification 4BA-300, characterized by a combined liquid/vapour valve located a the top. A recovery cylinder should be painted gray with a yellow shoulder.

Recovery Index of a material's ability to recover from deformation in the compressibility and recovery test, the deformation under load and the plastometer test. In the compressibility and recovery test, it usually is reported with compressibility and given as %. It is calculated by dividing the difference between recovered thickness and thickness under load, by the difference between original thickness and thickness under load. In the de-formation under load test, it indicates the extent to which a non rigid plastic recovers from prolonged compressive deformation at an elevated temperature. It is given as %, and is calculated by dividing the difference between height recovered 1.5 hours after load is re-moved and height after three hours of loading, by the change in height under load. In the plastometer test, it indicates the extent to which an elastomer recovers from compressive loading at an elevated temperature. It is equal to plasticity number minus recovered height.

Recovery time Refers to the length of time required for a system to return to its specified level of performance or accuracy after a significant external disturbance.

Recovery/recycle unit A term used to identify the complete unit used to recover and/or recycle refrigerant from the air conditioning system.

Reduction catalyst The section of a three-way catalytic converter that breaks nox (oxides of nitrogen) down into harmless nitrogen and oxygen through a reduction reaction.

Reduction of area Measure of the ductility of metals obtained in a tension test. It is the difference between original cross sectional area of a specimen and the area of its smallest cross section after testing. It is usually ex-pressed as % decrease in original cross section. The smallest cross section can be measured at or after fracture. For metals, it usually is measured after fracture and for plastics and elastomers, it is measured at fracture.

Redundancy Power supplies connected in parallel operation so that if one fails, the others will continue delivering enough current to supply the maximum load. This method is used in applications where power supply failure cannot be tolerated.

Reed Relay Consists of two thin magnetic strips (reeds). When a coil close to the reeds is energised, they are magnetised and drawn together making a connection between leads attached to the reeds.

Reed valve 1. A thin leaf of steel located in the valve plate of automotive compressors, serving as suction and discharge valves.

2. Used to control exhaust backpressure in a pumpless AIR system

Reengineering The process of examining and altering an existing system to reconstitute it in a new form. May include reverse engineering (analyzing a system and producing a representation at a higher level of abstraction, such as design from code), restructuring (transforming a system from one representation to another at the same level of abstraction), recommendation (analyzing a system and producing user and support documentation), forward engineering (using software products derived from an existing system, together with new requirements, to produce a new system), and translation (transforming source code from one language to another or from one version of a language to another).

Reentry interface An altitude 400,000 feet; the point at which reentering spacecraft are considered to enter the Earth's atmosphere.

Reformate Hydrocarbon fuel that has been processed into hydrogen and other products for use in fuel cells.

Reformer Device used to generate hydrogen from fuels such as natural gas, propane, gasoline, methanol, and ethanol for use in fuel cells.

Reforming A chemical process in which hydrogen containing fuels react with steam, oxygen, or both to produce a hydrogen-rich gas stream.

Reformulated gasoline (RFG) Gasolines which have had their composition and/or characteristics altered to reduce vehicular emissions of pollutants. Specifically, those gasolines which meet the RFG requirements of the 1990 Clean Air Act Amendments.

Refraction The deflection or bending of electromagnetic waves when they pass from one kind of transparent medium into another.

Refractometer An Instrument used to measure substances dissolved in water and certain oils. It works on the principle of light refraction through liquids. As light passes from air into a liquid it slows down and this phenomenon is what gives a "bent" look to objects that are partially submerged in water. Refractometers use this principle to determine the amount of dissolved solids in a liquid by passing light through a sample and showing the refracted angle on a scale. In automotive applications reflectometers are used to measure the specific gravity of battery electrolyte and antifreeze concentrations.

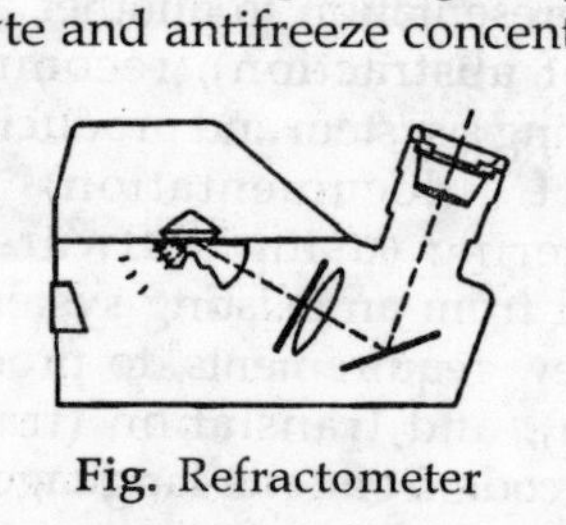

Fig. Refractometer

Refrigerant CFC-12 (R-12) The refrigerant, dichlorodifluoromethane, used in automotive air conditioners, as well as other air-conditioning and refrigeration system.

Refrigerant lines Specially designed vapour-barrier hoses reinforced with woven nylon mesh and fabric capable of withstanding the high temperatures and pressures of the system.

Refrigerant The chemical compound used in a refrigeration system to produce the desired cooling effect.

Refrigerant-containment device A device introduced on some car lines to guard against high pressure resulting in refrigerant loss by controlling the compressor and/or condenser fan motor.

Refrigeration cycle The complete cycle of the refrigerant back to the starting point, evidenced by temperature and pressure changes.

Refrigeration lubricant A mineral oil or synthetic oil-like lubricant, such as PAG and ESTER, that is formulated for specific use and application in designated refrigeration systems.

Refrigeration oil A highly refined (specially formulated), non-foaming organic mineral oil free from all contaminants, such as sulphur, moisture, and tars, used to lubricate the air-conditioner compressor.

Refrigeration To use an apparatus to cool; keep cool; or keep chilled under controlled conditions by natural or mechanical means, as an aid to ensure personal safety and comfort.

Refundable security deposit Money to be held until lease-end as a security method for all lease obligations.

Regeneration system A system in a gas turbine that converts some of the heat, usually wasted, into useable power.

Regenerative cooling Circulation of a propellant through a jacket around the combustion chamber in order to cool the chamber wall, the propellant subsequently being injected into the combustion chamber.

Regenerative Fuel Cell 1. Several fuel cell types in which fuel and oxidants are regenerated from the oxidation product.

Regenerative fuel cell A fuel cell that produces electricity from hydrogen and oxygen and can use electricity from solar power or some other source to divide the excess water into oxygen and hydrogen fuel to be re-used by the fuel cell.

Regenerator A device placed on a gas turbine to take the heat of exhaust and put it into the intake of the engine.

Register A high-speed device used in a central processing unit for temporary storage of small amounts of data or intermittent results during processing.

Regular production option (RPO) Items of new car equipment available to any buyer for a price.

Regular-duty coil spring A coil spring supplied to handle average loads to which the vehicle is subjected, having a small wire diameter as compared to a heavy-duty spring.

Regulated power supply A device that maintains within specified limits a constant output voltage or current for specified changes in line, load, temperature or time.

Regulation The process of holding constant selected parameters, the extent of which is expressed as a percent.

Regulator The power supply circuit that controls or stabilizes the output parameter at a specified value

Reheat principle A principle that is used in automotive air-conditioning systems to control in-car relative humidity by first cooling, then reheating the air.

Reinforced insulation As defined by Underwriters Laboratories, Inc., an improved basic insulation with such mechanical and electrical qualities that it, in itself, provides the same degree of protection against electrical shock as double insulation. It may consist of one or more layers of insulating material. It is acceptable in place of double insulation.

Relative Compression Testing Measuring the engines compression pressure relative to the amount of amperage required to turn the crankshaft during a cranking event. Using an oscilloscope in conjunction with an inductive amps probe.

Relative pressure The difference between intake manifold pressure and the output pressure in a fuel-injection system.

Relative wheel weights The weight on each of the vehicle's wheels as measured by individual matched scales under each wheel.

Relaxation Rate of reduction of stress in a material due to creep. An alternate term is stress relaxation.

Relay A magnetic component or solid state device that opens or closes an isolated switch(es) when a voltage is applied to the control terminals.

Relay/quick-release valve A valve, similar to a remote control foot valve, used on trucks with a wheel base of 254 inches (6.45 meters) or longer attached to an air tank to main supply line to speed the application and release of air to the service chambers.

Release fork Disengages the clutch disc from the flywheel by pressing on the pressure plate release springs.

Release levers A lever-type, pivoting pressure-plate device in the clutch assembly that is moved by throw out bearing movement, causing the clutch-spring pressure to be relieved so the clutch is released from the flywheel.

Reliability analysis A predictive tool used to estimate the "life" of a product. This is usually expressed in terms of hours as "mean time between failure".

Reliability prediction Reliability prediction is the analysis of parts and components in an effort to predict the rate at which an item will fail. A reliability prediction is one of the most common forms of reliability analyses.

Reliability The probability that an item will continue to function at customer expectation levels at a measurement point, under specified environmental and duty cycle conditions.

Reliability-centered maintenance (RCM) A methodology used to define a maintenance program while having reliability as an input to the decision making process. Originally developed by the aviation industry and known as MSG-3 (Maintenance Steering Group), Reliability Centered Maintenance or RCM has since been adopted by many different industries.

Relieved An engine in which the intake and exhaust passages have been cleaned of all obstructions and ridges.

Reluctance (R) In a magnetic path, the opposition to the flow of magnetic lines of force.

Reluctance sensor A magnetic pulse generator or pick-up coil that sends a voltage signal in response to a varying reluctance of its magnetic field and its reluctor core.

Reluctor A gear-like part of an electronic ignition system, having the same number of teeth as there are cylinders, near the top of the distributor shaft so that as a tooth passes a pickup coil, resistance to magnetic flux is reduced and the magnetic field is strengthened, which triggers the electronic control unit.

Remanufactured A term used to describe a component that is disassembled, cleaned, inspected, and reassembled using new or reconditioned parts. According to the Automotive Parts Rebuilders Association APRA), this same component is also referred to as being rebuilt.

Remote back-pressure transducer An exhaust back-pressure sensing device mounted in the vacuum line leading to the EGR valve rather than on the valve itself, to bleed off the vacuum signal to prevent recirculation.

Remote bulb A sensing device connected to an expansion valve or thermostat by a capillary tube, sensing temperature and transmitting pressure to the control for its proper operation.

Remote instrument An instrument that does not have access to the ATST light beam. Note that this definition encompasses both local instruments that are not in the active instrument set as well as instruments located at other sites (e.g., in orbit).

Remote sensing A technique for regulating the output voltage of a power supply at the load by connecting the error-sensing leads to the load. Remote sensing compensates for voltage drops in cables and connectors between the power supply and the load. Since the remote sense leads usually directly control the power supply's internal feedback loop, the system should not be operated with the remote sense leads disconnected. Also, when connecting remote sense leads, make sure the + lead goes to the + end of the load. A very common mistake is to reverse the leads, which can damage the system.

Remove and replace (R&R) To remove a component and replace it with a reconditioned, rebuilt, good used, or new component on a vehicle.

Rendezvous A place of meeting at a given time, for example, a spaceship with a space station.

Renewable energy A form of energy which is never exhausted because it is renewed by nature (within short time scales; e.g. wind, solar radiation, hydro power).

Renkin scale Used to measure low temperatures in the Fahrenheit scale. Similar to Kelvin which is used to measure low temperatures in the Centigrade system.

Rent charges The amount charged in addition to the depreciation and any amortized costs.

Replacement springs New springs having the same characteristics, shape, and type of ends as the springs in the vehicle, all four of which should be replaced at the same time.

Replacement tire A tire purchased to replace an original tire that was supplied by the vehicle manufacturer, having the same size and temperature range as the original tire.

Replication The industrial process for copying optical storage disks (CD, DVD) from a template master.

Repulsion A mechanical force that tends to separate like magnetic poles or electrical charges.

Required sidewall information Information that is required by the Department of Transportation to be included on each tire such as size, load rating, maximum inflation pressure, generic name of materials used in construction, actual number of plys, and so on.

Required voltage 1. The low voltage required by a device or system.

2. The highest voltage required to fire a spark plug.

Requirement A function, feature or capability that a product must provide or meet to satisfy the customer's needs and enterprise's objectives for a new product.

Requirements allocation matrix Requirements Allocation Matrix is a matrix showing the allocation of a requirement (e.g., reliability, weight, cost) to various subsystems or subassemblies so that requirement can be accurately flowed-down and the satisfaction of the overall requirement can be tracked and managed.

Requirements analysis The determination of product-specific performance and functional characteristics based on analyses of: customer needs, expectations, , and constraints; operational concept; projected utilization environments for people, products, and processes; and measures of effectiveness.

Requirements Characteristics that identify the accomplishment levels needed to achieve specific objectives for a given set of conditions

Requirements engineering Requirements engineering can be defined as the systematic process of developing requirements through the process of analyzing the problem or need, documenting the resulting requirements to solve the problem or meet the need in a variety of representation formats, and checking the accuracy of the understanding gained. Requirements engineering focuses on "what" needs to be designed. Requirements engineering is not a one time activity but instead should be revisited at every stage of the development process to find out if the requirements have changed at all and if not are they being met.

Requirements flowdown The process of deriving and allocating requirements to all levels of system decomposition.

Requirements management Requirements management is the process of managing the initial development of requirements and the subsequent changes to requirements to assure that they address only what is needed or required of the product and that adequate consideration is given to tradeoffs in product cost, development cost, development schedule, and competitor actions. Requirements management exercises control over the project scope to avoid scope creep and unnecessary or deferrable nice-to-have features/capabilities.

Requirements traceability The evidence of an association between a requirement and its source, its implementation, and its verification.

Research and development (R&D) As the term implies, to research and develop a concept, system, or product.

Residual A conservative projection of the market value of a vehicle at the end of a lease. Residual values are provided by major auto manufacturers or independent companies that specialize in auto valuation. A higher residual value will lower the monthly payment. However, the vehicle must be resalable at the residual amount or the lessor will lose money. Conversely, adjusting the residual too low will increase the monthly payment and may make the lease payment unattractive. Adjustments to residual may be made for excess mileage or wear and tear.

Residual elongation Measure of ductility of plastics. It is the elongation of a plastic specimen measured 1 minute after rupture in a tension test.

Residual flux The flux that exists in a magnetic core after the H field is returned to zero.

Residual magnetism The magnetism that remains in an electromagnetic material after the current is interrupted.

Residual pressure check valve A check valve used in the outlet port for drum brakes in some master cylinders. The valve retains a slight pressure in the lines to the drum brakes when the brakes are not applied.

Residual stress Stress present in a joint member or material that is free of external forces or thermal gradients.

Residual value The estimated value of the vehicle at the end of the lease. Often expressed as a percentage in decimal form. It is based on a vehicle's msrp. For example, the residual value of a vehicle may be 0.65, or 65% of its msrp at the end of the lease term. In this example, a car with an msrp of $20,000 is estimated to be worth $13,000 at the end of the lease term. Residual value may be a negotiable figure. This should be clearly stated in the lease. It is one of the elements used in determining the monthly lease payment and also in deciding whether to buy the vehicle at term end. Also called the walk-away value or guaranteed value.

Resistance Temperature Device (RTD) Resistance temperature devices (or detectors) rely on the principle that the resistance of a metal increases with temperature. When made of platinum, they may be known as platinum resistance thermometers (PRTs).

Resistance welding 1. A term used for spot welding.

2. A welding process that unites the surfaces with heat, obtained from the resistance of the work pieces, and the application of pressure.

Resistor spark plug A spark plug having a resistance of 5,000 to 10,000 ohms inside the upper part of the insulator, increasing the electrode life and suppressing radio interference.

Resolution Ability to distinguish visual detail, usually expressed in terms of the size (in kilometers) of the smallest features that can be distinguished.

Resonance 1. The state in which the natural response frequency of a circuit coincides with the frequency of an applied signal, or vice versa, yielding intensified response.

2. The state in which the vibration frequency of a body coincides with an applied vibration force, or vice versa, yielding rein-forced vibration of the body.

Resonant circuit A circuit in which inductive and capacitive elements are in resonance at an operating frequency.

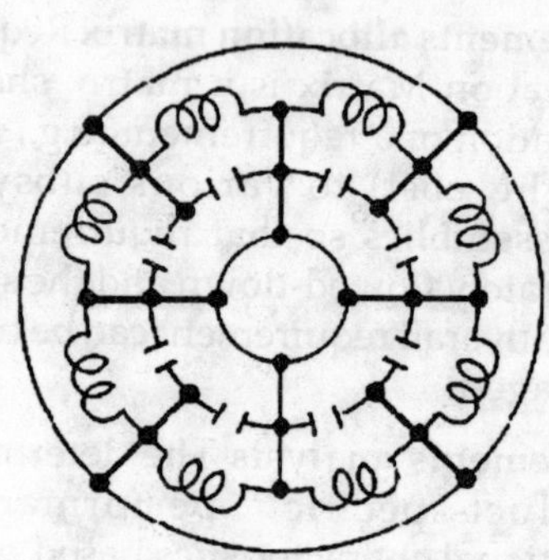

Fig. Resonant circuit

Resonant converter A class of converters that uses a resonant circuit as part of the regulation loop.

Resonant frequency The natural frequency at which a circuit oscillates or a device vibrates. In an L circuit, inductive and capacitive reactances are equal at the resonant frequency.

Resonator A small auxiliary muffler, after the main muffler, used to help reduce engine noise without restricting the flow of exhaust gas.

Resource activation Using a resource regardless of whether throughput is increased.

Resource utilization Using a resource in a way that increases throughput . **Response time** Refers to the allowable length of time between sending a signal commanding a system to execute an action and completion of the action, including settling time required for the system to return to its specified level of performance or accuracy.

Responsiveness to change Changes to system and program requirements in response to directed changes by the procuring activity, or problem solutions identified shall be evaluated for total program impact with respect to performance, cost, and schedules.

Rest Pressure Fuel pressure maintained within the system after the engine has been shut down

Retarded timing When the spark is late arriving to the combustion chamber and the flame front reaches the top of the piston when the piston is to far down in the cylinder to push the piston with any strength

Retarder 1. A mechanical, electrical, or hydraulic device used to slow a diesel engine when the throttle is lifted.

2. A chemical compound that reduces the speed of paint drying.

Retracting spring A spring that is used to pull the brake shoes away from the drum when the brake pedal is released, pushing the wheel-cylinder piston back into its bore and thus returns the brake fluid to the master cylinder.

Retread A used tire having had its old tread removed and a new tread applied.

Return hose 1. The low-pressure hose in a power-steering system through which fluid returns to the pump reservoir after leaving the steering-gear assembly or control valve.

2. A non-pressurized fluid or air system.

Return on investment Return on investment is a financial analysis technique which compares the expected return to the outlay or investment to determine a percentage of return.

Return spring 1. Springs used on drum brakes to pull the brake shoes away from the drums when the brakes are released.

2. A spring that is used to return a mechanism to its original position.

Return The name for the common terminal of the output of a power supply; it carries the return currents for the outputs.

Returnable cylinder 1. A cylinder of gas, such as oxygen, nitrogen, propane, or acetylene that can be reused.

2. A DOT approved refrigerant-recovery cylinder.

Rev limiter 1. A governor.

2. A mechanical or electrical device that limits engine speed.

Reveal file A small file that is "pulled, not pushed," available in numerous shapes, generally curved to fit tightly crowned areas such as around windshields, wheel openings, and other panel edges for final fitting of reveal molding.

Reveal moulding The metal trim that outlines an opening, depression, or raised area of an auto body.

Revenue release The point in the product life cycle that products are released for first sale to customers. Production is usually still at low rate, the customers that the product is sold to may need to meet certain requirements, and there may be special support capabilities provided to the product at this point.

Reverse 1. The transmission position enabling the vehicle to back up.

2. To increase wheel track or tread by installing the rims backwards.

Reverse balance A common type of hydraulic balance used on outside mounted mechanical seals.

Reverse bias 1. A dc voltage applied to a semiconductor junction, with a polarity opposite to that for forward bias.

Reverse bleeding A method of purging air from a hydraulic system by forcing fluid into the system through a bleeder valve and allowing the air to escape at the master cylinder or reservoir.

Reverse Elliot axle A solid-beam front axle on which the steering knuckles span the axle ends.

Reverse engineering 1. Reverse engineering is the process of capturing the geometry of existing physical objects and then using the data obtained as a foundation for designing a duplicate of the original or an entirely new adaptation.

2. Reverse engineering refers to the procedure of carefully dismantling and inspecting a competitor's product to look for design features that can be incorporated into one's own product.

Reverse flushing A method used to clean an engine or cooling system by flushing it in the opposite direction of normal coolant flow.

Reverse gear A transmission gear arrangement obtained from a planetary gear set when the planet-gear carrier is held and power is applied tot he sun gear.

Reverse idler gear 1. A gear in most standard transmissions that must be meshed to obtain reverse, and that idles in all forward gears.

2. A gear that transfers power in reverse only. Rotating on a separate shaft, it reverses the direction of the counter-gear motion.

Reverse voltage protection A circuit or circuit element that protects a power supply from damage by a voltage of reverse polarity applied at the input or output terminals.

Reversed polarity A condition that exists when battery cables are connected to the wrong terminals of a battery.

Reversion The interval during valve timing overlap when both intake and exhaust valves are open and a small mist of air/fuel is forced out back through the carburettor.

Revolution The movement around the center or axis such as the circling of Earth around the sun.

Revolutions per minute (rpm) 1. The number of times a member, such as a shaft, makes a complete revolution in one minute.

2. The rate of speed of a rotating part, such as a crankshaft.

Reynolds number A dimensionless parameter in fluid dynamics, which indicates transition from laminar to turbulent flow. The Reynolds number is a very important parameter in fluid dynamics. Flows at the same Reynolds number are similar, which make it possible to use smaller models (length l) for tests at higher velocities (velocity v) to achieve the same Reynolds number.

Right ascension With declination, one of the coordinates used to designate the location of an object on the celestial sphere. Right ascension is measured in hours, minutes, and seconds and is similar to longitude on Earth.

Right-hand thread A common thread for a bolt or nut that is tightened by turning it clockwise, or to the right.

Right-size Matching tooling and equipment to the job and space requirements of lean production.

Robotics The study of the design and use of robots, particularly for their use in manufacturing and related processes.

Robust design Design of the product in a manner to desensitize the product to variation including misuse and increase the probability that it will perform as intended.

Robustness The condition of a product or process where its operating parameters remain relatively stable with a minimum of variation even though factors which influence operation or usage , such as wear or environment, change.

Rocker arm 1. A pivoting part in a mechanical fuel pump, one end of which rides on a camshaft eccentric or pushrod, and the other end of which pulls on the diaphragm via a mechanical linkage, causing the diaphragm spring to be compressed.

2. A pivoting lever in the valve train of an overhead valve system which applies motion, directly or indirectly from the camshaft to open an intake or exhaust valve.

Rocker panel The sheet metal along the bottom of an auto body, beneath the doors, and between the front- and rear-wheel openings.

Rocker-arm ratio The difference between the upper movement of the pushrod, acting on one end of the rocker arm, and the downward movement of the rocker arm acting on the valve stem.

Root cause analysis (RCA) A technique for uncovering the cause of a failure by deductive reasoning down to the physical and human root(s), and then using inductive reasoning to uncover the much broader latent or organizational root(s.)

Root mean square (RMS) value The equivalent DC voltage that produces exactly the same amount of heating as the AC waveform in a simple (non-inductive) resistor. The RMS value of a sine wave is 0.707 x Peak Value.

Root sum of squares Root sum of squares (rss) is a tolerance method that makes use of rss to determine the best tolerance limits. Rss assumes that the print tolerance equals +/- 3 standard deviation limits and part nominal equals print nominal. This analysis exploits the manufacturing probability that a part is not always at its minimum or maximum value. It does not take into account process mean shifts (tool wear) and assumes the process is always centered.

Rotary vacuum valve That part of a vacuum control used to divert a vacuum signal for operation of doors, switches, and/or valves.

Rotary valve A semi-circular disc in a two-stroke engine that opens and closed the intake port.

Rotating seal When the spring loaded or moveable portion of the seal rotates with the shaft.

Rotation 1. A term used to indicate that a gear, shaft, or other device is turning.

2. A term used for tire rotation.

Rotational motion Movement that occurs when a shaft turns (rotates) on its axis. A dynamic seal is required to contain fluids were rotational motion is present.

Rotator A device in the cylinder head that causes the valve to rotate to reduce carbon build-up.

Rotor 1. Disc-shaped component that revolves with hub and wheel. The lining pads are forced against the rotor to provide a friction surface for the brake system, so as to slow or stop a vehicle.

2. In a conventional ignition system, the part that transfers the secondary voltage from the center terminal to the outer terminals of the distributor cap as it rotates inside the cap.

3. The rotating part of an alternator that provides the necessary magnetic field to create a current flow.

Rotor face The flat, parallel surface on each side of the rotor contacted by the pads (linings) during stopping.

Rotor oil pump A type of oil pump in which a pair of rotors produce the pressure required to circulate oil to the engine parts.

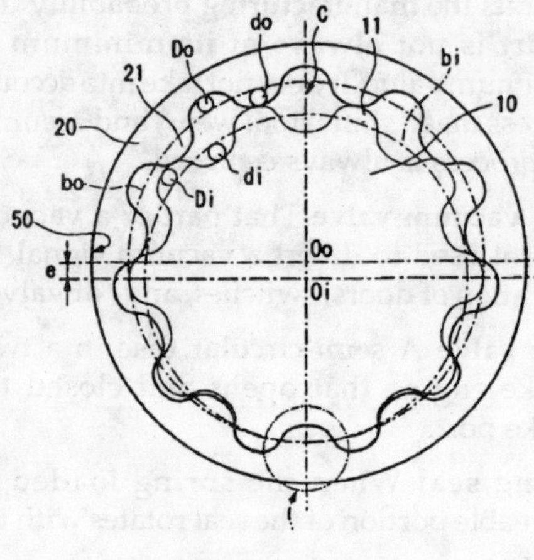

Fig. Rotor oil pump

Rotor pump A positive-displacement pump that uses an inner drive rotor and an outer drive rotor to produce an oil flow. Lobes on the rotors create a fluid chamber of varying volumes and eliminate the need for a crescent as used in a gear pump.

Rotunda Test equipment manufactured for Ford Motor Co. By many different manufacturers.

Round-trip light time The elapsed time is takes for light, or a radio signal, to travel from Earth, be received and immediately transmitted or reflected, and return to the starting point.

Route guidance database The detailed information required for a computer to generate a high-quality driving route between two locations. Information includes geometry, street names, addresses, speed limits, turn restrictions, one-way restrictions, road levels, and roadway connections.

Rubber bumper 1. Rubber stop on the car frame that prevents metal-to-metal contact when the rear-axle housing moves to its maximum upward position.

2. Rubber stops positioned on or near a control arm to limit the maximum upward or downward travel of the arm.

Rubbing block The insulated section of the movable breaker-point arm that contacts the breaker cam of a conventional ignition distributor.

Rule-based system A functional system in which knowledge is stored in the form of simple if-then or condition-action rules.

S

S&O Support and Operations.

Saddle valve 1. A term used for saddle-clamp access valve.

2. A two-part accessory valve that may be clamped around the metal part of a system hose to provide access to the air conditioning system for service.

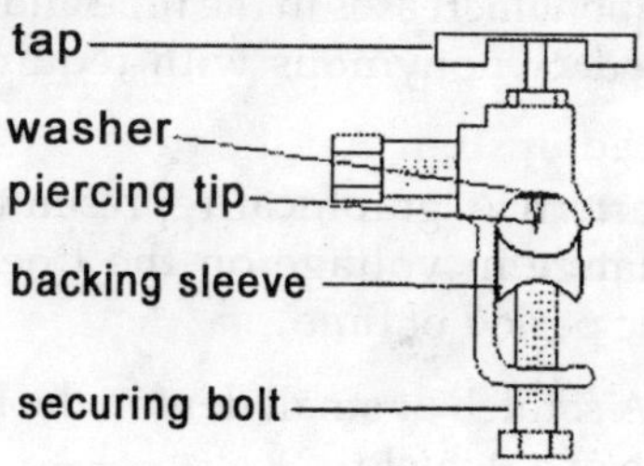

Fig. Saddle valve

Saddlebag Air chambers or openings in the left- and right-front corners of the car body between the kickpads and the exterior of the car. The evapourator is sometimes located in the right saddlebag.

Saddle-clamp access valve A two-port accessory valve that may be clamped around the metal part of a system hose to provide access to the system for diagnostics and service.

Safety factor The amount of load that can be absorbed by and through the chassis-frame members of a vehicle.

Safety glasses Eye glasses with or without side shields to protect the eyes when working in a hazardous environment and to be worn at all times when in the automotive shop.

Safety goggles For those who wear prescription glasses, eye protection from all sides fitting over the glasses and against the face and forehead to seal and protect the eyes from outside hazards.

Safety ground A conductive path from a chassis, panel or case to earth to help prevent injury or damage to personnel and equipment.

Safety lap An extra lap taken by the winner of a race to ensure that an error was not made by the official lap scorer.

Salt flats The dry lake beds in the desert, such as Bonneville, used by hot roders for top-speed runs.

Salvage yard A modern term for junkyard, an establishment that sells used auto parts.

Sample and hold A component of a type of analogue-to-digital converter. The analogue signal is frozen in a sample and hold circuit to prevent it changing during digitisation.

Sandbag 1. To legally secure equipment without having to pay for it.

2. To hold back in the staging area of a drag race in an attempt to select a specific opponent during eliminations.

3. To hold speed down in drag racing in an attempt to fall into a favorable bracket during final eliminations.

4. To deliberately run slower in road- or oval-track racing in an effort to conserve fuel and/ or to psych the competitors.

5. A bag filled with sand that is used to help shape metal panels.

Saturated desiccant A drying agent that contains all of the moisture it can hold at a given temperature.

Saturated drier A drier, accumulator-drier, or receiver-drier having a saturated desiccant.

Saturated point The point at which matter mush change states at any given temperature and pressure.

Saturated Switch Injector A fuel injection system that produces a spiked square wave signal with variable pulse width or on time

Saturated temperature The boiling point of a substance at a particular pressure.

Saturated vapour A term that indicates that the space holds as much vapour as possible and no further vaporization is possible at that particular temperature.

Saturation (Bipolar transistor) A condition when an increase in base current will not produce an appreciable change in the collector-emitter voltage.

Saturation (Magnetic) A condition in a magnetic material in which an increase in (B) will no longer produce an appreciable increase in (H).

Scheduled Downtime A period of time when the equipment is not available to perform its intended function due to planned downtime events. These include maintenance delay (delay after an interrupt is reported, but before anyone arrives to repair it); production test; preventive maintenance; change of consumables; setup; and facilities-related downtime.

Schematic A map-like drawing of the electrical system that gives the colours and shows the terminal points; used to trace the circuit for troubleshooting.

Schmitt Trigger An electronic circuit used to convert analogue signals to digital signals or vice-a-versa

Schottky diode A diode device that exhibits a low forward voltage drop and fast recovery time relative to a standard silicon diode.

Schrader valve A spring-loaded valve, similar to a tire valve, located inside the service-valve fitting and some control devices to hold vapour or fluid in the system. It requires special adapters for access to the system.

Scope creep The tendency for project requirements to grow over time, usually resulting in huge, unmanageable projects. As some projects progress, especially through development, requirements continuously change incrementally, causing the developer to add to the work scope with consequent increases in the time and budget required. Synonymous with requirements creep.

Scope pattern A graphical representation of the change in voltage on the line over a specific period of time

Score 1. A scratch or small dent in the finished surface of a vehicle.

2. (SCORE) An acronym for Short Course Off-Road Enterprises

Scoring The grooves worn into the friction surface on a brake drum or rotor which may be machined away; if the depth exceeds specified limits, the drum or rotor must be replaced.

Scrap yard 1. A firm selling used parts.

2. A firm that accepts scrap metal for recycling.

Scraper ring The second ring from the top of a piston used to scrape oil from the cylinder wall.

Scrappage A vehicle registered in the previous year but not re-registered in the current year.

Scratch 1. To spin the drive wheels, usually enough to leave a mark on the road surface.

2. A mark on the finish of a body surface made by a scribe or other sharp object.

3. To make a mark on a finished surface using a sharp object, such as a key.

Screen A metal mesh located in the receiver, expansion valve, and compressor inlet to prevent partials of dirt from circulating through the system.

Screening Screening is the process of evaluating and selecting new product ideas or concepts for development. These evaluation criteria include fit with company strategy, fit with other products/product lines, fit with customers and markets, profitability, growth, risk, investment requirements, technical capabilities, core competencies, etc.

Screw-thread pitch gauge A thin material with V-shaped notches that, when matched with a thread of a bolt or nut, indicates the number of threads per inch or millimeter, as well as the thread pitch.

Scuff in To run a new set of racing tires long enough to bring them up to temperature, and wear the manufacturer's protective coating off the tread area. Also known as scuff off.

Scuffing A type of wear between two parts where there is a transfer of material from one part to the other.

S-curve A graphical representation of costs, hours, technological progress and other factors. The name is derived from the s-like shape of the curve that is flatter at the beginning, accelerates sharply, and then tails-off. As it relates to technology, the s-curve is flat when the technology is first invented (technology performance improves slowly and incrementally). Then, as experience with a new technology accrues, the rate of performance increase grows and technology performance increases by leaps and bounds. Finally, some of the performance limits of a new technology start to be reached and performance growth slows.

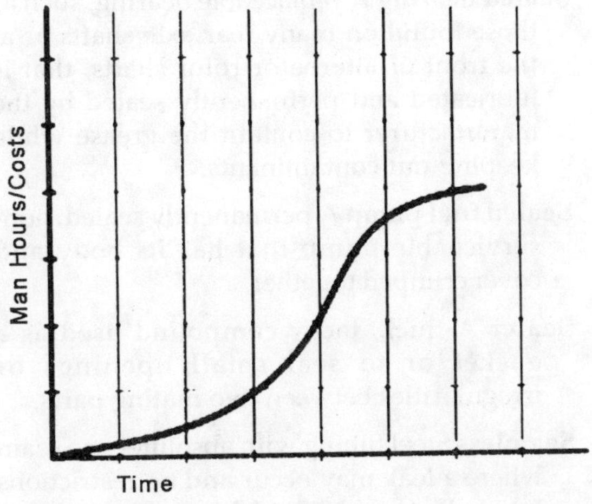

Fig. S-curve

Seal 1. A device used around a rotating shaft to prevent fluid or vapour leaks.

2. A gasket-like material used between two or more parts to prevent fluid or vapour leaks.

Seal faces The lapped faces that that provide the primary sealing in a mechanical seal.

Seal life Seals should run leak free until the sacrificial face (usually carbon/ graphite) is worn away. More than 85% of the mechanical seals in use today fail prematurely.

Seal only pump The pump does not have a conventional stuffing box so it could not be supplied with soft packing.

Seal weld A weld designed primarily to provide a specific degree of tightness against leakage.

Sealed battery A term used for maintenance-free battery.

Sealed beam headlamp A one piece, hermetically sealed headlamp in which the filament is an integral part of the unit and the lens itself is the bulb. Sealed beams are relatively inexpensive and when one burns out or the lens cracks, the whole unit is replaced.

Sealed bearing A replaceable bearing, such as those found on many rear axle shafts or at the front of alternator rotor shafts, that is lubricated and permanently sealed by the manufacturer to contain the grease while keeping out contaminants.

Sealed fuel pump A permanently sealed, non-serviceable pump that has its body and cover crimped together.

Sealer A thick, tacky compound used as a gasket or to seal small openings or irregularities between two mating parts.

Seamless Steel tubing with absolutely no seam where a leak may occur and no restrictions internally to cause fluid flow to be restricted.

Seasoned part A part that has been stress relieved.

Seat 1. The surface on which another part rests.

2. To wear to a good fit.

3. The bench-like unite one sits on in a vehicle.

Seat adjuster A device that permits fore and aft and/or up and down movement of a vehicle's front seat.

Seat back The upright portion of a vehicle seat.

Seat belt warning system A warning device to alert the driver and occupants to fasten their seat belts and/or shoulder harness.

Seat of the pants (SOP) Driving by sheer instinct where instruments cannot be used to determine speed or distance.

Security deposit May be required in some lease contracts. It is similar to a security deposit on an apartment. This money is held to cover any payments that might be missed. This deposit may or may not be refundable. It is usually equivalent to one month's lease payment.

Seize Two moving parts that suddenly bind together, usually due to lack of lubricant. Also known as freeze-up.

Seizing The stiffening or "freezing" of a chain joint or shaft as a result of roughness and high friction caused by galling or lack of lubrication.

Select low The program used by the microprocessor to determine which wheel is beginning to lock up.

Selective laser sintering Selective laser sintering - a rapid prototyping technology that uses a laser to trace a pattern on a powdered material to fuse it into a solid on layer after another to form a solid object.

Self sintered silicone carbide Same as Alpha sintered. No free silica present. Can be used in caustic and higher pH liquids.

Self test A computer function that scans systems and subsystems to provide data for troubleshooting.

Self-adjuster A mechanism on a drum brake that compensates for shoe-lining wear and keeps the shoe adjusted close to the drum.

Self-adjusting clutch A mechanism that automatically takes up the slack between the pressure plate and clutch disc.

Self-Calibrating A data acquisition card or module with a stable on-board reference voltage that software can use for automatic recalibration.

Self-diagnostics In automotive computers, especially those for engine control, a program which assesses the condition of the system, including the sensors and the computer itself, and communicates its findings to the technician by means of trouble codes.

Self-directed work teams Nearly autonomous teams of empowered employees, including hourly workers that share a common workspace and/or responsibility for a particular process or process segment. Typically such teams have authority for day-to-day production activities and many supervisory responsibilities, such as job assignments, production scheduling, maintenance, materials purchasing, training, quality assurance, performance appraisals, and customer service.

Self-discharge The discharging of a battery due to chemical action, although there is no electrical demand.

Self-energizing The placing of brake shoes so that the drum tends to drag the lining around with it, resulting in a wedging action between anchor and drum.

Self-inductance Inductance that produces an induced voltage in itself as the result of a change in current flow.

Self-induction The induction of a voltage in a current-carrying coil of wire, such as an ignition coil.

Self-locking screw A screw configured do that it locks itself in place when tightened.

Semiautomatic welding Manual welding with equipment that automatically controls one or more of the welding conditions.

Semicentrifugal clutch A clutch having weighted components in the pressure plate, such as rollers or release levers, that apply additional force against the pressure plate to hold the disc tighter during high engine rpm.

Semiconductor A solid material, usually germanium (Ge) or silicon (Si), that integrated circuits are made of, with an electrical conductivity between the high conductivity of metals and the low conductivity of insulators used to control the flow of electricity, containing 4 electrons in its valance shell a semiconductor is nether an insulator nor a conductor

Semielliptical Refers to the amount that the ends are higher than the center arch of a leaf-spring configuration.

Semielliptical spring 1. A leaf spring.

2. A set of leaf springs of regressive lengths stacked with the longest at the top to the shortest at the bottom.

Semifloating axle A popular automotive system in which the axle shaft provides three functions: transfers torque to drive the vehicle, supports the car weight, and retains the wheel.

Semitrailer A load-carrying vehicle equipped with one or more axles and constructed so that its front end is attached to, and supported by the fifth wheel of a semi-tractor that pulls it.

Sending unit An electrical or mechanical sensing device that sends information relative to some physical property such as temperature or pressure to a gauge or light. Also known as sensor unit.

Sense line return The conductor which routes the voltage on the output return to the control loop.

Sense line The conductor which routes output voltage to the control loop.

Sensi An outside master or teacher that assists in implementing lean practices.

Sensible heat Heat that causes a change in the temperature of a substance, but does not change the state of the substance that can be felt or measured with a thermometer.

Sensing voltage A condition that provides a means for the voltage regulator to monitor and control the battery voltage charge rate.

Sensitivity A measure of the minimum change in an input signal that an instrument can detect.

Sensor A device that can detect a change in a physical quantity (light or pressure for example) and produce a corresponding electrical signal.

Sensor algorithm An algorithm is a mathematical formula or series of formulas used by an on-board computer or processor to make a decision. In an airbag system, a crash-sensor algorithm determines whether the change in velocity indicates an impact of great enough force to require airbag deployment, based on pre-programmed parameters. If the change in velocity is great enough, the processor sends a signal to the device that inflates the airbag.

Sensor plate A plate used in the air intake of a continuous-flow, fuel-injection system to measure airflow.

Separator 1. A tank-like device used to remove liquids from a vapour, such as oil from refrigerant.

2. A nonconductive divider placed between the positive and negative plates of a battery.

Separator disc The metal plates in the multi-plate clutch that separates the friction discs.

Separator Pulse A pulse code consisting of a signal half-second pulse between the last service code and the first continusus code of a data stream

Sequence control The control of a series of machine movements, with the completion of one movement initiating the next. The extent of movements is typically not specified by numerical input data.

Sequencer A mechanical or electrical device which may be set to initiate a series of events and to make events follow a sequence.

Sequential fuel injection Similar to multi-port fuel injection, but the injectors spray fuel into the individual intake ports exactly at the beginning of each cylinder's intake cycle. The precise fuel control provides better engine performance.

Sequential Injection A fuel injection system where the injectors fire from individual drivers, in the same sequence as the engines firing order

Sequential Logic Circuits A flip-flop circuit in which the output is determined by the sequence of inputs. A given input affects the output produced by the next input

Sequentially Fired A type of fuel-delivery system in which the fuel injectors are fired individually according to ignition firing order and valve timing.

Sequential-port fuel injection (SPFI) A type of multi-port injection system where individual fuel injectors are pulsed sequentially, one after another in the same firing order as the spark plugs, rather than being pulsed simultaneously. This allows more precise fuel control for lower emissions and better performance. Also known as sequential fuel injection.

Serial Data The pulse train data signal sent on the communication line between the computer and the scantool.

Series 1. The interconnection of two or more power sources such that alternate polarity terminals are connected so their voltages sum at a load.

2. The connection of circuit components end to end to form a single current path.

Series circuit A circuit having only one path through which the current or fluid can flow, having to pass through one component before going on to another.

Series operation The pumps are connected with the discharge of the first pump discharging to the suction of the other. The speed and impeller widths must be the same or the difference in capacities could cause cavitation or over heating problems.

Series parallel system An arrangement where two 12-volt batteries are connected in such a manner as to provide 24-volts for starting and 12-volts for accessories and charging.

Series pass A controlled active element, such as a transistor, in series with a load that is used to regulate voltage.

Series regulator A regulator in which the active control element is in series with the dc source and the load.

Series wound In a starter motor, the field coils (windings) and the armature, are wired in series. All current flows through the field coils, through the hot brushes, through the armature, then to the ground through the ground brushes.

Serpentine belt A wide, flat belt having multi-V-grooves to provide friction contact with the pulleys that winds through all of the engine accessories to drive them off the crankshaft pulley.

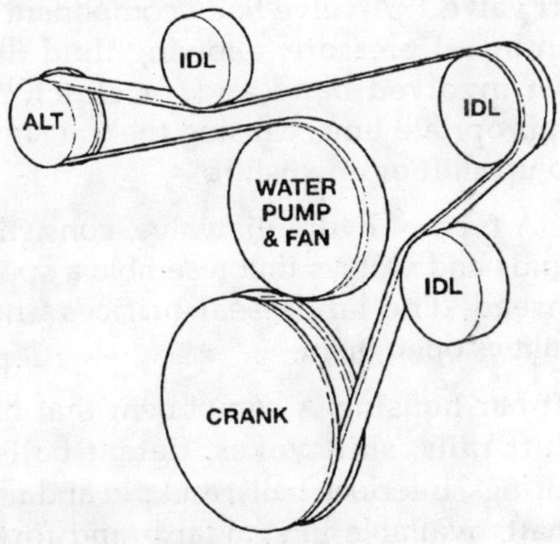

Fig. Serpentine belt

Serrated nut A self-locking type of nut having serrations on its contact side to prevent it from loosening when tightened.

Serrated rod cap A connecting rod cap that has serrated parting edges to help maintain alignment with the rod.

Server A software entity that responds to requests from other software entities (called clients).

Serviceability The characteristics of a product to make it more readily serviceable. These characteristics would address features related to fault identification, diagnosis, disassembly, repair, replacement, and re-assembly. A common serviceability metric is mean time to repair.

Services interfaces Interfaces used by Components to access ATCS services.

Servo 1. A device that converts hydraulic pressure to mechanical movement, such as a brake-wheel cylinder.

2. A vacuum operated unit that attaches to the throttle linkage to move the throttle on a cruse control system.

Shackle A swinging support for the rear end of a spring that permits it to vary in length as it deflects.

Shackle assembly This assembly is attached to the front spring eye and bushing an is then mounted through a shackle bushing to the frame allowing the leaf spring to pivot up and down.

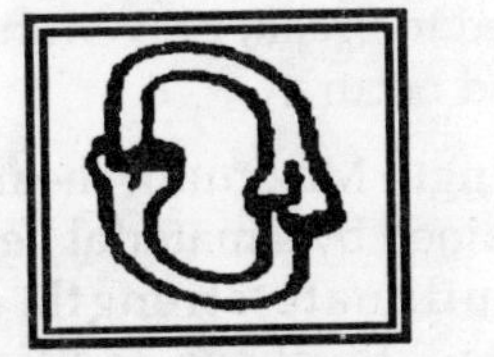

Fig. Shackle assembly

Shackle bushing Insulated bushings to help prevent the transfer of noise and road shock from the suspension to the chassis and vehicle interior.

Shade-tree mechanic 1. An amateur mechanic.

2. A mechanic with little or no knowledge of the trade.

3. A rural mechanic that literally works under a shade tree.

Shadow graph A scale using reflected light to indicate a difference in the weight of two parts.

Shaft packing The soft packing supplied by pump manufacturers. Most of these leaking packing are being replaced by mechanical seals.

Shaft seal An assembly that prevents vapour or fluid from escaping around the a rotating shaft.

Shaved 1. A vehicle body with the factory chrome trim removed and the holes filled in and painted.

2. A cylinder head that has been resurfaced.

Shear modulus of elasticity Tangent or secant modulus of elasticity of a material subjected to shear loading. Alternate terms are modulus of rigidity and modulus of elasticity in shear. Also, shear modulus of elasticity usually is equal to torsional modulus of elasticity. A method for determining shear modulus of elasticity of structural materials by means of a twisting test.

Shear pin A pin passing through two or more parts, as in securing a gear on a shaft, designed to break, preventing damage if an overload occurs.

Shear strength Maximum shear stress that can be sustained by a material before rupture. It is the ultimate strength of a material subjected to shear loading. It can be determined in a torsion test where it is equal to torsion strength. The shear strength of a plastic is the maximum load required to shear a specimen in such a manner that the resulting pieces are completely clear of each other. It is reported in Mpa (or N/mm^2) based on the area of the sheared edge.

Shielding gas Protective gas used to prevent or reduce atmospheric contamination while welding.

Shift forks A Y-shaped component located between the low/reverse, first/second, and third/fourth gears on the main shaft of a transmission that causes the gear to engage or disengage via the sliding clutches.

Shift interlock On a vehicle with automatic transmission, a safety device that prevents the driver from shifting out of park unless the brake pedal is depressed.

Shift kit Parts required to provide high performance of an automatic transmission shifter.

Shift lever 1. A lever used to change gears in a transmission.

2. A lever that moves the starter drive pinion in and out of mesh with the flywheel in some applications.

Shift rail A series of grooves in the forks to guide the shift forks, tension balls, and springs to hold the shift forks in gear, allowing them to interlock the rails to prevent the transmission from being shifted into two gears at the same time.

Shift tower The main interface between the drive transmission, a gearshift lever, pivot pin, spring, boot, and housing.

Shift valve 1. A valve body component acted on by oil pressure, allowing fluid flow to the involved band and/or clutch at the appropriate time, causing the transmission to up shift or downshift.

2. A type of hydraulic valve, consisting of lands and valleys that resemble a spool for thread. The lands seal orifices and the valleys open them.

Shift-bar housing A component that houses shift rails, shift yokes, detent balls and springs, interlock balls, and pin and neutral shaft; available in standard- and forward-position configurations.

Shifter A floor- or steering column-mounted lever on a motor vehicle used to select and/or shift the transmission gears.

Shim 1. A thin metal spacer used to align the clearance of a part.

2. A slotted strip of metal used to adjust the front-end alignment on many vehicles.

Shock absorbers Suspension device near each wheel that dampens the up-and-down movement of the vehicle. Inside a shock absorber, a piston rides up and down in a cylinder filled with thick fluid or compressed gas. The shock absorber counteracts the up-and-down movement allowed by the springs.

Shock compression The shock absorber in its shortened position, which occurs when the wheel moves upward.

Shock fluid Specially formulated hydraulic fluid used inside of shock absorbers.

Shock foaming The mixing of air and shock fluid, due to rapid movement of fluid between the chambers, causing the shock absorber to develop a lag because of piston is moving through an air pocket that offers up resistance. A gas-filled shock absorber is designed to reduce oil foaming.

Shock hydraulic principles Fluid is forced through orifices and/or valves at a controlled rate to provide the desired dampening effect.

Shock mounting position The direction and/ or angle at which a shock absorber is mounted: vertical, horizontal, or slanted inward at the top.

Shock mounts The rubber isolating bushing or grommets attached to the upper shock-mounting piston rod and the lower mounting cylinder tube in which the piston operates.

Shock operational check A method used to check shock efficiency that includes bouncing the vehicle bumper vigorously and observing the shock dampening action, or pumping the brakes slowly at low speed to see if the vehicle "rocking" motion is set up.

Shock piston The component attached tot he bottom of the piston rod containing the rebound valve that moves back and forth inside the inner cylinder.

Shock rebound The rebound travel when the shock absorber is in its lengthened position, which occurs when the suspension or spring moves downward.

Shock-absorber lag The incorrect operation of a shock absorber because of aeration due to the mixing of air with oils, causing the shock absorber to produce a poor ride.

Shock-absorber ratio A rating of shock-absorber extension control compared to the amount of compression control, varying from 50/50 to 80/20.

Shock-absorber strut assembly In a McPherson strut, the independent rear-suspension system that includes a rubber isolated top mount, upper and lower spring seal, coil spring insulator, and coil spring.

Shoe anchor The large pin, or post, or block against which a drum brake shoe pivots or develops leverage.

Shoe box Any car manufactured in the U.S. having a non-aerodynamic, box-like shape built from the late 1940's through the 1950's.

Short-circuit A direct connection that provides a virtually zero resistance path for current.

Short-circuit current (SCC) The initial value of the current obtained from a power source in a circuit of negligible resistance.

Short-circuit protection A protective feature that limits the output current of a power supply to prevent damage.

Should cost The most likely cost of developing and producing a product/system in accordance with specifications, with only nominal design changes resulting from maturation rather than requirements changes, and with normal allowances for scrap and rework.

Shrink wrap A shrink-fit tubing used to protect wires, wire splices, and terminals of an automotive electrical system.

Shrink-fit tubing An insulated tubing that shrinks to about half its original diameter when heated.

Shrinking hammer A special hammer that is used to shrink spots that have been stretched by excessive hammering.

Shroud 1. A metal or plastic duct that directs ambient air to the radiator cooling fan.

2. A hood-like device placed around an engine fan to improve air flow.

Shrouding An obstruction, such as carbon buildup, around a valve in the combustion chamber that interferes with the proper air flow.

Shunt 1. A parallel conducting path in a circuit.

2. A low value precision resistor used to monitor current.

Shunt regulator A linear regulator in which the control element is in parallel with the load, and in series with an impedance, to achieve constant voltage across the load.

Siamese 1. To exhaust pipes joined together.

2. Cylinders in an engine block that are cast so closed together that there is no room for a coolant passage between them.

3. Two adjacent valves in a cylinder head that is served by a single port.

Signal generator A test instrument for generating standard signals, used to analyze circuit performance.

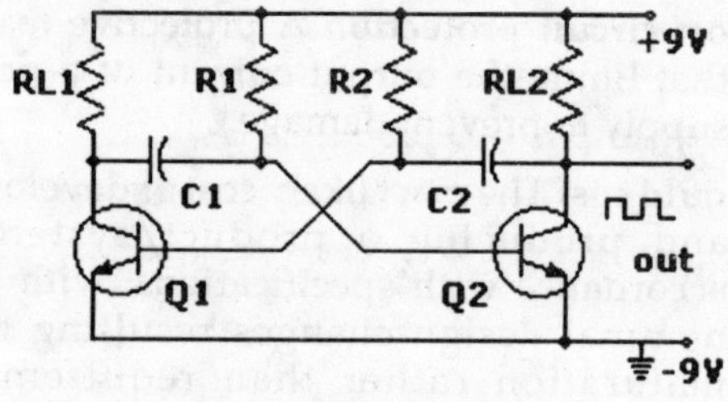

Fig. Signal generator

Signal ground The common return or reference point for analogue signals.

Signal In electronics, the information contained in electrical quantities of voltage or current that forms the input, timing, or output of a device, circuit, or system.

Signal integrity The accurate reconstruction of a signal, determined by the systems and performance considerations of an oscilloscope, in addition to the probe used to acquire the signal

Signal Source A test device used to inject a signal into a circuit input, The circuit output is then read by an oscilloscope.

Signal to noise ratio Signal to noise ratio as used in the design of experiments (doe) is the mathematical equation that indicates the magnitude of an experimental effect divided by the experimental error due to chance fluctuations. This metric is used in Taguchi methods, which is a form of doe.

Silica gel A drying agent used in many automotive applications, such as an air-conditioner desiccant because of its ability to absorb large quantities of water.

Silicon A element with the atomic number of # 14

2. one of the elements known as semiconductors that have exactly 4 electrons in there valance shell used to create semiconductor devices.

Silicon-controlled rectifier (SCR) A uni-directional, four-layer (PNPN) junction device in which conduction is initiated by the application of a gate current. Conduction will continue until the current is reduced to some minimum value.

Silverline A tool manufacturer producing the most popular Mity-Vac vacuum tool.

Similar window Pertaining to engine operation in which RPM and load are within predetermined percentages, Conditions in which certain monitors must pass to extinguish the MIL and erase DTC"s

Simple planetary gear set Gear set with a sun gear, planetary pinions, planetary carrier, and ring gear.

Simpson gear set A gear set having two simple planetary gears mounted on a common sun gear.

Simulation 1. A device, system, or computer program that represents certain features of the behaviour of a physical or abstract system. Vendors of planning and scheduling, forecasting and demand management, and other types of decision-support systems make growing use of simulation to compare the consequences of alternative courses of action.

2. A CAD/CAM program that simulates the effect of structural, thermal, or kinematic conditions on a part under design.

3. Within MRP II, utilizing a mathematical model to test various design configurations.

Simulator A device that mimics the operational conditions of equipment or vehicles.

Simultaneous Sampling When all analogue signals are read simultaneously. This is achieved by providing each input with its own A-D converter, and initiating sampling from a single clock. It ensures that there is no reduction in sampling rate when more signals are connected.

Simultaneously Fired A type of fuel delivery system in which the fuel injectors are fired in groups once per engine revolution.

Sine wave Waveform of a single frequency, indefinitely repeated in time. In practice there must be a transient at the start and finish of such a wave.

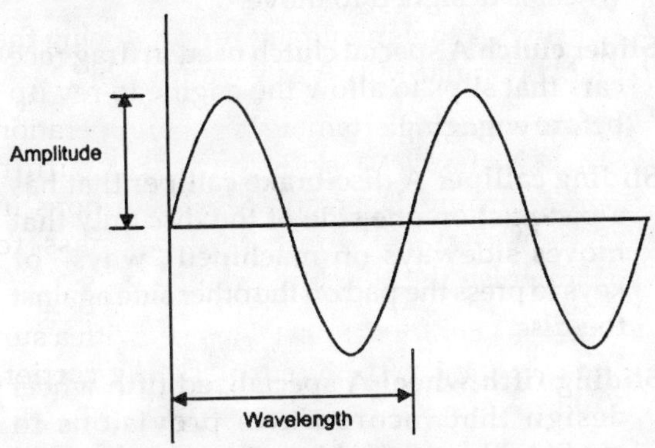

Fig. Sine wave

Single chamber capacity The measure of the maximum volume or displacement of the rotor chamber of a Wankel engine.

Single leaf spring A spring having one leaf that may be tapered thinner and wider toward the ends, allowing a variable flexing rate.

Single minute exchange of dies (SMED) The reduction in die set-up time. Set-up in a single minute is not required, but used as a reference.

Single overhead cam (sohc) An engine with a single overhead cam generally has one intake and one exhaust valve per cylinder; the single cam opens and closes both valves.

Single wrap A transmission brake with a circular steel strap that is lined internally with friction material.

Single-board engine controller (SBEC) A single microprocessor used to control engine functions.

Single-ended input An analogue input that is measured with respect to a common earth. Single ended inputs are only suitable for signals that are of good size - 100 mV full scale or above.

Single-plane crankshaft A crankshaft having throws on the opposite side of the same plane, at 180 degrees.

Single-plane manifold An intake manifold having a single plenum between the carburettor and intake ports.

Single-wire circuit A circuit using a single wire that relies n the metal structures of the vehicle, such as the frame or body, as a ground.

Sintered A somewhat porous metallic bearing formed by pressing particles of powdered metal to a temperature at which the adhere to each other.

Six-stroke cycle The stroke cycle of an experimental engine by Toyota where combustion occurs every third revolution of the crankshaft.

Sixty foot time The time it takes to cover the first sixty feet (18.3 meters) from the starting line in a drag race.

Skid control system A system designed to respond to a locking wheel by relieving hydraulic pressure to the locking brake.

Skid pad A flat area of pavement with a painted circle 300 feet (91 meters) in diameter used to determine a vehicle's lateral grip and lateral acceleration.

Skid plate A shield under the power train of an off-road vehicle to protect the engine and transmission.

Skidding A term used in describing how an impeller slides on the solids in the bottom of a mixer.

Skin 1. The surface hardening of RTV or some body fillers.

2. The outer sheet metal of a vehicle.

Skin effect 1. The tendency of current to crowd toward the outer surface of a conductor; increases with conductor diameter and frequency.

2. A thin, unburned layer of air/fuel mixture next to the combustion camber surface.

Skunk works A collocated project environment to shorten communication paths and to keep functional contributors close to one another and to the prototype or production center. Skunk works often have streamlined procedures and work processes that rely on informal communication and coordination to reduce bureaucratic overhead.

Slalom A contest of speed and manoeuvrability through a tight course marked off with pylons.

Slam the door 1. Beating another vehicle in competition.

2. Preventing a competitive vehicle from passing.

Slant six A six-cylinder engine design offered on some Chrysler car lines in the 1960's and 1970's.

Slave A power supply which uses the reference in another power supply, the master, as its reference.

Slave Cylinder 1. A device in the hydraulic clutch system which in activated by hydraulic force disengaging the clutch

2. In the brake system it is a device for transferring hydraulic energy from the master cylinder to the brake pads or shoes

Slave valve An air-pressure operated device that helps to protect gears and components in the transmission's auxiliary section by permitting range shifts to occur only when the transmission's main gearbox is in neutral.

Sleeve bearing A non precision bearing. Usually manufactured from carbon, Teflon. brass etc., Allows too much axial and radial movement for most mechanical seal applications.

Slewing rate The maximum rate of change a power supply output can produce when subjected to a large step response or specified step change.

Slide hammer A metal rod with a tool at one end, and a heavy weight that slides on the rod to pull, or push, the tool.

Slide travel The distance that a sliding fifth wheel is designed to move.

Slider clutch A special clutch used in drag race cars that slips to allow the engine to rev up before engaging.

Sliding calliper A disc-brake calliper that has a piston(s) on one side of the disc only that moves sideways on machined "ways" or keys to press the pad on the other side against the disc.

Sliding fifth wheel A specialized fifth wheel design that incorporates provisions to readily relocate the kingpin center forward and rearward, which affects the weight distribution on the tractor axles and/or overall length of the tractor and trailer.

Sling A loop of rope, cable, or chain used in hoisting heavy material.

Slinger A metal disc attached to the crankshaft to keep engine oil away from the front seal.

Slip angle The difference in the path the wheels follow during a turn compared to the actual direction they are pointing, caused by centrifugal force at higher speeds.

Slip joint A term used for lip yoke.

Slip stick An alternating slipping and sticking of the seal faces caused by a poor lubricant between the faces. Will cause vibration problems at the seal face unless the vibration is dampened.

Slip stream A partial vacuum that is created behind a vehicle travelling at a high speed.

Slipout A condition that occurs with a tractor-trailer when pulling with full power or decelerating with the load; tapered or worn clutch teeth tend to "walk" as they rotate, causing the sliding gear and clutch to slip out of engagement.

Slipper skirt A term used for slipper skirt piston.

Slipper skirt piston A piston that has a cutaway skirt so that the piston can come closer to the counterweights reducing the overall size of the engine. Also known as slipper skirt.

Slow reverse A simple planetary gear combination with sun gear as input, planetary carrier held, and ring gear as output.

Slow start A feature that ensures the smooth, controlled rise of the output voltage, and protects the switching transistors from transients when the power supply is turned on.

Slugging 1. The return of liquid refrigerant or oil to the compressor.

2. The act of adding a separate piece (s) of material in a joint before or during welding that results in a welding joint not complying with design, drawing or specification requirements.

Smog Air pollution, especially the photochemical variety formed when sunlight causes a chemical reaction between NOx and HC, resulting in the formation of ozone and other compounds smog is a conjunction of smoke and fog.

S-N diagram Plot of stress (S) against the number of cycles (N) required to cause failure of similar specimens in a fatigue test. Data for each curve on an S-N diagram are obtained by determining fatigue life of a number of specimens subjected to various amounts of fluctuating stress. The stress axis can represent stress amplitude, maximum stress or mini-mum stress. A log scale is almost always used for the N scale and sometimes for the S scale.

Snap ring A circular retaining clip used inside or outside a shaft or part to secure a shaft, such as a floating wrist pin.

Snapshot A feature of OBD-II that shows, on various scanners, the conditions that the vehicle was operating under when a particular trouble code was set.

Sneak analysis Analysis of modes of operation unanticipated during design that result in unexpected system behaviour and potential failure.

Snell Foundation An organization that sets safety standards for racing helmets that are adhered to by most race-sanctioning bodies.

Snipe A pipe that is placed on a wrench to increase leverage; generally an unsafe practice.

Snorkel tube A long, narrow tube attached to the air cleaner, used to direct air into the air filter.

Snotty A slippery surface.

Snow-and-mud tire A tire, identified with an "MS" suffix, having treads designed to provide traction when driving in mud or snow; available in various ply and belt designs.

Snubber An RC network used to reduce the rate of rise of voltage in switching applications.

Snubber network A circuit that uses a RC network and a diode in unipolar switching applications.

Sodium hydroxide A caustic soda that makes a good parts cleaner when mixed with water.

Sodium silicate A white, grey, or colourless compound sometimes used to seal small cracks or leaks in the cooling system.

Sodium-cooled valve A partially hollow valve containing metallic sodium that melts at a low temperature (208°F) 97.8°C and when in its liquid state at operating temperatures splashes around inside the valve, transferring heat away from the valve head.

Soft foot The tendency for one of the pumps feet to rise when the hold down bolt is loosened. Very bad for alignment.

Soft plug A term used for core plug.

Soft solder 95/5 A lead-free metallic alloy of 95% tin (Sn) and 5% antimony (Sb); used to repair or join ferrous metal parts for temperatures below 350°F (176°C).

Soft systems methodology Soft systems methodology (ssm) assumes differing viewpoints between individuals regarding a problem. It tries to move toward consensual action between these conflicting views. Ssm is a goal-driven, iterative process with a philosophy of continual improvement for which process is more important than the result. Ssm requires a facilitator to provide an unbiased viewpoint and is indicated for use when the facing a complex, organizational problem. Ssm is useful for providing a structure for understanding complex programs.

Solder A filler metal used in joining two or more parts that has a liquid state not exceeding 840°F (450°C).

Soldering A welding process that produces consolidation of materials by heating them to the proper temperature and using a filler metal having a liquidus not exceeding 840°F (450°C) and below the solidus of the base metals.

Sole Society of logistics engineers

Solenoid 1. An elecrtomechanical device similar in operation to a relay, however, the movement of the armature or iron core results in changing electrical energy into mechanical energy.

2. An electro-mechanical device used to effect a push-pull mechanical operation using electrical current.

3. A rely with an external mechanical movement.

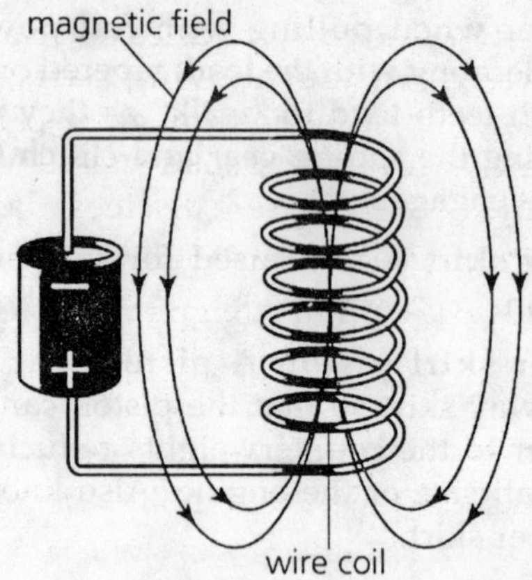

Fig. Solenoid

Solenoid relay A relay that connects a solenoid to an electrical circuit, such as a starter-motor solenoid relay.

Solenoid switch An electrical switch that is opened and closed mechanically by the movement of a solenoid core.

Solenoid valve A mechanical valve that is operated by a solenoid to control the flow of a liquid or a gas. Used in braking systems, air conditioning systems, emission control systems, cruise control systems and even suspension and steering systems.

Solid modelling Three-dimensional modelling in which solid characteristics of an object are built into the database so that complex internal structures can be realistically represented.

Solid Oxide 1. A solid combination of oxygen and zirconium that serves as an electrolyte for a fuel cell.

Solid propellant A rocket propellant in solid form; usually consisting of a mixture of fuel and oxidizer.

Solid rocket booster A rocket, powered by solid propellants, used to launch spacecraft into orbit.

Solid rotor A rotor that is a solid piece of metal with a friction surface on each side.

Solid state device Any device that can control current without moving parts, heated filaments, or vacuum gaps. Thus, all semiconductors are solid-state devices but some solid-state devices are not semiconductors (for example, transformers)

Solid state ignition An ignition system using diodes and transistors to control spark timing.

Solid-axle suspension Suspension system in which the wheels are mounted at each end of a solid, or undivided, axle or axle housing.

Solids modelling A geometric modelling method that completely and unambiguously describes both the exterior and interior of a part or assembly in three dimensions (geometry, topology and mass properties).

Solid-state device An electronic component constructed from any type of semiconductor material

Solvent 1. A liquid substance, such as water, in which other substances can dissolve.

2. A petrochemical liquid that will dissolve oil and grease.

3. A paint cleaner and thinner.

Sonic testing A procedure for testing the integrity of engine blocks, using sound waves.

SOP 1. An acronym for standard operating procedure.

2. An acronym for seat of the pants.

SOP rally A competitive event in which instruments cannot be used by the contestant to check time or distance.

Sorbent Material that sorbs another; i.e., has the capacity or tendency to take it up either by adsorption or absorption.

Spacer A device, such as a shim or washer, that is used to increase the space between to mating surfaces for parts.

Span The width of the metal bellows. The distance from the inside diameter to the outside diameter.

Spanner The British term for wrench.

Spare 1. Not in regular use or immediately needed.

2. Extra or reserve.

3. A term used for spare tire.

Spare tire A full-sized replacement tire or a compact space-saver tire, generally for emergency service, but available for use when needed.

Spark advance curve The rate at which ignition timing advances as plotted on a graph; the line rises from some initial amount of advance and levels off at the maximum advance.

Spark advance The moving ahead of the ignition spark in relation to the piston position.

Spark Amplitude The greatest voltage created during a spark event (the height of the firing line)

Spark decel valve A vacuum valve, located in the line between the distributor and carburettor, to advance the spark during deceleration, to reduce emissions.

Spark delay valve (SDV) A vacuum valve acting like a restrictor, used in the vacuum line between the distributor and carburettor, to delay vacuum-timing advance under certain driving conditions to reduce NOx (oxides of nitrogen) emissions.

Spark duration The time a spark is established across the gap of a spark plug.

Spark ignition (SI) An engine-operating system where the air/fuel mixture is ignited by an electrical spark.

Spark line The line on an oscilloscope that indicates the voltage required to fire the spark plug and the number of degrees the distributor turns while the spark exists.

Spark Output The output signal from the processor which triggers the Ignition module to fire the ignition coil

Spark plug An ignition component threaded into the cylinder head that contains two electrodes extending into the cylinder that form a gap across which high-voltage electricity arcs to ignite the compressed air-fuel mixture.

Spark plug fouling An accumulation of deposits on the lower, exposed end of the spark plug that act as an electrical conductor, thereby creating a path for electricity to leak to ground rather than jump across the electrode gap.

Spark plug heat range The temperature limits, hot, normal, or cold, within which a spark plug is designed to operate. It is determined by varying the length of the exposed lower section of the plug ceramic insulator.

Spark plug well The recess in a cylinder head for a spark plug.

Spark plug wire A special high-voltage wire from the distributor cap to the spark plug.

Spark plug, resistor type A spark plug having an electrode resistance of 5,000 to 10,000 ohms to increase electrode life and suppress radio interference.

Spark Retard Causing less spark advance to be added, resulting in a spark which is introduced later on the compression stroke

Spark test A quick check of the ignition system made by carefully placing a metal end of one of the spark plug wires close to the engine while cranking the engine to see if there is a spark gap and to determine its intensity.

Spatter paint The technique of applying two separate colours of paint simultaneously to provide a speckled finish.

Spatter The metal particles expelled during fusion welding that do not form a part of the weld.

SPC Statistical Process Control is a quality-control method focusing on continuous monitoring of the process rather than the inspection of finished products, with the intent to achieve control of the process and eliminate defective product.

Spec car 1. A term used for specification car. 2. One car in a group of identical specification cars that form their own racing category with a focus on driving skills.

Special A high-performance, individually built car, such as a prototype.

Special causes (of variation) Special causes are causes of variation in output from a manufacturing process or system of procedures that is not due to the inherent operation of the process or system itself (common causes), but is due to the intrusion into the system of a one-time or external cause of variation. One-time or external causes do not spring from the system, and so are preventable - i.e., their occurrence can be prevented. Consequently, the reason for each special cause must be investigated and steps then taken to see that it does not occur again. The presence of a special cause of variation must be determined statistically. This is done by knowing that variation in output due to common causes follows a regular pattern corresponding to the normal curve - i.e., with an average and a deviation on either side of the average within three standard deviations. Variation due to a special cause results in a performance outside these statistical limits.

Special characteristics Product and process characteristics designated by the customer; governmental, regulatory or safety agencies; and/or the supplier through knowledge of the product or process.

Special services A category that describes plants which add value to automotive parts manufactured by others by using process such as painting, plating, heat treatment, etc.

Specialty service shop A repair shop that specialized in certain vehicle components, such as engine rebuilding, brake repair, radiator repair, and so on. Also known a specialty shop.

Specialty tire Any of several types of tires such as, all-season tire, puncture-sealing tire, snow-and-mud tire, or studded tire.

Specific Gravity A measure of the weight of a liquid. Fresh water at 4°C (39°F) is given a value of one. If the liquid you are questioning will float on water the specific gravity is less than one. If it sinks, it is higher than one. Density is a better term.

Specific heat The quantity of heat required to change one pound of a substance by 1o F.

Specific impulse Parameter for rating the performance of a rocket engine. Indicates how many pounds or kilograms of thrust are obtained by consumption of a pound or kilogram of propellant in one second.

Specific speed A formula that describes the shape of a pump impeller. The higher the specific speed the less N.P.S.H. required.

Specification car One car in a group of identical cars that form their own racing category with a focus on driving skills. Also known as spec car.

Specifications 1. The document that prescribes the requirements with which the product or service has to conform.

2. As used with qfd, specifications are the particular measures or metrics to define a product requirement. Synonymous with target value in this context.

3. Specifications are boundaries, usually set by management, engineering, or customers, within which a system must operate. They are sometimes called engineering tolerances.

Spectrometer An optical instrument that splits the light received from an object into its component wavelengths by means of a diffraction grating; then measuring the amplitudes of the individual wavelengths.

Spectroscopy The study of the production, measurement and interpretation of electromagnetic spectra.

Speed Density A method of calculating airflow by sensing engine load and intake manifold pressure and other engine sensors.

Speed flare up An operating condition where the engine speeds up without an increase in vehicle speed.

Speed of light Light waves travel at a speed approximately equal to 186,000 miles per second, or 10 ft. Per nanosecond. It is the limiting speed of the universe

Speed of sound In air it is 1090 feet per second, in water it is 4800 feet per second.

Speed rating A tire rating that indicates the maximum safe vehicle speed that a tire will withstand.

Speed ratio A comparison of the difference in speed between two moving parts such as impeller speed and turbine speed.

Speed sensor An electrical device that can sense the rotational speed of a shaft or member and transmit this information to another device, such as a readout.

Speed shift 1. To up shift a manual transmission without releasing the accelerator.

2. The art of up shifting a manual transmission without using the clutch.

Speedometer An instrument, usually dash mounted, used to measure the speed of a vehicle.

Spice Simulation program with integrated circuit emphasis - one of the most widely used analogue circuit simulation programs.

Spider A set of gears in the differential that allow the rear wheels to rotate at different speeds as the vehicle is cornering.

Spike Short pulse of voltage or current - usually undesirable.

Spin on filter An oil filter having a threaded attachment for the ease and speed of replacement.

Spin stabilization Spacecraft stabilization accomplished by rotating the spacecraft mass, thus using gyroscopic action as the stabilizing mechanism.

Spin To skid out of control by 180 degrees or more.

Spindle 1. A shaft or stub axle upon which the wheel hub and bearing rides.

2. A shaft which carries either cutting tools or work that is being machined.

Spinner A term used for knock-off hub.

Spinning balancer A balancer that rotates the tire and wheel to determine the amount and location of any unbalanced condition.

Spiral bevel gears A differential ring gear with helical gear teeth.

Spiral development model The spiral development model combines the waterfall development model and the prototype approach. It consists of a series of partial implementations or releases of the product. This approach is useful when the risks are significant, there is a need/opportunity to field a partial system in a short amount of time, and the requirements are not completely understood or can change over time. Key assumptions with the spiral development model are a) the initial release is sufficient to key system stakeholders that they will continue to participate in its evolution; b) the architecture of the initial release is scalable to accommodate the full set of system life cycle requirements; c) and the users/customers are sufficiently flexible to adapt to the pace of system evolution.

Spiral spring Springs formed from flat strip of wire wound in the form of a spiral, loaded by torque about an axis, normal to the plane of the spiral.

Spiral-grooved shock absorber An arrangement that is used to reduce the mixture of air with the fluid as it passes through the valves by breaking up the air bubble and, at the same time, reducing lag.

Splash lubrication A non-pressurized system of lubrication, same as splash-feel oil system.

Splash shield A stamped sheet-metal deflector plate located behind the disc-brake rotor providing component protection from dirt and water and improving the flow of air over the rotor.

Splash-feel oil system A type of engine lubrication system in which oil is splashed onto the engine parts to be lubricated.

Splined discs Multiple-disc clutch components splined internally to the side gears or over a special hub on single-pack types.

Splined yoke A consideration to allow the driveshaft length to increase and decrease to accommodate the movement of the rear axle.

Splines The internal or external longitudinal grooves in a gear or shaft that mesh when assembles, causing the gear to turn with the shaft, but allow lateral movement.

Split bobbin winding The method of winding a transformer whereby the primary and secondary are wound side-by-side on a bobbin with an insulation barrier between the two windings.

Split brake system A service-brake system having tow or more separate fluid, electrical, mechanical, or other circuits. If one circuit fails, full or partial brake actuating capability is retained.

Split crankcase An oil pan split horizontally, in the same plane as the crankshaft, into two or more pieces to simplify service.

Split guide ring Part of a torque converter assembly designed to reduce fluid turbulence and improve efficiency.

Split point The pressure at which a proportioning valve closes during brake application and reduces the rate at which further pressure is applied to the rear drum brakes.

Split sump A term used for split crankcase.

Split torque converter A design having a simple planetary gear set to divide engine torque between mechanical and hydraulic operation.

Split valve guide A two-piece valve guide used in some older flathead engines with mushroom valves.

Spoiler 1. An aerodynamic device that "spoils" the airflow over the vehicle.

2. A device attached below the front bumper to reduce drag by deflecting air away from the vehicle.

3. A device mounted on the rear deck to provide a downward force.

Spoke The wire-like bracing between the hub and the rim of a wire wheel.

Sponge lead Lead with many small holes used to make a surface porous or songlike for use in battery negative plates.

Spongy brake pedal A condition where the brake pedal is not solid when depressed, but bounces softly. This is caused by air in the hydraulic lines, distortion or stretching of the connecting parts, or swelling of the hydraulic hoses.

Spongy pedal A term used for spongy brake pedal.

Spontaneous combustion The process by which a material ignites and burns by itself.

Spook To psych, distract, or unnerve a competitor at the start of a race.

Spool 1. A final drive without any differential action.

2. A term used for locked rear end.

Spool piece Usually refers to a seal and bearing combination that is supplied for some mixer applications. It depends upon the application, but this is not a good idea if there are already two anti-friction bearings already installed on the shaft

Spot weld A weld made between or upon overlapping members in which bonding may start and occur on the mating surfaces or may proceed from the outer surface of one member.

SPOUT Spout is the abbreviation for Spark Output which is the output line from the Electronic Control Unit to the Ignition Module.

Sprag 1. A figure-eight-shaped locking element of a one-way sprag clutch.

2.A one-way clutch used in an automatic transmission.

Sprag clutch 1. A device containing numerous oblong parts, called sprags, that operate by tilting between an inner and outer race to lock up when the outer race is turned in one direction and to slip when it is turned in the opposite direction.

2. A one-way clutch having cam-profiled locking elements that engage cylindrical outer and inner races.

Spread tandem suspension A two-axle assembly in which the axles are spaced to allow maximum axle loads under existing regulations, usually more than 55 inches (140 cm).

Spring 1. Small steel coils that are used to close the intake and exhaust valves when the cam lobes release pressure on the valve stem.

2. A steel or composite elastic leaf- or coil-like device that compresses as it absorbs energy and returns to its original position when it releases that energy.

Spring back The tendency of a material to return to its original shape or near its original shape after being bent.

Spring bind A term used for coil bind.

Spring break chamber 1. A device used with brake-foundation assemblies as a fail safe unit that automatically applies the truck's service brakes in the event of an air loss.

2. A device used as a parking brake and is used with cam and wedge-type foundation brakes.

Spring break valve 1. A device that limits the hold-off pressure to the spring-brake chambers via a relay valve or quick release valve.

2. A device that provides a modulated spring-brake application to the front axle proportional to the service braking pressure whenever a loss of pressure occurs.

Spring energized A one-way roller clutch that has the rollers held in contact with the cam and race by individual springs, or by means of a spring-actuated cage.

Spring eye The main spring-leaf end that is formed in a O-shape for placement of the rubber spring-mounting bushings.

Spring hanger The vehicle-frame bracket for the eye at the front of the rear leaf spring.

Spring leaf A long, flat section of spring steel making up all or part of a leaf spring.

Spring leaf insert Small, replaceable composition pads placed between the spring leaves near the ends to aid in slippage between the leaves while they are flexing.

Spring load A measure, in pounds or kilograms, of how much weight a spring can support, generally at the installed height.

Spring oscillation Continued compression and rebound motion of a spring, after the wheel has encountered a bump or hole on the road surface, which diminishes gradually, depending on the condition of the shock.

Spring pin A small piece of metal rolled in a pin-like manner that is used to dowel small parts that may occasionally have to be disassembled and reassembled.

Spring rate The relationship of spring deflection to lead applied, such as the amount of weight, in pounds per inch or newtons per millimeter, required to deflect the rear spring.

Spring retainer The element which locates and provides reaction for the spring in a one-way roller clutch.

Spring sag The loss of spring load due to overloading and/or metal fatigue.

Spring shackle A small, swing-arm attachment at the rear of the leaf spring to allow the spring to flex.

Spring walk The tendency of a valve spring to bounce around in its seat at high-engine speed.

Spring windup The slight S-shape assumed by the leaf spring during extreme acceleration and braking that may be controlled by traction bars on high-performance vehicles.

Springs, torsion bar A long straight bar that is fastened to the frame at one end and to a control arm at the other. Spring action is produced by a twisting of the bar.

Springy thingy A dragster that has a light, flexible structure to allow for maximum weight transfer.

Sprocket A sheet-like disk with teeth around its outer perimeter that mesh with a belt or chain.

Sprocket pitch The dimension between the centers where the rollers would be bedded against the bottoms of adjacent tooth spaces.

Sprocket-pitch diameter The pitch diameter of a sprocket used in a synchronous belt drive that coincides with the belt pitch line and is always greater than the sprocket outside diameter.

Sprung weight The mass of the vehicle that is supported by the springs, including the body, engine and transmission.

Spun bearing Any bearing on the crankshaft that has seized on the journal and turned in the housing bore.

Spur gear A transmission or differential gear having teeth cut straight across its face, parallel to the rotational axis.

Spyder 1. A light, two-person, horse-drawn carriage.

2. A light, two-person roadster.

Square wave A repetitive waveform whose shape is essentially square or rectangular, usually with equal on and off times.

Square-cut piston seal A fixed seal for a calliper piston that has a square cross section.

Squeegee 1. A flexible rubber block used to apply glazing putty and light coats of body filler.

2. A metal-backed rubber blade having a handle used to clean windshields.

Squib A pyrotechnic term used for a fire cracker that burns but does not explode. The squib starts the process of air bag deployment

Squirt hole 1. A hole in the side of a connecting rod in an OHV engine which squirts oil toward the camshaft.

2. A hole in the pin end of a connecting rod to squirt oil to the underside of a piston for cooling.

Squish The action where some compressed air/fuel mixture is pushed out of a decreasing space between the piston and cylinder head of the combustion chambers in some engines.

Sr 1. Strontium An element with the atomic number of 38

2. An acronym for street roadster

Stability 1. The percent change in output parameter as a function of time, with all other factors constant, following a specified warm-up period.

2. The ability to stay on a given frequency or in a given state without undesired variation.

Stabilize bar link A device that connects the lower control arm to the stabilizer bar.

Stabilizer A device that uses the torsion resistance of a steel bar to reduce the roll of a vehicle and prevent too great a difference in the spring action at the two front wheels.

Stabilizer bar A long, spring-steel bar attached to the cross member and interconnects the lower control arm that twists like a torsion bar during turns to transmit cornering forces from one side of the vehicle to the other to help equalize wheel loads and prevent excessive leaning.

Stage A portion or phase of the product development process with a clear objective of milestone that ends with a stage-gate review before authority is granted to proceed with the next stage or phase.

Staged turbos turbocharger in series, one feeding the other.

Stage-gate process A widely employed product development process that divides the development effort into distinct time-sequenced stages or phases separated by management decision gates. Product teams must successfully complete a prescribed set of related activities in each stage prior to obtaining management approval to proceed to the next stage of product development. The framework of the stage-gate™ process includes work-flow and decision-flow paths and defines the supporting systems and practices necessary to ensure the process's ongoing smooth operation.

Staggered timing camshaft A camshaft ground so as to provide a longer duration for the cylinders farthest away from the main intake tract to equalize the amount of fuel mixture reaching all of the cylinders.

Staging area The area between the pits and starting line in drag racing where the cars are lined up.

Staging lane 1. Lanes within the staging area where cars are grouped according to class.

Staging lights The light beams used in drag racing to guide the front wheels of a car into proper position.

Stainless steel Alloy steels containing a high percentage of chromium.

Staking 1. Securing a part, such as a bearing, in place by exerting pressure around the opening by punching or peening.

2. To punch or chisel an indentation into two touching parts so that one of the parts cannot turn against the other.

Stall 1. A work area.

2. To come to a stop because the engine failed.

3. The disruption of air flow around the air foil.

4. A condition where the engine is operating and the transmission is in gear, but the drive wheels are not turning because the turbine of the torque converter is not moving.

Stall speed The maximum possible engine and torque converter impeller speed, measured in RPM, with the turbine held stationary and the engine throttle wide open.

Stall test A starter-motor test of the current draw made when the motor is stalled.

Stall-torque test A starting system test that determines pound-feet or newton-meters of torque developed by the armature when restrained from rotating by a special measuring instrument attached to the drive pinion gear.

Stamping A process technology which manufactures automotive parts by shaping rolled sheet metal or by bending or stretching it in a sequence of purpose-built tools fitted to a general purpose press.

Stand off The fine mist of air/fuel mixture that may be forced back into the carburettor during timing overlap when the intake and exhaust valves are open.

Stand on it 1. To hold the accelerator to the floor.

2. A popular novel based on actual events credited to a fictitious driver, Stroker Ace, and believed to be hilarious by most race enthusiasts.

Stand-alone Refers to a device that is self-contained, one that does not require any other devices to function.

Standard bolt-and-nut torque specifications A chart showing the standard torque recommendations for standard English and metric sizes of bolts and nuts.

Standard cost Standard cost is the predetermined or planned cost of manufacturing a single unit or of providing a single unit of service. It represents a goal or baseline that is used to project cost, based on experience and/or analysis.

Standard Mandatory requirements employed and enforced to prescribe a disciplined uniform approach to software development, that is, mandatory conventions and practices are in fact standards.

Standard operating procedures (SOP) 1. The way things are generally done.

2. The routine of things.

Standard temperature and pressure (STP) A set of engineering reference conditions for dealing with gases with temperature equal to 273.15°K and pressure equal to 101.325 Pascal.

Standard transmission A transmission that must be manually shifted into the various gears.

Standard work Specifying tasks to the best way to get the job done in the amount of time available while ensuring the job is done right the first time, every time.

Standard work week Average number of hours worked normally scheduled in a work week.

Standardization Standardization or parts, materials, modules and assemblies makes possible the interchangeability of these items among products, resulting in higher volume production and purchasing, lower investment in inventory, easier purchasing and material handling, fewer quality inspections, and less difficulties in production.

Standby Time A period of time, other than non-scheduled time, when the equipment is in a condition to perform its intended function, facilities are available, but it is not operated. This includes time when no operator is available, time when no product is available (no boards or components), and waiting on upstream or downstream equipment.

Standing kilometer The international standards for acceleration records for a kilometer taken from a dead stop over a measured distance.

Standing mile The international standards for acceleration records for a mile taken from a dead stop over a measured mile.

Standing quarter mile 1. The length of a typical drag-race course.

2. The acceleration from a dead stop to the end of a measured quarter mile.

Standpipe A type of check valve that prevents a reverse flow of the lubricant that becomes liquid due to heat generated during operation. It prevents the loss of lubrication at start up when the lubricant has cooled and is solidus.

Star connection The interconnection of 3 electrical-equipment windings in star or wye fashion, as in an alternator.

Star wheel adjusting screw A toothed wheel that is manually or automatically rotated to expand the brake shoes outward as the lining wears.

Starter An electric motor powered by the battery that turns the crankshaft before the pistons begin operating.

Starter circuit The electrical system that carries the low and high current power to the starter components required to start the vehicle.

Starter Clutch An overrunning clutch assembly used to prevent starter damage when the flywheel begins to turn faster than the pinion gear

Starter drive The part of the starter motor that engages the armature-mounted pinion gear to the engine-flywheel ring gear.

Starter motor A small electric motor used to turn the engine for starting.

Starter relay A magnetic switch, generally operated by the ignition switch, that uses low current to close a circuit to control the flow of very high current to the starter.

Starter ring gear A relatively large diameter starter-driven gear attached over the torque converter or flywheel flex-drive plate.

Starter safety switch An electrical switch that prevents the starting of a vehicle with an automatic transmission while in gear.

Starter solenoid A magnetic switch, energize by the ignition switch, that shifts the starter-drive pinion gear into the flywheel ring gear, then closes the electric circuit to the starter.

Starving A condition where: 1. The evaporator does not get enough refrigerant to properly function.

2. The engine does not get enough fuel and/or air to function.

State of charge The degree or the amount in percentage that a battery is charged. A fully charged battery would be 100% charged

State of the art 1. A phrase that implies being up-to-date in technology.

2. Pertaining to the latest technology.

3. Using the latest scientific or engineering developments.

Statement of work (SOW) A document primarily for use in procurement, which specifies the work requirements for a project or program. It is used in conjunction with specifications and standards as a basis for a contract. The SOW will be used to determine whether the contractor meets stated performance requirements.

Static 1. The state in which a quantity does not change appreciably within an arbitrarily long period of time.

2. A body at rest.

Static balance A process of checking and/or correcting wight distribution of interrelated parts while they are not in motion.

Static head The maximum height the liquid is being pumped as long as you take into consideration the siphon affect in some piping systems.

Static pressure seal No movement between the sealing surfaces, such as the oil pan to engine.

Static pressure The pressure of a vapour or liquid in a container or system under ambient temperature conditions with no load imposed.

Static seal A seal that prevents fluid passage between two parts that are in fixed positions relative to each other.

Static system pressure The pressure of a vapour or liquid in a system under ambient temperature conditions with no load imposed.

Static tension The tension in the belt when the drive is at rest, a factor that determines drive efficiency, service life, or other operating characteristics.

Statically balanced A tire or other rotating device that is balanced at rest so that it will not turn when free to do so because its weight mass is distributed evenly around the circumference of the tire or the axis of rotation.

Stationary fifth wheel A non-movable fifth wheel whose location on the tractor frame is fixed.

Stationary seal The spring loaded or moveable portion of the seal does not rotate with the shaft. Must be used when the seal surface speed exceeds 5000 fpm. or 25 meters per second.

Statistic Filtering The control module internally charts the reselts of diagnostic testing over a perion of time and creates a baseline for testing.

Statistical fluctuations Kinds of information that cannot be precisely predicted.

Statistical process control (SPC) Use of variation analysis, with manual or computerized control charts, to detect irregular variations in a process as quickly as possible. Often, SPC charts display upper and lower limits for part characteristics or process parameters and show trends over time, indicating when the limits are exceeded (or are about to be exceeded) and corrective actions are needed. In some closed-loop systems, adjustments are made automatically when readings indicate that a control limit is being approached.

Stator 1. A reaction device or member unit inside the torque converter, mounted on a one-way roller or sprag clutch, having vanes to redirect the fluid flow that discharges from the turbine center back toward the impeller at an angle most advantageous to torque multiplication.

2. A stationary field having three windings to provide overlapping, three-phase electrical pulses to smooth the current flow.

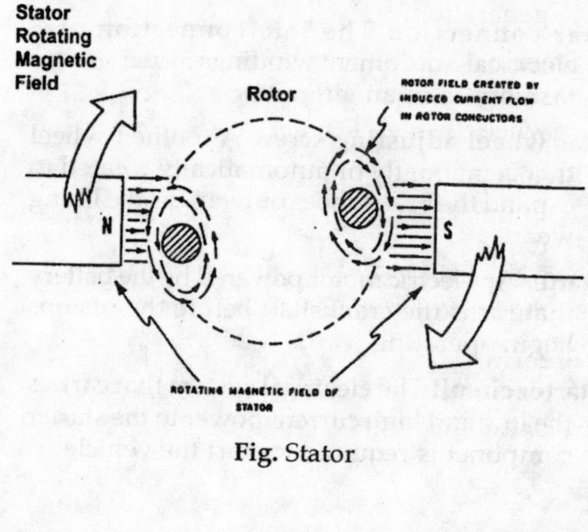

Fig. Stator

Stator assembly The reaction member or torque multiplier that is supported in a free-wheel roller race splined to the valve and front support assembly.

Stator Gain The increase in voltage, amperage, or wattage in the alternators stator

Stator support A part of the transmission pump housing in which the stator assembly in a torque converter locks.

Status accounting The process of documenting the correct, approved status of the system, including a historical record of the development of specifications and all approved changes.

Status signals Logic signals that indicate normal or abnormal conditions of operation, including: ac low, dc low, ac ok, dc ok, overtemperature, overvoltage, undertemperature, overcurrent.

Steady rest A journal support on a camshaft or crankshaft grinder to prevent vibration.

Steam engine An external-combustion engine operated by steam generated from a boiler.

Steam holes Passages designed to permit the flow of steam in an engine cooling system in hot spot areas where steam is expected to collect.

Steam Reforming The process for reacting a hydrocarbon fuel, such as natural gas, with steam to produce hydrogen as a product. This is a common method for bulk hydrogen generation.

Steering arm An arm that is attached to the steering knuckle that turns the knuckle and wheel for steering.

Steering axis inclination (SAI) The angle of a line through the center of the upper strut mount and lower ball joint in relation to the true vertical centerline of the tire, viewed from the front of the vehicle.

Steering axis The vertical line through the centerline of the upper and lower pivot- or ball-joints on a steered wheel.

Steering box A general term used for the steering mechanism at the end of the steering column.

Steering brake A braking system that provides separate control for the left and right rear wheels as an aid in steering around curves and/or correcting a pull to either side in some off-road vehicles.

Steering column shift An arrangement where the transmission shifter is mounted on the steering column.

Steering column Tubing through which the steering shaft mounts and rotates, providing a surface for the mounting of the hazard switch, turn signal switch, ignitions switch, and transmission selector or shifter.

Steering damper A device that reduces or eliminates road shock and vibration.

Steering drift The tendency of the steering to gradually drift to either side when the vehicle is driven straight ahead on a smooth, level road surface, generally caused by improper caster or camber or an under-inflated tire. Also known as steering pull.

Steering gear The assembly that converts motion from the steering column to the pitman arm.

Steering geometry The relationship of the steering linkage and the wheels to the road affected by caster, camber, scrub radius, steering offset, toe in , and toe out.

Steering kickback The shock felt in the steering wheel as the front wheels encounter obstructions in the road.

Steering knuckle The part around which each front wheel pivots as it is steered.

Steering linkage damper A shock-absorber type device that connects the steering linkage to the framework of the vehicle and absorbs some of the road shock and dampens most of the vibrations.

Steering linkage The assembly of tie rods, idler arms, and links that transfer steering motion from the steering gear box to the steering spindles with the rods, pivoting joints, and supporting parts that transfer steering motion from the pitman arm to both knuckle arms.

Steering lock A locking device on the steering column that prevents steering wheel rotation and/or selector lever motion unless unlocked using an ignition key.

Steering pull The tendency of the steering to gradually pull to the right or left when the vehicle is driven straight ahead on a smooth, level road surface, generally caused by improper caster or camber or an under inflated tire. Also known as steering drift.

Steering ratio 1. The ratio of the worm to the sector.

2. The ratio of the rack to the pinion.

Steering shaft A shaft that extends from the steering wheel toward the gearbox through the steering column.

Steering system The mechanism that permits the driver to change vehicle direction by turning a wheel inside the vehicle.

Steering terminology Terms that apply to steering, such as bump steer, memory steer, steering pull, steering wander, and torque steer.

Steering wander The tendency of the steering to pull to the right or left when the vehicle is driven straight ahead on a smooth road surface that may be caused by improper caster adjustment.

Steering wheel centering The procedure of turning both tie-rod couplings equally in the proper direction to correctly position the steering-wheel spokes, and placing the steering gears on their high (center) position.

Steering wheel The wheel, located at the top of the steering shaft, which the driver uses to steer the vehicle.

Steering wonder The tendency of a vehicle to pull to one side when driving straight ahead.

Steering-and-ignition lock A device that locks the ignition open and, at the same time, locks the steering wheel in position so it cannot be turned.

Stemming A condition where the valve radius section has corroded to the extend that it has a smaller diameter than the stem.

Stem-type service valve A service valve requiring a special wrench be affixed to a stem for opening and closing.

Step Standard for the exchange of product model data - an international product data standard to provide an complete, unambiguous, computer-interpretable definition of the physical and functional characteristics of a product throughout its life cycle.

Step-down transformer A transformer that produces a lower voltage at its secondary output than is applied to its primary input.

Step-down transformer A transformer with a turns ratio more than one. The output voltage is less than the input voltage.

Stepless transmission A transmission without gears that goes from low gear to overdrive without meshing gears.

Stepped flywheel A flywheel having a ledge to which a pressure plate is attached.

Stepped resistor 1. A resistor having two or more fixed-resistance values.

2. A resistor assembly having a switch that is wired in series to increase/decrease the circuit resistance thereby controlling an electrical motor speed.

Stepper motor A motor that moves a specific amount of rotation with each voltage application.

Steptronic A transmission strategy from BMW that allows clutch less shifting

Stereolithography A form of rapid prototyping involving the building of solid plastic objects from a CAD file by projecting a laser-generated beam of ultraviolet radiation onto the surface of a vat of photosensitive resin.

Stiffness A measure of the dynamic elongation of a belt under tension.

Stinger A slightly conical pipe used as an exhaust resonator to which individual headers feed, such as at the top of a high-performance Volkswagen Beetle engine.

Stirling engine A type of internal combustion engine where the piston is moved by changes in pressure of the alternately heated and cooled working gas.

Stock A factory-manufactured vehicle having standard design, parts and colour.

Stoichiometric ratio The ideal air to fuel ratio of 14.7:1, in terms of mass, to achieve the most complete combustion possible in an internal combustion engine.

Stoicíomitry The science of providing the correct air / fuel ratio to the combustion chamber of a gasoline internal combustion engine.

Stokes parameters The parameters, relative to polarized light and the Poincaré sphere, that are usually represented as: I, the intensity of the light beam; M, that part of the beam polarized in the horizontal plane; C, that part polarized in the +45° direction; and S, that part circularly polarized.

Stone Poor performing engine or vehicle.

Stop bit The last, or final bit of a digital data communications sequence.

Stop solenoid A term used for idle-stop solenoid.

Stoplight switch An electrical device used to illuminate the stoplights when the brakes are applied.

Stoplights Lights at the rear of a vehicle that are illuminated when the driver applies the brakes to slow or stop the vehicle. Also known as brake lights.

Stopping distance The distance required to stop a vehicle based on speed, reaction time, and road conditions.

Storage battery A device that stores electrical energy in chemical form and produces that electrical energy when required.

Storage Cell A cell which stores potential for electricity in the form of a chemical charge.

Storage temperature The range of ambient temperatures through which an inoperative power supply can remain in storage without degrading its subsequent operation.

Straight 1. Moving or continuing in one direction without turning.

2. Not curved or bent.

3. A straight stretch in a closed race course.

Straight in damage 1. Damage caused by one vehicle hitting another directly or straight on.

2. Damage caused by hitting the wall nose first in closed-course racing.

Straight in Hitting the wall nose first in closed-course racing.

Straightaway A straight stretch in a closed race course such as the front and back straightaway at Indy.

Straightedge 1. A percussion metal bar, ground so that one or more edges are in a very straight line; used in measuring and checking straightness or alignment

2. A metal bar used to check the engine block deck and cylinder head for warpage.

Straightness A condition where all elements of a surface or an axis are in a straight line.

Strain Change per unit length in a linear dimension of a part or specimen, usually expressed in % Strain, as used with most mechanical tests, is based on original length of the specimen. True or natural strain is based on instantaneous length, and is equal to ln × l lo , where l is instantaneous length and *lo* is original length of the specimen. Shear strain is the change in angle between two lines originally at right angles.

Strain energy Measure of energy absorption characteristics of a material under load up to fracture. It is equal to the area under the stress-strain curve, and is a measure of the toughness of a material.

Strain gauge A transducer that converts information about the strain placed upon an object into a change of resistance.

Strain hardening exponent Measure of increase in hardness and strength caused by plastic deformation.

Stranded wire A conductor made up of several small wires twisted together.

Stratified charge A type of combustion having a small amount of rich air/fuel mixture near the spark plug with a leaner mixture throughout the remainder of the combustion chamber.

Stratified To layer or to have in layers.

Stratified-charge engine An engine in which each cylinder has two combustion chambers connected by a small passage; the smaller prechamber contains the spark plug and receives a rich mixture while the main chamber receives a lean mixture which is ignited by a flame front from the prechamber.

Stratosphere An upper portion of the atmosphere that extends 10 to 30 miles (16 to 48 km) above the surface for the Earth.

Stratospheric ozone layer A layer extending from 6 to 15 miles (9.7 to 24.1 km) above Earth's surface, protecting the Earth from ultraviolet (UV) rays from the sun.

Streamliner 1. A specific class vehicle with a fully enclosed body, including the wheels, for dry-lakes racing.

2. A racing car with an aerodynamically enclosed body.

Streamlining The shaping of a vehicle body so as to minimize air resistance.

Strehl ratio The ratio of the illuminance at the peak of the diffraction pattern of an aberrated point image to that at the peak of an aberration-free image as formed by the same optical system.

Strength reduction ratio Alternate term for fatigue notch factor.

Stress Load on a specimen divided by the area through which it acts. As used with most mechanical tests, stress is based on original cross-sectional area without taking into account changes in area due to applied load. This sometimes is called conventional or engineering stress. True stress is equal to the load divided by the instantaneous cross-sectional area through which it acts.

Stress amplitude One-half the range of fluctuating stress developed in a specimen in a fatigue test. Stress amplitude often is used to construct an S-N diagram.

Stress concentration factor Ratio of the greatest stress in the area of a notch or other stress raiser to the corresponding nominal stress. It is a theoretical indication of the effect of stress concentrators on mechanical behaviour. Stress concentration factor usually is higher than the empirical fatigue notch factor or strength reduction ratio, because it does not take into ac-count stress relief due to local plastic deformation.

Stress ratio Ratio of minimum stress to maximum stress in one cycle of loading in a fatigue test. Tensile stresses are considered positive and compressive stresses negative.

Stress relaxation Decrease in stress in a material subjected to prolonged constant strain at a constant temperature. Stress relaxation behaviour is deter-mined in a creep test. Data often is presented in the form of a stress vs. time plot. Stress relaxation rate is the slope of the curve at any point.

Stress relieve To take residual stress out of an object. This is very important with lapped seal faces; especially those that have been inserted into a metal holder.

Stress riser An area of a part that is most likely to crack due to mishandling and/or misuse.

Stress rupture strength Alternate term for creep strength.

Stress strain diagram Graph of stress as a function of strain. It can be constructed from data obtained in any mechanical test where load is applied to a material, and continuous measurements of stress and strain are made simultaneously. It is constructed for compression, tension and torsion tests.

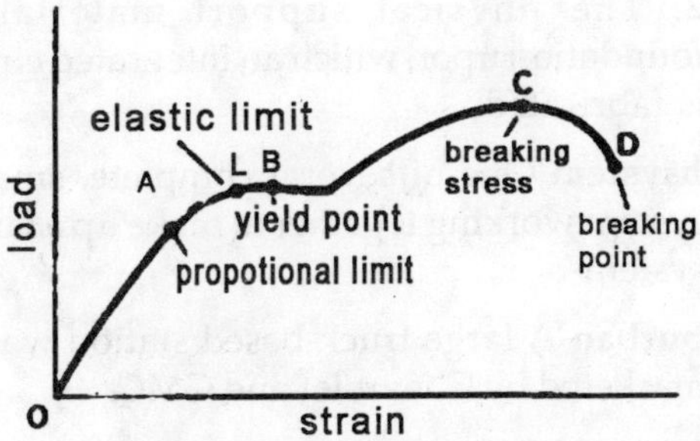

Fig. Stress strain diagram

Stress-strain ratio Stress divided by strain at any load or deflection. Below the elastic limit of a material, it is equal to tangent modulus of elasticity. An alternate term is the secant modulus of elasticity.

Stretchout A luxury sedan that has been lengthened and made into a limousine.

Stroboscope An instrument used in the study of the rapid revolutions or vibrations of a body by rendering it visible at frequent internals with a flash of light.

Stroker 1. An engine that has been stroked.

2. A camshaft that used to stroke and engine.

Stroker kit A special crankshaft and connecting rod kit used to increase the displacement of an engine by lengthening the stroke of the pistons.

Stroking seal A seal, similar to a lip seal that is installed at the rear of a calliper piston and moves with it.

Structured analysis and structured design Structured analysis and structured design is composed of two parts - structured analysis and structured design. Structured analysis is composed of an essential model, an environmental model, a behavioural model and lastly an implementation model. The essential model is a model of what the system must do, the environmental model defines the scope and interaction between the system and the world. The behavioural model specifies the required behaviour of the system so that it can interact with it's environment. Lastly, the implementation model implements the system. The structured design section is divided into three levels. The processor model assigns processes to processes. The task model assigns processes and data to tasks. Lastly, the program implementation model is an internal definition of individual tasks. Structured design breaks up the program into a hierarchy of modules with a computer program as the result.

Structured systems analysis Structured systems analysis uses process and data perspective to analyze, develop and document the requirements of a system. Structured systems analysis uses dataflow diagrams, entity relationship diagrams, data dictionaries to communicate with designers and describe the requirements.

Strut 1. Components connected from the top of the steering knuckle to the upper strut mount that maintain the knuckle position and act as shock absorbers to control spring action in a vehicle's suspension system. This is used on most front wheel drive cars and some rear wheel drive cars.

2. A device that transfers servo apply force to the transmission band. A strut may be located either between the servo or lever and the band, or between the anchor and the band.

Strut actuator An electronically controlled actuator that controls strut firmness in a computer-controlled suspension system.

Strut adjusting knob In an adjustable strut, an eight-position adjusting knob which can vary the strut orifice opening, generally accessible without having to raise the vehicle.

Strut cartridge A self-contained unit with a pressure tube and a piston rod assembly, factory sealed and calibrated.

Strut rod A heavy steel rod in the suspension system located ahead or behind a lower control arm. A strut attached between car framework and lower control arm outer end on many vehicles to determine fore-and-aft position of the outer arm.

Studded tires Tire that provides improved traction on ice, but are prohibited by law in many states because their use resulted in road-surface damage.

Stud-mounted rocker arms Individually mounted rocker arms with the use of a stud and ball.

Stuffing box pressure Between suction and discharge pressure but closer to suction pressure.

Stuffing box That portion of the pump that held the packing and now holds the mechanical seal.

Stumble 1. The tendency of an engine to falter and catch resulting in a hesitation.

2. A momentary abrupt deceleration during acceleration.

Subcontractor An individual, partnership, corporation, or an association that contracts with an organization (i.e., the prime contractor) to design, develop, and/or manufacture one or more products.

Subcooler A section of liquid line used to ensure that only liquid refrigerant is delivered to the metering device; may be a part of the condenser or may be placed in the drip pan of the evaporator.

Subcooling The cooling of a liquid below its condensing temperature.

Subframe A partial front or rear chassis frame often used in unibody design to support engine or suspension assemblies.

Sublimator An exposed metal plate, located on the outside of a spacesuit, that functions as a cooling coil to control suit temperatures.

Submerged arc welding(Saw) A process where a weld is made by feeding a bare wire into a blanket of granular fusible flux.

Substrate 1.The supporting structure in a catalytic converter where the catalyst is applied.

2. The physical support material or foundation upon which an integrated circuit is fabricated.

Subsystem One of several complete, smaller system working together to make up a larger system

Suburban A large truck-based station wagon marketed by Chevrolet and GMC.

Subvented lease A special lease, subsidized by an auto manufacturer, that features a low money factor rate or high residual value, making the monthly payments extremely attractive. Automakers increasingly are using subvented-lease specials instead of rebates to boost sales of particular models.

Successive approximation A technique used in A-D converters to measure an analogue signal. It compares the signal with progressively smaller values, each step getting nearer the actual voltage. More details are in *Monitor* Issue 4.

Suction A negative pressure.

Suction accumulator A term used for accumulator.

Suction accumulator/drier A term used for accumulator.

Suction head The head on the suction side of the pump. You subtract it from the discharge head to determine the head being produced by the pump. It is a sum of the static, pressure and friction heads.

Suction line 1. The line connecting the evaporator outlet to the compressor inlet.

2.A low-pressure line in a system usually used as a return line.

Suction manifold The point where vapour enters a compressor.

Suction pressure Compressor inlet pressure.

Suction recirculation Piping from the bottom of the stuffing box back to the suction side of the pump. Used to lower pressure and circulate liquid in the stuffing box.

Suction service valve A term used for low-side service valve.

Suction side That portion of a system that is under low pressure.

Suction specific speed A formula that will predict one of the types of cavitation. Pumps should be purchased with a number below 8500 (10,000 metric)

Suction throttling valve A back-pressure regulated device that prevents freeze-up of the evapourator core.

Suction-line regulator A device used to regulate the pressure in the suction line.

Sulphated battery A battery condition where sulphate has built up on the plates to the extent that it can no longer be charged.

Sulphating The lead sulphate that builds up on a battery as a result of battery action that produces electric current.

Sulphur oxides Small amounts of acids (SOx) that forms as a result of a reaction between the hot exhaust gas and the catalytic converter catalyst.

Sump 1. The bottom part of some compressors that contains oil for lubrication of the moving parts of the air conditioning compressor.

2. The reservoir for oil at the bottom of an engine.

3. The oil pan, or reservoir, that contains a supply of fluid for the transmissions hydraulic system

Sun gear Central gear of the planetary gearset around which the other gears rotate

Sun shade The portion of the enclosure that rotates about a horizontal axis that is coincident with the telescope altitude axis. The sun shade rotates up to close the opening formed created between the two arch girders of the carousel when the shutter has rotated toward zenith.

Sun synchronous orbit A walking orbit whose orbital plan processes with the same period as the planet's solar orbital period. In such an orbit, a satellite crosses periapsis at about the same local time every orbit.

Sunday afternoon rally One day rallies, usually run by a local car club. These may be run on public access roads.

Sunload Heat intensity and/or light intensity produced by the sun.

Sunload sensor A sensor placed on the dashboard to determine the amount of sun coming into the vehicle; a device that senses heat and/or light intensity.

Sunroof A window-type opening in the roof of the vehicle that can tilt or slide open.

Sunspot cycle The recurring, eleven-year rise and fall in the number of sunspots.

Sunspots Dark regions on the Sun which are the centers of large vortices and possess powerful magnetic fields. Maximum sunspot activity occurs in cycles with a period of about 11 years.

Supercharged, supercharger Serves the same function as a turbocharger but avoids lag time because it runs off an engine-driven pump. Both turbochargers and superchargers are used to produce more power without increasing engine displacement, but neither are particularly fuel efficient and both can require costly maintenance as vehicles age.

Supercharger This is a mechanism that pressurizes air to increase engine power, but unlike a turbocharger it is driven by a belt or gearing and thus though stronger, is more complex.

Superheat switch An electrical switch activated by a n abnormal temperature-pressure condition, such as a superheated vapour, used for system protection.

Superheat The heat intensity added to a gas after the complete evaporation of a liquid.

Superior conjunction Alignment between Earth and a planet on the far side of the Sun.

Super-stock eliminator Any vehicle running under the NHRA index/handicap system.

Supervisory control The use of microcomputers and workstations to accomplish operator interface, data acquisition, process monitoring, and some degree of production control.

Supplemental brake system An additional brake system used to assist the service-brake system in retarding a vehicle.

Supplementary insulation As defined by Underwriters Laboratories, Inc., an independent insulation provided in addition to the basic insulation to protect against electric shock in case of mechanical rupture or electrical breakdown of the basic insulation. An enclosure of insulating material may form a part of the whole of the supplementary insulation.

Supplementary payments The agreement or policy that an agency will pay defense costs, premiums, and interests.

Supplier certification A supplier becomes "certified" when it has delivered parts with perfect quality over a pre-specified time period (say six months). At that point, inspection is no longer needed.

Supplier cost reduction effort (SCORE) A United States program that urges suppliers to develop processes to cut costs. Suppliers are allowed to keep half the savings and pass on the other half to the automakers.

Supplier qualification A supplier is "qualified" when a customer when it has been determined that the supplier is capable of providing a part.

Supplier roadmap Technology roadmaps of the suppliers current and future product and process technology capabilities. These are typically represented in tabular or graphic form over time to aid in the selection of the appropriate product or process technology for a new product.

Suppliers The term 'suppliers' includes contractors, sub-contractors, vendors, developers, sellers or any other term used to identify the source from which products or services are obtained.

Supply Chain Management The procurement, stocking and distribution of components, subassemblies and products throughout the design, manufacturing, and distribution stages, ensuring that the correct components, subassemblies and products are delivered to their appropriate destination at the proper time, the lowest overall cost, and acceptable quality levels.

Supply voltage Voltage from the supply with the circuit in operation.

Surface filter A filter that traps contaminants on its surface. Screen and paper filters are types of surface filters.

Surface finish The roughness or smoothness of a surface.

Surface ignition The ignition of the air/fuel mixture in a combustion chamber by heated metal or carbon deposits.

Surface loading The transfer of pressure as one part acts on another.

Surface modelling A 3d modelling technique to describe geometry by its surfaces. This is typically used where surface shape is critical such as the design of auto body panels and aerostructures and industrial design.

Surface mount technology Surface Mount Technology (SMT) is a method of attaching electrical components directly to a board substrate rather than through a plated hole.

Surface mounted devices (SMD) A family of components intended to be mounted directly upon the surface of a substrate or circuit board.

Surface penetrator A probe designed to penetrate the surface of a body, surviving an impact of hundreds of g's, measuring and telemetering the properties of the penetrated surface.

Surface rover A semi-autonomous roving vehicle deployed on the surface of a planet or other body, taking images and soil analyses for telemetering back to Earth.

Surface speed A measure of the feet per minute the seal face or some other component is moving.

Surfacing The application by welding, brazing, or thermal spraying of a layer of material onto a surface to obtain desired properties or dimensions.

Surge 1. Electrically An instantaneous change in voltage, current or power.

2. A condition where the engine speed increases and decreases slightly with no throttle action.

Surge A large, momentary, increase in the voltage on a power line.

Surge bleeding A brake-bleeding technique designed to dislodge air from wheel cylinders by creating turbulence in the wheel cylinder.

Surge tank The reservoir of a cooling system where coolant condenses before being returned to the radiator.

Suspendability The ability of a fluid to suspend heavier dirt particles rather than letting them fall to the bottom.

Suspension arm An arm pivoted on the frame at one end and on the steering-knuckle support on the other end.

Suspension height The height of the vehicle at its four corners.

Suspension Springs, shock absorbers, struts, and links used to suspend the frame, body and engine above the wheels.

Suspension system Includes springs, shock absorbers/struts, and linkage used to suspend a vehicle's frame, body, engine and drivetrain above the wheels.

Sustainer engine An engine that maintains propulsion of a launch vehicle once it has discarded its boosters.

Swaged end A tube or pipe having an inside diameter that has been stretched to accept the outer diameter of another tube or pipe that is the same size.

Swaging To reduce or taper.

Swash plate 1. An angular plate attached to the bottom of the four pistons on a Stirling engine. A s the pistons move downward, the swash plate is turned.

2. A mechanical system that is used for pumping, having an angled plate attached to a center shaft, and pistons that are attached to the plate along the axis of the shaft. As the shaft rotates, the pistons move in and out of a cylinder, producing suction and pressure.

Swash-plate compressor A compressor in which the pistons are driven by an offset (swash) plate affixed to the main shaft, such as the six-cylinder air-conditioning compressor.

Sway bar A bar on the suspension system that connects the two sides together. It is designed so that during cornering, forces on one wheel are shared by the other.

Sway-bar link A connector from the lower control arm to the sway bar.

Swept area The total area of the brake drum or rotor that contacts the friction surface of the brake lining.

Swing attachment A leaf-spring shackle.

Swing pedals Pedals that are suspended from beneath the dash, such as clutch, brake and accelerator.

Swirl A cylinder-head design that causes the air/fuel mixture to enter the combustion chamber at a high rate of speed, increasing its atomization.

Switch An electrical device that controls the on and off of a sub-system or system.

Switching frequency The rate at which the dc voltage is switched in a converter or power supply.

Switching power supply "Switchers" are typically smaller (by about 10X), lighter (by about 10X) and generate less heat (by about 5X) compared to an equivalent linear supply. At output power ratings above about 25W they are usually less expensive. However, their transient response is slower, output regulation less precise, and have higher PARD (noise). These latter issues aren't usually a problem for most digital circuits, but must be considered when powering sensitive analogue circuits.

Switching regulator A switching circuit that operates in a closed loop system to regulate the power supply output.

Swivel foot A valve-adjusting screw having a ball that swivels when it contacts the valve stem.

Symmetrical camshaft lobes Camshaft lobes having identical opening and closing ramps.

Symmetrical Corresponding in size, form, and relative position on opposite sides of a line, plane, point, or axis.

Symmetrical rear-leaf spring A term used for rear-leaf spring.

Synchromesh A type of manual transmission where the synchronizer is used to bring a selected gear up or down tot he speed of the main shaft.

Synchromesh transmission A transmission having a synchronizer.

Synchronize To cause two or more events to operate at the same time and/or the same speed.

Synchronizer A device used in a manual transmission to bring a selected gear up or down to the speed of the main shaft.

Synchronous belt A belt having cogs or teeth that mesh in mating cogs or teeth of a pulley.

Synchronous Fuel Timing Fuel injectors firing in direct relationship to ignition reference pulses.

Synchronous motor Runs with out slippage. Used in elevators and compressors frequently, but seldom used in pumping. Pumps mostly use induction or "squirrel cage" motors.

Synchronous observing The observing of a common solar phenomenon by multiple instruments using a shared exposure timing sequence.

Synchronous rectification A rectification scheme in a switching power supply in which a FET or bipolar transistor is substituted for the rectifier diode to improve efficiency.

Synchronous The condition that exists when two events occur in a specified time relationship with each other, under the control of master timing signals.

Synthesis The translation of input requirements (including performance, function, and interface) into possible solutions (resources and techniques) satisfying those inputs. Defines a physical architecture of people, product and process solutions for logical groupings of requirements (performance, functions, and interface) and their designs for those solutions.

Synthetic aperture radar A radar imaging instrument which provides a penetrating illumination of radio waves, and is capable of imaging surfaces covered by clouds and haze. SAR images are constructed of a matrix where lines of constant distance or range intersect with lines of constant Doppler shift.

Synthetic fuel A fuel made by liquefying coal or by extracting oil from shale or tar sands.

Synthetic lining Brake friction material made from nonorganic, non-metallic, and nonasbestos materials. Typically fibreglass and aramid fibres.

Synthetic oil 1. A type of engine lubricant consisting of highly polymerized chemicals.

2. A non-mineral based lubricant for use in automotive air conditioners.

Syringe A tool with a rubber bulb and a tube or a plunger in a tube; used to suck out or force in a liquid

System Combination of several pieces of equipment to perform in a particular manner.

System curve A description of what the pump is required to perform. The pump will pump where the system curve intersects the pump curve.

System design authority The part of the developing contractors organization which is responsible for the overall design of the CIs under development.

System design The process of designing a system that comprises the interaction and integration of subsystems and subassemblies into a single system that performs an intended function. The sub-assemblies can consist of electrical, mechanical, optical, software, and other components to achieve overall functionality.

System elements A system element is a balanced solution to a functional requirement or a set of functional requirements and must satisfy the performance requirements of the associated item. A system element is part of the system (hardware, software, facilities, personnel, data, material, services, and techniques) that, individually or in combination, satisfies a function (task) the system must perform.

System head The head caused by friction in the piping. valves and fittings.

System integration team A system integration team is a higher-level IPT used in a larger program which flows down requirements and workscope in individual ipt's, monitors and coordinates their activities from a technical perspective, resolves interface and integration issues, and redirects technical activities when required to assure that the development work is accomplished to meet the overall system requirements.

System integration The successive combining and testing of hardware and software system components in a prescribed manner to prove compatibility and performance.

System overcharge A term used for an overcharged system.

System pressure The average pressure in a system, such as the fuel-injection system.

System protective valves A device that is used to protect the brake system against accidental loss of air pressure, build-up of extreme pressure, or backflow and reverse air flow in a truck braking system.

System requirement A condition or capability that must be met or possessed by a system or system component to satisfy a condition or capability needed by a user to solve a problem.

System requirements review System requirements review is a design review at which the system requirements document is reviewed and approved. This review determines which needs of the total user requirements statement will be satisfied by the proposed project.

System status (I/M readiness status) A signal for emission testing which states that all the vehicle's on-board emissions diagnostics have been run. System status is not concerned whether the emission system passed or failed the test, only that the test was run.

System-dependent recovery system Refrigerant recovery system that relies on system components, such as the compressor, to remove the refrigerant from the system.

Systems engineering group The collection of departments, managers, and staff who have responsibility for specifying the system requirements allocating the system requirements to the hardware and software components specifying the interfaces between the hardware and software components, and monitoring the design and development of these components to ensure conformance with their specifications.

Systems engineering management process A logical sequence of activities and decisions transforming an operational need into a description of system performance parameters and a preferred system configuration.

Systems engineering management process group A group of specialists who facilitate the definition, maintenance, and improvement of the systems engineering process used by the organization. In the key practices, this group is generically referred to as "the group responsible for the organization's systems and software engineering process activities".

Systems engineering Systems engineering is the application of scientific and engineering effort to: Transform an operational need into a description of system performance parameters and a system configuration through the use of an iterative process; for example, definition, syntheses, analysis, design, test and evaluation, etc.; Integrate related technical parameters and assure compatibility of all physical, functional, and program interfaces in a manner which optimizes the total system definition and design, and; Integrate reliability, maintainability, safety, human, and other such factors into the total engineering effort.

Systems security engineering An element of system engineering that applies scientific and engineering principles to identify security vulnerabilities and minimize or contain risks associated with these vulnerabilities. It uses mathematical, physical, and related scientific disciplines, and principles and methods of engineering design and analysis to specify, predict, and evaluate the vulnerability of the system to security threats.

Systems specification A system level requirements specification. A system specification may be a System/subsystem specification, Prime Item Development Specification, or a Critical Item Development Specification.

T

Tachometer An instrument for measuring the speed of the engine crankshaft in revolutions per minute (rpm).

Tack cloth A varnish-coated cheese cloth having a tacky surface used to clean metal prior to painting.

Tack weld A weld made to hold the parts of a work piece in proper alignment until the final weld can be made.

Taguchi methods A quality engineering methodology developed by Genichi Taguchi that includes off-line quality control, on-line quality control, and system of experimental design to reduce costs and improve quality. Taguchi methods are not just a statistical application of design of experiments. Taguchi methods include the integration of statistical design of experiments into a powerful engineering process. The goal is not just to optimize an arbitrary objective function, but also to reduce the sensitivity of engineering designs to uncontrollable factors or noise. This moves design targets toward the middle of the design space so that external variation affects the behaviour of the design as little as possible. This permits large reductions in both part and assembly tolerances, which are major drivers of manufacturing cost.

Tailoring The tailoring of requirements is the responsibility of the acquirer (customer), suggested tailoring may be provided by prospective and selected developers (prior to contract agreement).

Tailpipe 1. The pipe from the muffler or catalytic converter to the rear of the vehicle

2. The outlet pipe from the evaporator to the compressor.

Talk time Daily production number required to meet orders in hand divided into the number of working hours in the day.

Talker A device on the GPIB (general purpose interface bus) that sends information to a Listener on the bus.

Tall block An engine assembly consisting of a short block and a cylinder head.

Tall gears A final drive assembly with a high gear ratio.

Tall-deck engine An engine-block design height that permits a longer crank stroke.

Tandem booster A power brake vacuum booster with two small diaphragms in tandem to provide added vacuum force.

Tandem master cylinder A master cylinder that consists of a single bore with two pistons and separate fluid compression chambers used with "split" braking systems to ensure that there will be some braking power in one breaking circuit, even if there is a fluid loss in the other.

Tandem One directly in front of another and working together, such as the rear axles of a tractor.

Tandem seals The seals are facing in the same direction with a low pressure barrier fluid circulating between them.

Tandem-axle drive A type of drive that combines two single axles through the use of an interaxle differential or power divider and a short shaft that connects the two.

Tandem-axle suspension A suspension consisting of two axles with a means for equalizing weight between them.

Tang A projection peace of metal such as the peace on the end of the torque converter on an automatic transmission, used to rotate the oil pump

Tank A container for vapour or liquid such as an expansion tank of a fuel tank

Tank unit That part of a fuel-indicating system that is mounter in the fuel tank

Taper-bored piston pins Piston pins having tapered bores to reduce weight and add strength

Tapered roller bearing A specific kind of baring that is based on tapered steel rollers held together in a cage riding in a tapered seat.

Tapped resistor A fixed resistor having tow or more taps to provide different resistance values, as for fan speed control

Tappet A mechanical, solid-valve lifter located between the camshaft and valve stem, usually on an L-head engine

Target costing A market-driven strategy and process that begins with what price a product can sell for in the marketplace to achieve a desired sales volumes. Target cost is then calculated by subtracting the desired profit margin from this target price. The target cost is treated as an independent variable that must be satisfied along with other customer requirements rather than the result of design decisions (dependent variable). This cost would be considered the unit production cost that is expected to be achieved during a mature production stage. Depending on the definition, it may or may not include warranty costs and selling, general and administrative costs.

Target standard The target design standards for prototype Systems, toward which work during the development phase is directed in regard to prototypes.

Tariff A duty or tax imposed on imports.

Tarpaulin 1 A cover used to protect the passenger compartment of an open car when the seat is not occupied. 2 A cover used to protect the cargo bed of a pickup truck

Task 1. A sequence of instructions treated as a basic unit of work.

2. A well-defined unit of work in the software process that provides management with a visible checkpoint into the status of the project. Tasks have readiness criteria (preconditions) and completion criteria (postconditions).

Task manager PCM software designed to manage, regulate and perform various monitor testing.

Team A collection of people, often drawn from diverse but related groups, assigned to perform a well-defined function for an organization or a project. Team members may be part-time participants of the team and have other primary responsibilities.

Team charter A team charter is a brief written document used to define the mission and objectives of the team. The charter typically includes a statement of mission, objectives or statement of work; background; authority, boundary conditions (scope, constraints, resources, and schedule); membership; high-level requirements or specifications, and interface responsibilities.

Teambuilding The process of influencing a group of diverse individuals, each with their own goals, needs, and perspectives, to work together effectively for the good of the project such that their team will accomplish more than the sum of their individual efforts could otherwise achieve.

Tech 1 A scan tool recommended by GM for computer diagnosis

Technical (Formal) reviews A series of system engineering activities by which the technical progress on a project is assessed relative to its technical or contractual requirements. The formal reviews are conducted at logical transition points in the development effort to identify and correct problems resulting from the work completed thus far before the problem can disrupt or delay the technical progress. The reviews provide a method for the contractor and procuring activity to determine that the development of a CI and its identification have met contract requirements.

Technical architecture A view of the software architecture that focuses on the implementation and support structures

Technical data package A collection of product related engineering data comprised of the Engineering Drawing Package (EDP) and non-EDP data related to the design and manufacture of the item or system. The EDP contains all the descriptive documentation needed to ensure the competitive re-procurement of an item or system. The non-EDP consists of data such as system and development specifications, product specifications, concurrent repair list packaging data sheets, special production tool data, acceptance inspection, equipment data, project standards, repair manuals, supplementary quality assurance provisions, preparation for delivery requirements and other data as required.

Technical data Recorded information, regardless of medium or characteristics, of a scientific or technical nature. It may, for example document research, experimental, developmental, or engineering work; or be usable or used to define a design or process or to acquire, support, maintain, or operate materiel. Technical data does not include computer software or financial, administrative, cost and pricing, and management data, or other information incidental to contract administration.

Technical directive A form that passes technical information to a Contractor that does not change the contract or scope of work. If the information may result in a change to the contract or scope of work, then it must be submitted using a Change Order.

Technical effort A technical effort is any activity that influences system performance by defining, designing, or executing a task, requirement, or procedure. All the activities required to implement and execute the systems engineering process are technical efforts.

Technical inspection The examination of vehicles before an event to ensure that they conform to the rules.

Technical objectives Technical objectives shall be established for each program so that meaningful relationships among need, urgency, risks, and worth can be established.

Technical performance measurement The continuing prediction and demonstration of the degree of anticipated or actual achievement of selected technical objectives. It includes an analysis of any differences among the "achievement to date", "current estimate", and the specification requirement. "Achievement to date" is the value of a technical parameter estimated or measured in a particular test and/or analysis. "Current Estimate" is the value of a technical parameter predicted to be achieved at the end of the contract within existing resources.

Technical program planning and control The management of those design, development, test, and evaluation tasks required to progress from an operational need to the deployment and operation of the system by the user.

Technical service bulletins (TSB) The periodic information provided by the vehicle manufacturers regarding any problems encountered

Technician One involved in the design, operation, service or repair of a vehicle.

Technique A method of performing an analysis.

Technology roadmap A tabular or graphic representation of technology plans mapped against time to guide the selection and use of technology in new product development or represent the technology embodied in future products.

Technology Specification requirements shall be delineated in light of acceptable technological risks defined by risk assessment.

Technology transfer Technology transfer is the process of transferring research and technology from laboratories, government and outside organizations into the enterprise for practical application in new products.

Technology Validation Confirming that technical targets for a given technology have been met.

Tee pee exhaust An exhaust system for the Volkswagen Beetle engine.

Teething troubles Having difficulties with a new part or with a new vehicle.

Teflon pin bottom A Teflon piston-pin retainer that fits between the pin and cylinder wall.

Telecommunication Any process of communication over considerable distance.

Telematics control unit (TCU) The embedded vehicle control unit that communicates with the automobile controls, GPS satellite, and customer service centre to provide telematics features to a driver.

Telematics Telematics can be defined as a combination of wireless voice and data communication systems aimed at providing drivers with safety and information including automatic airbag deployment notification, vehicle tracking, personalized information, real-time traffic data, emergency aid, and entertainment from a central service center. Innovative features include Internet access, voice activation, and the ability to control the car audio and climate control.

Telemetry The system for radioing information, including instrument readings and recordings, from a space vehicle to the ground.

Telescope control system (TCS) The TCS is responsible for control of all telescope subsystems.

Telescope mount Also known informally as the Mount. The Mount is arranged in an altitude-over-azimuth configuration. The Mount is comprised of all the major structural elements and bearings that provide support and rotation of the ATST optical mirrors, cells, and related instrumentation packages in azimuth and altitude. The Mount can be subdivided into three main structural subassemblies: 1. the Optics Support Structure, or OSS;

2. the Mount Base; and

3. the Azimuth Track Assembly.

Telescope pier Also known informally as the Pier. The Pier is a monolithic concrete foundation poured in place at the site. It includes interfaces to the Coudé Rotator and the Mount assemblies. Contractor is responsible for the detail analysis and design of the Pier, but is not responsible for constructing it at the site.

Telescoping shock absorber A shock having certain working parts that move in and out during compression and rebound.

Temperature gauge A device that provides a visual indication of temperature.

Temperature glide The range of an evaporating or condensing temperature for a given pressure.

Temperature indicator A gauge or lamp to warn of an overcooling or overheating condition.

Temperature range, operating The range of ambient or case temperatures within which a power supply may be safely operated and meet its specifications.

Temperature range, storage The range of ambient temperatures within which a power supply may be safely stored, non-operating, with no degradation in its subsequent operation.

Temperature rating The temperature-resistance ratings that indicate a tire's ability to withstand heat.

Temperature sending unit A device in contact with the engine coolant that changes resistance, depending on its temperature, and sends a voltage signal to the gauge or lamp.

Temperature sensing bulb A term used for remote bulb.

Temperature stick A stick-like crayon designed to melt at a specific temperature when rubbed on a heated surface.

Temperature-regulated valve A term used for hot-gas bypass valve.

Temporary strainer A device installed during the run-in period of a new or rebuilt system, intended to remove coarse debris particles present in the system. After a suitable period it is removed, cleaned, and stored for future use.

Temporary-use-only spare tire A term used for compact spare tire.

Tensile failure Failure in a synchronous belt drive caused by mishandling, misalignment, overloading, shock loads, or by large debris entering into the drive.

Tensile force The moving force that slides or pulls an object over a surface.

Tensile member The muscle of the belt whose purpose is to withstand the tension that is imposed to transmit the desired power.

Tensile modulus of elasticity Tangent or secant modulus of elasticity of a material subjected to tensile loading. Alternate terms are Young's modulus and modulus of elasticity in tension. Tensile modulus of elasticity is approximately equal to compressive modulus of elasticity within the proportional limit.

Tensile strength Ultimate strength of a material subjected to tensile loading. It is the maximum stress developed in a material in a tension test.

Tension impact test Method for determining energy required to fracture a specimen under shock tensile loading.

Tension-loaded ball joint A suspension component mounted in the lower control arm with the ball-joint stud facing upward into the knuckle so the vehicle weight tends to pull the ball out of the joint.

Terminal 1. A battery post or connector 2 An electrical connector.

3. The point where two or more conductors are connected

Terminal speed The maximum speed, in drag racing, that a vehicle achieves when passing through the traps at the end of the strip.

Termination fee An amount sometimes charged at the end of a lease.

Ternary Having three parts, elements, or divisions.

Terrestrial Of or pertaining to the Earth.

Terrestrial planet Any of the four planets closest to the Sun: Mercury, Venus, Earth, Mars

Tertiary winding A third winding on a transformer.

TES Thermal electric switch

Tesla (T) SI unit of magnetic flux density (magnetic induction) expressed as 1 weber/square meter.

Test and evaluation Measurement and evaluation of system performance to validate that the system meets its specifications. Test and evaluation often occurs in the field under actual operating conditions with actual users rather than in a simulated laboratory environment.

Test data Recorded information, regardless of the form or method of the recording, of a scientific or technical nature. The term includes computer software or data incidental to contractual administration, and usually does not include financial and/or management information.

Test lamp A term used for test light.

Test light A small, non-powered or self-powered light used to determine electrical continuity in a circuit.

Test plan A document that describes the approach and test steps for all developmental, integration, qualification/certification, factory acceptance testing, and customer acceptance testing.

Test requirement The stimulus, measurement, power, loads and any special test equipment or procedure essential to validate proper operation of a device or some predetermined design control or product specification definition.

Test wheel A controlled procedure by the National Highway Traffic Safety Administration (NHTSA) to perform laboratory tests for temperature-resistance testing of automotive tires as determined by how long the tire lasts on the test wheel. The ratings are A, B, or C, with an A rating having the best temperature resistance.

Testability The characteristic of a product's design that facilitates it's testing during development/qualification, in production, and in the field.

Testable A requirement or set of requirements is considered to be testable if an objective and feasible test can be designed to determine whether each requirement has been met.

Tetrachloroethylene A chlorinated cleaning solvent with four chlorine atoms per molecule.

Tetraethyl lead (TEL) A chemical, Pb, once added to gasoline to increase its octane and aid in lubrication of the valves. Lead damages catalytic converters and oxygen sensors and therefore cannot be used in vehicles designed to operate on unleaded fuel. Leaded gasoline is no longer sold in the United States.

Theory of constraints (TOC) A lean management philosophy that stresses removal of constraints to increase *throughput* while decreasing *inventory* and *operating expenses*.

Theory of Inventive problem solving Theory of Inventive Problem Solving (Russian acronym is TRIZ) is a structured methodology developed by Genrich Altshuller for problem solving and innovation based on analysis and codification of technology solutions from millions of patents.

Thermactor An air pump or air-aspirator, air-injection system by Ford.

Thermactor pump A device that injects ambient air into the exhaust system.

Thermal Coefficient of Resistance The change in resistance of a semiconductor per unit change in temperature, over a specific range of temperature.

Thermal conductivity A measure of the material's ability to conduct heat. This is a very important factor in the selection of mechanical seal faces.

Thermal cutting A cutting process that removes metal by localized melting, burning, or vapourizing of the work piece.

Thermal delay fuse A device once used in conjunction with the compressor protection switch that heats and blows a fuse to stop a compressor action during abnormal operation.

Thermal efficience The difference between potential and actual energy developed in a fuel measured in Btu's per pound or gallon (kilogram or litre).

Thermal energy Energy in the form of heat.

Thermal fuse A temperature-sensitive fuse link designed so that it melts at a certain temperature and opens a circuit.

Thermal imaging A troubleshooting piece of equipment that will let you see hot spots in the equipment.

Thermal limiter An electrical or mechanical device, similar to a fuse, that opens at 300 degrees F to control the intensity or quantity of heat.

Thermal protection A protective feature that shuts down a power supply if its internal temperature exceeds a predetermined limit.

Thermal reactor An obsolete early attempt of an emissions-control device comprising a large, heavy, exhaust manifold in which hydrocarbons and carbon monoxide that escape from the cylinders are oxidized.

Thermal stability The ability of a device to reach and maintain a specific temperature for a long period of time.

Thermal system Provides active cooling of the exterior surfaces of the carousel and the shutter, as well as controlling other thermal aspects of the enclosure. It includes the cooling plant (e.g., chiller, a/c units situated in a remote, stationary location next to the observatory), delivery system to provide cooling media to the enclosure, and temperature sensors. The control system used to maintain the required skin temperatures is included in the overall enclosure system active cooling of the exterior surfaces of the carousel and the shutter, as well as controlling other thermal aspects of the enclosure. It includes the cooling plant (e.g., chiller, a/c units situated in a remote, stationary location next to the observatory), delivery system to provide cooling media to the enclosure, temperature sensors, and control system to maintain the required skin temperatures.

Thermal tile Silica fibre insulation used to protect 70% of the exterior of the Space Shuttle orbiter against reentry temperatures of up to 1430°C. Surface heat dissipates so rapidly that an uncoated tile can be held by its edges with the bare hand while its interior glows red hot.

Thermic reaction The chemical reaction between metal oxide and aluminium that produces superheated molten metal and a slag containing aluminium oxide.

Thermistor 1. A resister that changes it's resistance in relationship to the temperature surrounding it

Thermocouple Thermoelectric couple used to measure temperature.

Thermoelectric meters A type of dash instrument that uses heat created by current flow through the gauge to deflect the indicator needle

Thermoelectric principle The production of a voltage potential created by heating the connection of two dissimilar metals.

Thermoplastic A plastic material that can be softened or melted repeatedly without change of properties. Injection moulded parts are manufactured with Thermo-plastic materials.

Thermosetting A plastic material that can be softened and moulded but cannot be re-used or reverted to its original state.

Thermosphere The Earth atmosphere between 120 and 250 to 400 km (depending on the solar and geomagnetic activity levels), where temperature has an exponential increase up to a limiting value Texo at the thermopause. The temperature Texo is called the exospheric temperature.

Thermostat 1. A driver-operated device used to cycle the clutch to control the rate of refrigerant flow as a means of temperature control

2. A temperature-sensitive component located in a housing at the coolant outlet of the engine that restricts coolant flow to the radiator to maintain the desired engine-operating temperatures. Often referred to as Ranco control.

Thermostatic air cleaner An engine air-cleaner assembly that controls the temperature of the intake air by blending relatively cool underhood or outside air with relatively hot air picked up fro a shroud over the exhaust manifold.

Thermostatic clutch control A method of temperature control using a thermostat to cycle a clutch on and off.

Thermostatic de-icing switch A thermostat that prevents the evaporator from icing up due to low temperature conditions.

Thermostatic expansion valve The component of a refrigeration system that regulates the rate of flow of refrigerant into the evaporator as governed by the action of the remote bulb-sensing tailpipe temperatures.

Thermostatic gauge A type of gauge used to indicate fuel level, oil pressure, and engine-coolant temperature.

Thermostatic pressure valve A valve that opens and loses in relation to temperature changes.

Thermostatic switch A term used for thermostat.

Thermostatic vacuum switch (TVS) A valve that controls the passage of vacuum according to temperature, blocking vacuum until a certain coolant temperature is reached, at which point it opens.

Thermostatic vacuum valve (TVV) A valve that is operational with a vacuum signal that is proportional to an applied temperature.

Thermostatically controlled air cleaner (TCAC) A device used to regulate the temperature of the air entering the engine air cleaner as an aid in reducing emissions.

Thermosyphon The flow of coolant in a heat exchanger, such as a radiator, that is possible by the difference in densities of heated and cooled fluid.

Thexton An aftermarket tool manufacturer

Thin film technology A Unaxis core competence. The technology is used to deposit a layer of material with as little as one-thousandth the thickness of a human hair onto a substrate. Such ultra-thin coatings can filter and reflect light, protect and enhance surfaces, insulate against heat or cold, conduct and control electric current, and are also vital to information storage.

Thin wall guide An insert that may be used to repair a damaged valve guide.

Thinner A solvent used to reduce the viscosity of a liquid.

Third gauge A low-side gauge used to check pressure drop across a suction-pressure regulator at the compressor inlet on some car lines.

Third member A differential.

Third party insurance Protection for the damage of property or bodies of others.

Thixotrophic fluid The viscosity of the fluid decreases with agitation. Non drip paint is an example of such a fluid.

Thou A term used for thousandths, as in a measurement or thousands, as in revolutions per minute.

Thread class A standard designation to indicate the closeness of fit between two threaded parts, such as a bolt and nut.

Thread series A standard designation of the number of threads and their pitch per unit, inch or millimeter.

Threaded fastener Any of several types of components that hold parts together in assembly by interference of screw threads with mating threads in another component, such as capscrews, self-threading screws, stud-and-nut, and bolt-and-nut assemblies.

Threaded insert A threaded coil that is used to repair a damaged internal thread.

Three on the tree A three-speed manual transmission having a column-mounted shifter.

Three phase A type of electrical power such as that produced by an alternator.

Three piece valve An intake or exhaust valve made of three different types of material in the head, stem, and tip.

Three terminal regulator A power integrated circuit in a 3-terminal standard transistor package. It can be either a series or shunt regulator IC.

Three way monolithic Catalytic Converter A converter that contains Rhodium in the first bed to reduce NOx emissions and Plutonium and Platinum in the second bed to change HC to H_2O and CO to CO_2. A monolithic Converter has a one piece construction.

Three-axis stabilization Stabilization accomplished by nudging a spacecraft back and forth within a deadband of allowed attitude error, using small thrusters or reaction wheels.

Three-mode cycle A quick test procedure to study the causes of high emissions performed on a dynamometer under load or in a service bay without a load.

Three-speed differential A type of two-speed axle arrangement with the capability of operating both axles at a different speed at the same time. The third speed is actually an intermediate speed between the high and low range.

Three-speed transmission A transmission providing three forward-gear ratios, one reverse gear ratio and neutral.

Three-way catalyst (TWC) A catalytic converter that oxidizes hydrocarbons and carbon monoxide and reduces oxides of nitrogen emissions. It has a chamber upstream handling reduction, and one downstream handling oxidation.

Three-way Coherent communications mode wherein a DSS receives a downlink whose frequency is based upon the frequency of an uplink provided by another DSS.

Three-window coupe A coupe without rear-quarter windows.

Threshold limit value The percentage, in parts per million (ppm), of refrigeration in atmospheric air above which a human can become drowsy or have loss of concentration.

Threshold Voltage Level The level a voltage must attain to activate a circuit or component

Throat That part of a rocket engine between the combustion chamber and nozzle.

Throttle A hand control to adjust engine speed 2. An accelerator pedal.

Throttle body The housing of a throttle-body injection system that contains the valve.

Throttle cut-out relay A term used for wide-open-throttle cut-out relay.

Throttle plate A term used for throttle valve.

Throttle position sensor A variable three-wire resistor-type electrical sensor which sends a signal to the electronic control unit relative to the throttle position.

Throttle pressure The transmission hydraulic pressure that is directly related to engine load. Throttle pressure, which increases with throttle opening and engine torque output, is one of the principle pressures used to control shift points.

Throttle return check A dashpot.

Throttle solenoid positioner An electric device that holds the throttle plate in the hot-idle position and closes it when the ignition switch is turned off.

Throttle valve A flap valve that controls the amount of air admitted into the induction system.

Throttle-body fuel injection A form of electronic fuel injection in which the injectors are centrally located in a throttle-body housing that contains a valve to regulate air flow through the intake manifold. Less efficient and precise than multi-port or sequential fuel injection.

Throttle-body Throttle-body fuel injection is a type of electronic fuel injection which positions the injector(s) centrally in a throttle-body housing. This housing contains a valve to regulate the airflow through the intake manifold.

Throttling valve A term used for throttle valve, suction throttling valve, or evaporator pressure regulator.

Throughput 1. The rate at which work proceeds through a manufacturing process.

2. The rate at which information is processed through a computer.

Throw 1. A connecting rod journal on a crankshaft

2. The number of output circuits on a switch.

Throw a rod 1. A loose connecting-rod bearing

2. A broken connecting rod that has been forced through the block or oil pan.

Throwout bearing The clutch-release bearing.

Thrust In a centrifugal pump it refers to the axial movement of the shaft. The thrust can be towards the wet or power end of the pump and at start up it thrusts in both directions.

Thrust angle The angle between the geometric centerline and the thrust line of a vehicle.

Thrust bearing A bearing or a part of the main bearing that limits end-to-end movement of the crankshaft.

Thrust chamber The area in a propulsion rocket in which force accumulates before ejection, e.g. the reaction chamber.

Thrust line The bisector of total toe on the rear wheels, or the direction in which the rear wheels are pointing.

Thrust load Load placed on a part that is parallel with the center of the axis.

Thrust plate A retainer that positions the camshaft in an OHV engine and limits its end-to-end movement.

Thrust surface the area of a crank or block that absorbs end-to-end thrust pressure.

Thrust The force that propels a rocket or spacecraft measured in pounds, kilograms or Newtons. Thrust is generated by a high-speed jet of gases discharging through a nozzle.

Thrust vector control Control of the thrust vector direction to steer a rocket or spacecraft during powered flight. Thrust vector control is most often achieved by hydraulically gimbaled engines.

Thrust washer A washer that is capable of supporting a thrust load while maintaining the proper spacing within the component.

Tie rod The linkage between the idler arm or pitman arm and the steering arm.

Tier 1 supplier Manufacturer to the vehicle assemblers who are responsible for delivery of the finished assembly, product development and continued technology renewal.

Tier 2 supplier Producer of parts providing value-added to minor sub-assembly.

Tier 3 supplier Supplier of engineered materials and special services, such as rolls of sheet steel, bars and heat treating, surface treatments.

Tie-rod coupling A threaded sleeve between the tie rod and the tie-rod end providing lengthwise adjustment to set front-wheel toe in.

Tie-rod end A pivoting ball and socket joint located near the outer end of the tie rod used to connect the tie rods to the center link and to the steering arms.

Tijuana chrome Silver, chrome, or aluminium spray paint.

Tilt steering wheel A type of steering wheel that can be tilted to various angles to provide for driver comfort and convenience.

Time Base Oscilloscope circuitry that controls the timing of the sweep. The time base is set by the operator in seconds/divisions

Time box or time-boxing Time box or time-boxing refers to a technique for setting interim or end-date goal for a project and the project scope (e.g., list of features in priority order to account for the time available), approach, and plan for achieving the deadline.

Time compression technologies Time compression technologies - technologies to support the product development process, that when effectively integrated into the process, offers opportunity for significant reductions in cycle time.

Time constant Time period required for the voltage of a capacitor in an RC circuit to increase to 63.2 percent of maximum value or decrease to 36.7 percent of maximum value.

Time for rupture Time required to rupture specimen under constant stress and temperature in a creep test.

Time guide A reference, providing average time required to perform specific repairs or service to a motor vehicle, that may be used as a labour charge based on an hourly rate.

Time, speed, and distance (TSD) A type of rally where the object is to cover the greatest distance in the shortest period of time without exceeding the speed limit.

Time-based competition There is a belief that the first producer and vendor of a product has a significant market advantage. Thus companies may be organized to minimize the time it takes to develop a product to exploit this advantage.

Time-to-Market 1. Time-to-Market is the cycle time of product development from conception of a new product to initial sale of the new product.

2. Time-to-Market is the dimension of strategy focused on getting products to market quickly as the basis of competition.

Timing 1. The combustion spark delivery in relation to the piston position

2. The procedure of marking the appropriate teeth of a gear set prior to installation and placement in proper mesh

3. The time relationship between reference clock signals in a computer and other control signals, e.g. Other inputs or outputs that control the action or response of a circuit.

Timing belt The belt through which the crankshaft drives the camshaft or camshafts in an overhead valve or overhead cam engine.

Timing chain The chain through which the crankshaft drives the camshaft or camshafts in an overhead valve or overhead cam engine.

Timing diagram A graphical method used to identify the time in which all of the events of the four-stroke engine operate.

Timing Disc Also Known as an armature, reluctor, trigger wheel, pulse wheel, or timing core. It is used to conduct lines of magnetic force and indicate piston position

Timing gear the gear or gears through which the crankshaft drives the camshaft or camshafts in an overhead valve or overhead cam engine.

Timing light a stroboscopic tool used to check the precise timing of a conventional ignition, adjusted by loosening the hold down bolt and turning the distributor.

Timing Timing refers to the crankshaft angles at which the valves open and close and at which time the ignition system fires the spark plugs.

Timing valve a device in the fuel-injector pump which times the delivery of fuel to the injectors.

Tin 1. An element (Sn)

2. A term often used for metal.

Tin knocker One who repairs auto bodies.

Tin top A vehicle with a fully enclosed body.

Tin work Auto bodywork.

Tin worker One that works on vehicle bodies.

Tinny A station wagon having an all metal body.

Tip insert A term used for leaf-tip insert.

Tip the can To increase the ratio of nitro in a racing-fuel mixture.

Tolerance Tolerance is the upper and lower limits of some dimension or parameter relating to a component part, material or assembly which an actual item must comply with in order for it to be acceptable in procurement or manufacturing. The difference between the upper and lower tolerance is the tolerance spread.

Tolerance analysis Tolerance analysis - an analysis of the dimensional tolerances of manufactured parts. In general, the component tolerances are all known or specified and the resulting assembly tolerance is calculated. Tolerance analysis methods include worst case, root sum of squares, statistical Monte Carlo analysis and other techniques. Analysis can be done in one, two or three dimensions. Economic compromises can also considered.

Tolerance design Tolerance design is a step in the design process (following parameter design) where the determination is made of how much variation is acceptable with a design parameter that will still allow the satisfactory functioning of the product to meet the customer's needs. Often tolerance design is not adequately considered, and the designer merely specifies standard tolerances which may be inadequate or overstated.

Tonneau cover Automotive classic jargon used for tarpaulin or bed cover.

Tool Device for use in, or attachment to, production machinery that is for the assembling of materials or the working of materials by turning, milling, grinding, polishing, drilling, punching, boring, shaping, shearing, pressing or planning.

Tooling A set of required standard or special tools needed to produce a particular part, including jigs, fixtures, gauges, and cutting tools, but excluding machine tools.

Tooth A projection on a gear rim or synchronous belt that meshes or engages with another component.

Tooth cracks In asynchronous belt drive, tooth cracks and eventual tooth separation can be caused by under tensioning, over tensioning, or using a backside idler with too small a diameter or as a result of an under-designed belt drive.

Tooth form The shape of the working surface of a sprocket tooth from the bottom of the seating curve up through the working faces to the tip of the tooth.

Tooth separation A tooth crack.

Top dead center (TDC) The piston position at the top of its stroke.

Top eliminator (TE) the overall winner in a series of drag races.

Top end 1. High engine rpm; a point that horsepower is the greatest

2. The far end of a quarter-mile drag strip.

Top fuel The hottest category in drag racing having cars capable of a quarter mile in less than five seconds, and a top speed of over 300 mph (482.7 km/h).

Top inch The first downward inch of a piston stroke where most wear occurs.

Top loader A Ford four-speed transmission of the mid 1960s to early 1970s.

Top ring The top ring of a piston.

Top time (TT) The speed at which a vehicle passes through the traps at the end of a quarter mile.

Top U-bolt plate A plate located on the top of the spring and held in place when the U-bolts are tightened, clamping the spring and axle together.

Topcoat Usually the final paint film applied to a surface.

Top-down design Top-down design is a design methodology whereby an entire design is decomposed into its major components, and then these components are further decomposed into their major components, etc. The constraints are established early in the design flow, and then are passed on and adhered to by the back-end processes.

Top-end power The engine output at high speed.

Top-fuel dragsters The fastest of drag racing vehicles, these have a characteristicly long body and use top fuel which accounts for the tremendous speeds these vehicles can attain.

Top-mount battery A battery having terminals located at the top of the case.

Topocentric apparent place Topocentric apparent right ascension and declination (α, δ) will be the interface to the telescope control system. The pole is the precessing-nutating celestial intermediate pole (CIP); the zero point of right ascension will be the true equinox of date and/or the celestial intermediate origin (CIO).

Torch English term for Flashlight

Torching 1. The cutting or burning of a valve face caused by excessive detonation

2. To flame-cut metal.

Toroid A round magnetic core with a hole in the middle. A doughnut shaped core.

Torque A measure of twisting force, given in foot-pounds (abbreviated as lb.-ft.) Or newton-meters (n-m). In the case of an automobile, it is the twisting or rotational force the engine exerts on the crankshaft. Vehicle specifications often include the maximum torque an engine produces at a specific number of revolutions. An engine that produces 200 lb.-ft. Of torque at 3,000 revolutions per minute, or 200 lb.-ft.@ 3,000 rpm, accelerates better at low speeds than an engine that provides 200 lb.-ft.@5,000 rpm.

Torque and twist The forces developed in the trailer and/or tractor frame that are transmitted through the fifth wheel when a rigid trailer, such as a tanker, is required to negotiate bumps, such as street curbs.

Torque converter 1. Unit that transfers power from the engine to the transmission input shaft by directing and redirecting fluid flow.

2. A type of fluid coupling used to connect the engine crankshaft to the automatic transmission input shaft. Torque converter multiply the available engine torque under certain operating conditions.

Torque converter capacity The ability of a torque converter to absorb and transmit engine torque in relation to the amount of slippage in the converter.

Torque converter clutch A clutch located inside the torque converter that locks the turbine and the impeller together to prevent any slippage. Also called a lockup torque converter.

Torque curve A graphic depiction of the amount of torque available at different engine speeds.

Torque multiplication phase The period of torque converter operation when the vortex flow is redirected through the stator to accelerate impeller flow to the turbine and increase engine torque.

Torque multiplication Torque increase as a result of converter action that allows the turbine to revolve slower than the impeller during acceleration and heavy-load conditions at a ratio as much as 2:1.

Torque plate A thick, metal plate bolted to the deck of a block during cylinder boring and honing to reproduce the stress caused when the heads are installed and the bolts are properly torqued.

Torque rating A measure of the engine's power capability, whereby the amount of twisting or rotating effort being exerted on the crankshaft is expressed in lb.-ft. Of force. Torque is the force that gets the weight of the vehicle moving, making it an important consideration.

Torque rod shims Thin metal shims used to rotate the axis pinion to change the operating angle of the U-joint.

Torque sequence The order in which a series of bolts or nuts should be tightened.

Torque spec The force required to tighten a nut or bolt.

Torque steer An outside influence, such as uneven front tire-tread wear, causing the steering wheel to turn right or left during hard acceleration, relating to the dual torque and steering requirements that are applied to the drive axle of front-wheel-drive vehicles.

Torque test A starting motor test to determine torque produced and current required while the specified voltage is applied.

Torque tube A drive-shaft system that is enclosed in a tube extending from the transmission tailshaft to the rear-axle housing.

Torque wrench A wrench that indicates the amount of torque being exerted when used, to enable threaded parts to be tightened a specified amount.

Torque, engine Engine torque is the amount of twisting effort exerted at the crankshaft by an engine expressed in foot-pounds of force. A foot-pound represents the force of one pound acting at the right angle to the rotating crankshaft at distance of one foot in length.

Torqueflite A three-speed, plus reverse, automatic transmission used by Chrysler.

Torque-limiting clutch brake A system designed to slip when loads of 20-25 pounds-feet are reached, to protect the brake from overloading and the high heat damage.

Torque-to-yield bolt A bolt that has been tightened to a specified yield or stretch point.

Torr One millimeter of mercury. Used in both the imperial and metric systems a a unit of measurement in vacuum service.

Torsion bar A long straight bar fastened to the frame at one end and to a suspension part at the other. In effect, a torsion bar is merely an uncoiled spring, and spring action is produced by twisting the bar. The main advantage of the torsion bar over the coil spring in the front suspension is the ease of adjusting the front suspension height.

Torsion test Method for determining behaviour of materials subjected to twisting loads. Data from torsion test is used to construct a stress-strain diagram and to determine elastic limit torsional modulus of elasticity, modulus of rupture in torsion, and torsional strength. Shear proper-ties are often determined in a torsion test.

Torsional balance A term used for vibration damper.

Torsional deformation Angular displacement of specimen caused by a specified torque in torsion test. It is equal to the angular twist (radians) divided by the gage length (mm).

Torsional modulus of elasticity Modulus of elasticity of material subjected to twist loading. It is approximately equal to shear modulus and also is called modulus of rigidity.

Torsional rigidity The resistance of a structure to twisting and flexing.

Torsional rings Rings that have a slight twist when placed within the cylinder wall, made by adding a chamfer or counterbore on the ring.

Torsional stabilizer bar A stabilizer bar made of spring steel reduces the body roll, thereby reducing the effect of centrifugal force when a vehicle rounds a corner.

Torsional stiffness A vehicle body's resistance to twisting motions.

Torsional strain Strain corresponding to a specified torque in the torsion test. It is equal to torsional deformation multiplied by the radius of the specimen.

Torsional strength Measure of the ability of a material to withstand a twisting load. It is the ultimate strength of a material subjected to torsional loading, and is the maximum torsional stress that a material sustains before rupture. Alternate terms are modulus of rupture and shear strength.

Torsional stress Shear stress developed in a material subjected to a specified torque in torsion test.

Torsional vibration A rotary vibration that causes a twist-untwist action on a rotating shaft.

Torsion-bar spring A term used for torsion bar.

Torsion-bar steering gear A rotary valve power-steering gear.

Torsion-bar suspension A suspension featuring a torsion bar rather than a coil or leaf spring.

Torus Solid geometrical figure with the shape of a doughnut or innertube.

Torx A star shaped driver used to increase torque and for reset headed fastners.

Torx fastener A six-point fastener that is easy to grip and drive without slippage. Sometimes called a "star" fastener, this relatively new fastener is used on most late model cars in luggage racks, headlights, tail-light assemblies, mirror mountings, and exterior trim.

Total effect The change in a stabilized output produced by concurrent worst case changes in all influence quantities within their rated range.

Total head The amount of head produced by the pump. Discharge head minus suction head. If suction head is a negative number it is added to the discharge head.

Total heat load The amount of heat to be removed or added, based on all conditions.

Total loss Any loss of sufficient size where there is nothing left of value of the property.

Total pedal travel The complete distance a clutch pedal must move to disengage the flywheel.

Total quality management (TQM) A management technique to improve the quality of goods and services, reduce operating costs and increase customer satisfaction.

Total regulation band The range of combined regulation tolerances such as the effects of input voltage variation, output load variation, temperature variation, drift and other specified variables. It is expressed as a plus/minus percent from nominal. Also called accuracy limits.

Total solar irradiance The amount of radiant energy input by the Sun into the Earth's atmosphere.

Total Time All time during the period being measured (at the rate of 24 hours per day, seven days per week.) Also equal to the sum of time spent in all six equipment time categories.

Total volume Volume of space in the cylinder and combustion chamber above the piston at bottom dead center (BDC).

Totaled the condition of a vehicle that has been damaged beyond repair.

Tow truck A truck set up to pull a trailer or other vehicle.

Tow vehicle A vehicle set up to pull a trailer or other vehicle.

Towing & labour Pays for towing and labour at the site of an automobile breakdown.

Towing capacity The amount of weight a vehicle can tow behind it.

Town car A four-door, chauffeur-driven sedan with an open cockpit and an enclosed passenger compartment.

Toxicity A measure of how poisonous a substance is.

Toxics 1. A chemical harmful to human life

2. As definer in the 1990 clean air act amendment, toxics include benzene, butadiene, formaldehyde, acetaldehyde, and polycyclic organic mater.

Trace 1. A coloured dye introduced into a system to detect leaks

2. To follow logically from beginning to end

3. A bar hinged to two parts to transfer motion between them.

4. A graphical representation of the change in voltage on the line over a specific period of time.

Traceability The evidence of an association between items, such as between process outputs, between an output and its originating process, or between a requirement and its implementation.

Track 1. A term used for tread

2. The roadway or pathway a vehicle travels during a competitive event

3. To follow

4. A belt or tread that guides lines of motion.

Track bar A steel bar attached to the axle housing on one side and he frame on the other side of the vehicle to maintain sideways alignment between the axle housing and frame on some coil-spring equipped models.

Track Vehicle width, measured from the centre of one tire's contact patch to the center of the opposite tire's contact patch.

Tracking Rear wheels following directly behind the front wheels.

Tracking station A station set up to track an object through the atmosphere or space, usually by means of radar or radio.

Tracking The science of monitoring satellite locations by means of radio antennas at ground stations or by using other satellite systems in space.

Traction bar A device used on a rear-wheel-drive to prevent axle twist and spring wind up during hard acceleration.

Traction Control System A system that attempts to control wheel spin during acceleration on slick road surfaces.

Traction rating A tire rating that indicates the traction capabilities of the tire to the consumer.

Traction The amount of friction between the tire and the ground.

Traction-Lock A limited-slip differential by Ford.

Tractive effort The pushing force exerted by the vehicle's driving wheels against the road's surface.

Tractor 1. A term used for truck tractor

2. A motor vehicle having a fifth wheel used for pulling a semitrailer

3. A truck or short wheel-base vehicle used for pulling a trailer or semitrailer

4. A self-propelled vehicle having tracks or wheels.

Trade-off analysis Trade-off Analysis is the process of making decisions when each choice has both advantages and disadvantages. In a simple trade-off, it may be enough to list each alternative and the pros and cons. For more complicated decisions, list the decision criteria and weight them. Determine how each option rates on each of the decision score and compute a weighted total score for each option. The option with the best score is the preferred option. Decision trees may be used when options have uncertain outcomes.

Trade-off study An objective evaluation of alternative requirements, architectures, design approaches, or solutions using identical ground rules and criteria.

Trading down Buying a less expensive vehicle than the one currently owned.

Trading up Buying a more expensive vehicle than the one currently owned.

Traffic message channel (TMC) The TMC transmits (on FM subcarriers) digital codes representing standardized traffic information messages to be decoded and displayed (or spoken) in any given language by in-vehicle receiver.

Trail braking A driving technique where the brakes are applied after entering a turn to help rotate the vehicle.

Trailer 1. A platform or container on wheels pulled by a truck or tractor

2. A non-powered vehicle used to haul cargo

3. To put on the trailer.

Trailer charge To fill the trailer air tanks with air by means of a dash control valve, tractor protection valve, and a trailer relay emergency valve.

Trailer hand-control valve A device located on the dash or steering column used to apply only the trailer brakes; primarily used in jack-knife situations.

Trailer slider A movable suspension frame that is capable of changing the trailer wheelbase by "sliding" and locking into different positions.

Trailer tongue An extension at the front of a trailer that attaches to a hitch ball on a tow vehicle.

Trailing arm A rear suspension element consisting of a lengthwise member that pivots from the body at its forward end and has a wheel hub rigidly attached to its trailing end.

Trailing Edge 1. In a Hall-effect switch, the edges of the rotating blades that exit the switch

2. The descending edge of a waveform

Trailing shoe A shoe, toward the rear of a non-servo brake assembly, which pivots around a fulcrum in a direction opposite to normal drum rotation.

Trailing side For a satellite that keeps the same face toward the planet, the hemisphere that faces backwards, away from the direction of motion.

Trajectory The flight path of a projectile, missile, rocket or satellite.

Tramp a form of wheel hop where a pair of wheels on an axle hop in opposite phase due to axle windup.

Tranny A term for transmission.

Trans A term for transmission.

Trans-Am An annual road-racing series, the oldest in the United States, for pony cars and small sedans.

Transaxle This mechanism combines the duties of a differential and a transmission, and is commonly used in front wheel drive (FWD) automobiles. Having a transaxle means that there is no need for the drive shafts which are used to connect the differential to the transmission in rear wheel drive (RWD) cars.

Transceiver An abbreviation for Transmitter-Receiver; a device that both transmits and receives analogue or digital signals.

Transducer 1. A transducer is a device that creates an electrical signal in response to a physical stimuli.

2. a device that converts an input of one form into an output of another form, such as a

motor that converts an electric signal to a rotary motion or a device that converts vacuum signals into electrical signals.

Trans-Earth injection The firing of a spacecraft's engines to increase speed and break out of a orbit around the Moon, or another planet, and begin it on a trajectory to Earth.

Transfer case A small transmission used on four-wheel-drive vehicles to transfer engine torque to the front and/or rear axles.

Transfer gears Gears used to transfer engine torque from one level to another in a transaxle.

Transformer A set of coils wound on an iron core in which a magnetic field couples energy between two or more coils or windings.

Transient A short surge of current or voltage, often occurring before steady-state conditions have become established.

Transient A short surge of voltage or current, often occurring before steady-state conditions have become established.

Transient recovery time The time required for the output voltage of a power supply to settle within specified output accuracy limits following a transient.

Transient response time The interval between the time a transient is introduced and the time it returns and remains within a specified amplitude range.

Transient The instantaneous change of voltage or current produced when a charge is made from one steady-state condition to another.

Transient Voltage Voltage produced most likely by self-induction. Transient voltage travels randomly throughout the vehicle's electrical system

Transistor A semiconductor device for controlling the flow of current between two terminals, the emitter and the collector, by means of variations in the current flow between a third terminal, the base, and one of the other two.

Transistorized ignition An ignition system having conventional breaker points with transistor regulation of voltage.

Transition The point where a condition changes such as the point when a digital signal changes from off to on

Transitional spring coil Coils that become inactive when compressed to their maximum load-carrying capacity.

Trans-Lunar injection The firing of a spacecraft's engines to increase speed and break out of a parking orbit around Earth and begin it on a trajectory to the Moon.

Transmission A gearing device of a vehicle that provides variable ratios between the engine output and the differential input.

Transmission band A flexible steel band lined with friction material that is clamped around a circular drum to hold it from turning.

Transmission controlled spark (TCS) An emissions-control system to prevent distributor vacuum advance at normal operating temperature until the transmission has shifted into high gear by using a transmission-mounted electric switch controlling a solenoid-actuated vacuum valve.

Transmission oil cooler Heat exchanger located in the radiator outlet end section through which transmission fluid flows for cooling purposes on most automatic transmission cars.

Transmission-regulated spark system (TRS) A system that allows distributor vacuum advance only when the transmission is in high gear.

Transmitter An electronic device that generates and amplifies a carrier wave, modulates it with a meaningful signal, and radiates the resulting signal from an antenna.

Transplant 1. An automobile factory operated by an automaker in a country other than its own.

2. An engine swap.

Transponder A device that transmits a response signal automatically when activated by an incoming signal.

Transportation expense Pays for car rental expenses if the vehicle is not driveable.

Transportation planning Optimizes assignments from plants to distribution centres, and from distribution centres to stores. The systems combine moves to ensure the most economical means are employed on a scale that no human planner could match.

Transverse crack A crack with its major axis oriented approximately perpendicular to the weld axis.

Transverse Mounted side-to-side or latitudinally.

Transverse vibration A vibration caused by an unbalanced drive line.

Traps 1. The measured section at the end of a drag strip

2. The fastest part of an oval track or road racing course.

Trash box 1. An engine built from junk parts

2. A car built from junk parts.

Travel 1. A term used for rebound travel

2. To move or go from one place to another.

Travel trailer A trailer with living accommodations.

Travel-sensitive strut A strut with the capability to adjust its firmness in relation to the amount of piston travel within.

Traverse engine An engine mounted sideways in a vehicle, such as in most front-drive vehicles.

Traverse leaf spring A leaf spring mounted sideways, such as in the Model T, or Corvette.

Tread compound A blend of synthetic rubber used in the tire tread material.

Tread distortion Tread shape change on certain tires as rotating contact is made with the road surface; less severe in belted tires.

Tread groove The recessed portion of the tire tread between the ribs.

Tread rib The raised portion of the tire tread between the grooves.

Tread rubber A blend of up to 30 different synthetic, and eight natural rubbers used in a tire tread.

Tread That portion of a tire that comes into contact with the surface of the road and is designed to allow the air flow to cool the tire and to channel water during wet weather.

Treadle A dual, heavy-truck brake valve that 1. Releases air from the service reservoir to the service lines and brake chambers

2. Opens ports to service lines to vent air in the primary and secondary systems.

Tree diagnostics chart A flow chart that suggests logical procedures to follow when troubleshooting a problem.

Triac A bi-directional silicon-controlled switch.

Triangulation The arrangement of frame members for greater durability and rigidity.

Triaxle suspension A suspension consisting of three axles with a means of equalizing weight between them.

Tri-Chevy 1955, 1956, and 1957 Chevrolets.

Trichloroethylene A chlorinated cleaning solvent often used in aerosol brake cleaner.

Trickle charger A battery charger with a very low output.

Tri-coat A painting system having three layers of topcoat paint.

Trigger A trigger is something that causes a data acquisition system to start collecting data. It may be as simple as pressing a software button or a set of conditions which when met trigger data capture (internal triggers), or an externally generated, hardware signal (an external trigger).

Trim Level Vehicle designation assigned by vehicle manufacturers that represent specific equipment packages.

Tri-oval track A racing course with three high-banked turns such as at the Daytona International Speedway.

Trip 1. A key cycle (key on, run, key off/power-down) where the enable criteria for a particular diagnostic are met and the diagnostic test runs.

2. Ignition cycle and engine run sequence in which the PCM runs emission control monitors.

Triple evacuation The process of evacuating a system that involves three pump downs and two system purges with an inert gas, such as dry nitrogen (N).

Tripod The central part of certain CV joints having three arms or trunnions with needle bearings and rollers running in grooves or races in the assembly.

Tri-power A triple two-barrel carburettor setup.

Tri-Y header An exhaust manifold setup that pairs cylinders according to their firing order to provide better exhaust scavenging.

Trophy An award for achievement, such as winning a race.

Trophy dash a short match, at the start of the event, among the fastest qualifying cars in an oval track race.

Trophy run A match for an individual class trophy in drag racing.

Troposphere A division of the Earth's atmosphere extending from ground level to altitudes ranging 5-10 miles.

Trouble code A number generated by a computer to indicate a failure in a sensor, circuit, or the computer itself. The number may be communicated to the technician by the flashing of a dash light when the diagnostic mode is entered. Also known as failure code.

Trouble diagnosis The work required to find the cause of a problem.

Troubleshooting The location of a problem or malfunction by a process of elimination.

Truck A large vehicle generally used for hauling cargo or for pulling a trailer.

Truck tractor A large truck with a fifth wheel used to pull a semitrailer.

Truck-Frame and Axle Repair Association (TARA) A trade association of those engaged in repair of truck and heavy equipment frames.

True anomaly The angular distance of a point in an orbit past the point of periapsis, measured in degrees.

True Downtime Cost (TDC) A method of recording and analyzing all significant cost metrics associated with equipment downtime in a building or manufacturing facility. TDC provides a way to assign time and/or monetary value to previously considered "non-tangible" cost of downtime. Also TDC includes downtime factors commonly overlook to arrive at a more true value for the cost of downtime.

True power Actual power generated or consumed in a circuit.

True strain Instantaneous % of change in length of specimen in mechanical test. It is equal to the natural logarithm of the ratio of length at any instant to original length.

True stress Applied load divided by actual area of the cross section through which load operates. It takes into account the change in cross section that occurs with changing load.

Trueing 1. Straightening bent shafting or parts.

2. Dressing a grinding wheel.

3. Setting up a flywheel or pulley so that it runs evenly without sideways motion at its rim.

4. Making an edge straight or bringing two parts square with each other.

Truncation Rejection of the final digits in a number, thus lessening the precision but not necessarily the accuracy.

Trunk A storage compartment at the rear of most passenger cars.

Trunk unit An early automotive air-conditioning evaporator that mounted in the trunk compartment and was channelled through the package tray.

Trunnion The ends of a universal joint cross where the bearings ride.

Truth in leasing Also known as the consumer leasing act of 1976, this act was designed to protect consumers against inadequate and misleading lease information.

Truth table A chart used to show the output condition of a logic gate or circuit as compared to different conditions of input.

Tub 1. A term used for phaeton.

2. A phaeton body.

3. The external structure of an auto body.

Tubbed A standard passenger-car body with cut-out rear wheel wells to accommodate oversize rear wheels.

Tube-and fin condenser A type of heat exchanger that consists of tubes to which fins are attached to facilitate the transfer of heat from the fluid passing through it to the air passing over it.

Tube-and-fin evaporator A type of heat exchanger that consists of tubes to which fins are attached to facilitate the transfer of heat from the air passing over it to the fluid passing through it.

Tube-and-fin radiator A type of heat exchanger that consists of tubes to which fins are attached to facilitate the transfer of heat from the fluid passing through it to the air passing over it.

Tubeless tire A tire having an integral inside rubber liner to retain air; mounted directly on the rim.

Tubular core A radiator or heater core made up of tubular cells.

Tubular pushrod A term used for hollow pushrod.

Tucked and rolled A type of upholstery pleating.

Tudor An early Ford designation for a four-door sedan.

Tules The boondocks or the back country.

Tulip valve An intake or exhaust valve with a tulip-shape on the backside of the valve head.

Tune up A routine maintenance procedure that includes replacing the rotor and spark plugs; in older cars, replacing the rotor, points, capacitor, spark plugs and adjustment of the timing.

Tuned exhaust An exhaust system having equal length passages.

Tuned header A term used for tuned exhaust.

Tuned intake A term used for ram tuning.

Tuned intake and exhaust systems Intake and exhaust systems that increase the flow of intake charge into and out of the combustion chambers by varying the length, shape, or diameter of the component.

Tuned port injection (TPI) A multiport fuel-injection system used on 5.0 and 5.7 Liter General Motors V-8 engines featuring tuned intake runners from a common plenum.

Tunes Slang for a stereo sound system.

Tune-up A regularly scheduled maintenance to check normal operation of the vehicle.

Tune-Up Manufacturers Council (TMC) A trade association for the manufacturers of ignition parts.

Tune-up specifications Specifications used during a tune-up procedure on a vehicle.

Tungsten carbide A common hard face seal material available in several grades depending upon hardness and corrosion resistance. Cobalt and nickel are the two most common types.

Tungsten electrode A non-filler metal electrode used in arc welding, arc cutting, and plasma spraying, made principally of tungsten (W).

Tungsten inert gas welding A welding process using tungsten (W) with an inert gas, such as argon (Ar), as a shield.

Tunnel ram manifold An intake manifold having a large plenum and long runners to the intake ports, improving output at high rpm.

Turbine 1. The driven member inside a torque converter consisting of many vanes that receive fluid flow from the impeller.

2. A driven member that connects to the transmission input shaft and provides torque input to the planetary gearset.

Turbine engine An engine in which gas pressure is created by combustion to spin a turbine.

Turbine Machine for generating rotary mechanical power from the energy in a stream of fluid. The energy, originally in the form of head or pressure energy, is converted to velocity energy by passing through a system of stationary and moving blades in the turbine.

Turbo An acronym for turbocharger.

Turbo intercooler A term used for intercooler.

Turbo lag The time it takes the turbocharger to kick in after the driver accelerates; the lag results because a turbocharger compressor is spun by exhaust gases in the exhaust manifold.

Turbocharger (Turbo) This piece of equipment is driven by flowing exhaust gases, and works to increase the power produced by the engine by allowing it to burn up more fuel than it ordinarily could. The turbocharger achieves this feat by pressurizing the air taken in by the engine - more air means more combustion, of more fuel.

Turbocompressor Machine for compressing air or other fluid (reactant if supplied to a fuel cell system) in order to increase the reactant pressure and concentration.

Turbopump A pump driven by a gas turbine, generally used to pump propellant into a combustion chamber.

Turbulence Disturbed fluid. Can cause cavitations problems in a centrifugal pump. Often caused by an elbow located too close to the pump suction inlet.

Turbulent flow State of fluid flow, where the particles move in irregular, wavy paths. Shapes.

Turkey pan Flanges that are installed as deflectors in an engine to reduce oil splash in the lifter areas.

Turn 1. To change directions.

2. A specific speed as in "to turn 90 mph:.

Turn ratio The ratio between the number of turns between windings in a transformer, primary to secondary.

Turn signal Lights located at the four corners of a vehicle to signal a turn.

Turn-by-turn route guidance A method of providing route guidance by sequentially delivering upcoming maneuver information to the driver visually and/or audibly as the route is driven.

Turning circle A term used for turning diameter.

Turning diameter the diameter of a circle a vehicle would make if the steering were locked.

Turning torque Amount of torque required to keep a shaft or gear rotating.

Turns ratio Ratio of the number of turns on the primary winding of a transformer to the number of turns on the secondary winding.

U

U-bolt U-shaped bolts to fit around the rear-axle housing for clamping the leaf spring to the housing.

Ultimate elongation Alternate term for elongation of material at rupture under tensile loading.

Ultimate strength Highest engineering stress developed in material before rupture. Normally, changes in area due to changing load and necking are disregarded in determining ultimate strength.

Ultrahigh frequency (UHF) The band of signal frequencies falling between 300 and 3000 MHz.

Ultraviolet An invisible spectrum with wavelengths shorter than visible light b longer than X-rays about 200 to 4000 angstrom units.

Umbilical A cable conveying power to a rocket or spacecraft before lift-off. Also a tethering or supply line for an astronaut outside a spacecraft.

Umbrella liability policy A liability coverage that goes above and beyond the normal coverage.

Umbrella seal An umbrella-shaped valve stem seal used to direct oil away from the stem and guide.

Umpire Brought in if the agency and policyholder don't agree on the amount of loss or damage to a property.

Unbalanced seal Not hydraulically balanced to generate low heat at the seal faces. Typical of original equipment designs.

Unburned hydrocarbons HC is a major pollutant in exhaust gases as a result of the partially unburned fuel after combustion.

UNC An acronym for Unified National Coarse. Identifying a standard thread pitch and count.

Unclassified Not attracting a security classification but control may be required for other reasons. Therefore official information of a unclassified nature should not be disclosed without formal departmental/program approval.

Undercharge An air-conditioning system that is short of refrigerant, resulting in improper cooling.

Under-dash unit A term used for hang on unit such as for a CB radio or other aftermarket accessory.

Underfill A depression on the weld face or root surface extending below the adjacent surface of the base metal.

Under-hood lubrication services The work performed under the hood during a chassis lubrication such as lubricating the distributor, manifold heat-control valve, upper suspension, hood hinges and latch, checking the fluid in the engine crankcase, brake master cylinder, radiator, battery, and power steering reservoir, plus inspecting belts and safety-related items.

Underinflation A condition with respect to tire inflation that decreases the rolling diameter and increases the contact area, which results in excessive sidewall flexing and tread wear causing excessive heat buildup, ultimately leading to severe tire damage.

Undershoot A transient change in output voltage in excess of specified output regulation limits.

Undersize (US) A part of item that is smaller than that specified.

Underslung suspension A suspension system in which the spring is positioned under the axle.

Understaging A vehicle placed at the beginning of a drag race, behind the normal starting point.

Understeer The opposite of oversteer, this is when a vehicle turns much less than the driver desired it to. This is a state of affairs that is more commonly encountered when cornering aggressively in FWD than RWD vehicles because sudden power to the front wheels can cause them to lose traction and push onwards without turning well.

Undertread compound A blend of synthetic rubbers used in the material under the tire tread.

Under-vehicle lubrication services Work performed under the vehicle during a chassis lubrication such as changing oil and filter, checking or changing transmission fluid, and gear oil or differential gear oil, lubricating the suspension and steering system, universal joints, clutch/or transmission linkage, front-wheel bearings, and inspection of safety-related items.

Undervoltage protection A circuit that inhibits the power supply when output voltage falls below a specified minimum.

Unequal A-arms A suspension system where one arm is larger than the other.

Unequal length control-arm suspension Front suspension-control arms of unequal length to compensate for jounce and rebound.

UNF an acronym for Unified national fine. Identifying a standard thread pitch and count.

Unfair advantage Not really "unfair", the term given for that extra margin realized by a racing team with careful planning and attention to detail.

Unfilled carbon Containing carbon/ graphite and nothing else. Filled carbons contain inorganic that will be sensitive to some chemicals and temperature extremes. Unfilled carbons are the preferred seal faces.

Unglued An engine or transmission that has blown and is destroyed.

Unibody A short term for unit body.

Unibody construction a type of body construction that doesn't require a separate frame to provide structural strength or support for the vehicle's mechanical components. Also called "unitized."

Unicast A component or part cast as a single unit.

Unicast rotor Type of rotor used on man late-model cars having the rotor and hub cast as one integral component, often from gray iron rather than cast iron.

Unidirectional rotor A rotor with cooling fins that are curved or formed at a angle to the hub centre to increase cooling airflow. Because the fins only work properly when the rotor turns in one direction, unidirectional rotors cannot be interchanged from right to left.

Unidirectional tire Tire whose tread pattern is designed to get optimum traction only when the tire is mounted to roll in one direction.

Unidirectional tread pattern A tire tread pattern that can be rotated only in one direction. Thus left to right rotation cannot be done.

Uniform pitch spring A valve spring having uniformly spaced coils.

Uniform tire quality grading Information required by the Department of Transportation to be moulded into the sidewall of each tire to provide information relating to tread wear, traction, and temperature ratings.

Uninsured motorist coverage Insurance which pays for costs resulting from a hit-and-run or an accident with an uninsured motorist.

Uninterruptible power supply (UPS) A type of power supply designed to support the load for specified periods when the line varies outside specified limits. UPS systems generally filter the AC line voltage and provide battery-backup power in the event that the main AC power fails. UPS systems are separated into two major categories, standby and continuous operation systems. Within each of these categories are square, quasi-square and sinewave output systems. The simplest and least expensive UPS is a standby UPS with square wave output. In a standby UPS, input power flows through the UPS (usually through a triac) to the output. If there is no filtering, the output will include noise or transients present at the input. When the input AC voltage falls below a minimum value, the input is turned off, battery power is inverted to AC and delivered to the output. The transition or switching time (from mains power to battery power) must be shorter than the power supply's holdup time or the power supply will momentarily fail. In a continuous UPS, the DC/AC inverter is continuously supplying power to the load and the AC power serves to maintain battery charge. Of the three different types of waveforms, generating a square waveform is the easiest, least expensive and least desirable waveform. Not only does a square wave include a rich compliment of undesirable harmonics, it also charges the input energy storage capacitor of a conventional switcher (which is usually the UPS load) to a lower-than-normal peak voltage, increasing switching current, switch transistor temperature and reducing component lifetime. On the other hand, generating a sinewave output is far more difficult, expensive but more electrically desirable.

Unipolar A signal range that is always positive or always negative, for example 0 to +10 V.

Unit 1. One.

2. An assembly or device that can perform its function without outside interference or assistance.

Unit body A form of vehicle construction that includes the chassis, frame, and body in one unit.

Unit cable A cable having pairs of cables placed into groups (units) of a given quantity, then these groups form the core.

Unit distributor A breakerless ignition distributor used by General Motors containing the ignition coil, magnetic pickup coil, and timer core.

Unit testing Testing of individual hardware or software units or groups of related units.

Unitized distributor An electronic ignition-system magnetic-pulse distributor, that combines the distributor, electronic module, pickup coil, cap, and coil into one compact assembly.

Universal flexible hose A flexible radiator hose.

Universal input Power supply's ability to accept a wide input voltage range (90VAC to 264VAC) without the selection of input range, either manually or electronically. As opposed to autoranging or auto-select input.

Universal joint a joint that transmits rotary motion between two shafts that aren't in a straight line.

Universal time coordinated The world-wide scientific standard of timekeeping; based upon carefully maintained atomic clocks and accurate to within microseconds. The addition or subtraction of leap seconds, as necessary, keeps it in step with Earth's rotation. Its reference point is Greenwich, England; when it is midnight there, it is midnight UTC.

Universal time The mean solar time of the meridian of Greenwich, England. Formerly called Greenwich mean time.

Universe All matter and energy, including Earth, the galaxies and all therein, and the contents of intergalactic space, regarded as a whole.

Unleaded fuel Fuel that does not contain tetraethyl lead.

Unleaded gasoline Gasoline that contains less than 0.0018 ounces (0.05 grams) of lead per gallon (3.785 liters).

Unloader A device linked to the throttle valve to open the choke valve when in the wide-open throttle position.

Unloading solenoid An electrically controlled valve for operating the throttling valve or bypass valve in some applications.

Unobtanium A mythical alloy that is lighter, stronger, more flexible and, at the same time, more rigid than anything available.

Unscheduled deployment A term used when referring to air bag system that could be set off without the vehicle being involved in a crash.

Unscheduled Downtime A period of time when the equipment is not available to perform its intended function due to unplanned downtime events. These include maintenance delay, repair, change of consumables, out-of-spec input, and facilities-related downtime.

Unsprung weight the vehicle weight not supported by the springs, including the steering knuckle, brake assembly, tire and wheel.

Updraft A device or system having an upward airflow.

Updraft carburettor A carburettor having an updraft airflow.

UPI Undesired Portion of Image

Uplink The radio signal transmitted to a spacecraft from Earth.

Upper A-arm A term used for upper-control arm.

Upper control arm a front-suspension component bolted to the frame between the pivoting attachment point on the vehicle-frame crossmember and the upper ball joint, which is fastened to its outer end having bushings that are attached to the control arm.

Upper radiator hose The radiator hose from the engine-coolant outlet to the radiator inlet.

UPS Uninterruptible Power Supply. Used to keep critical equipment, including computers, running in the event of a mains power failure.

Upshift from a lower gear to a higher gear ratio.

Usability testing Usability testing focuses on understanding the user's experience with a product or process and gathering user feedback to improve product design. The immediate result of usability testing is a list of specific and general recommendations for improving the software/hardware, documentation, training and/or other collateral materials provided to end users. The longer-term benefit is a better understanding of how to design more usable and marketable products by increasing and managing user input throughout the product development cycle.

Utility body A replacement for the cargo bed of a pickup truck that includes compartmental storage for parts and tools.

Utility Services The electrical, pneumatic and hydraulic services that are supplied to the Primary Mirror Assembly.

Utilization equipment Equipment that uses electric energy for mechanical, chemical, heating, lighting, or other useful purposes, such as an electric motor.

V

Vacuum 1. Any pressure less than atmospheric

2. An enclosed space from which all air has been removed, having an absolute atmospheric pressure of near zero.

Vacuum advance The principle-advancing ignition timing by using the vacuum generated by the engine, accomplished through the use of a mechanism attached to the distributor which moves the breaker point or pickup coil plate when it receives vacuum.

Vacuum booster A power-brake actuating mechanism that uses vacuum on one side of a diaphragm as a power source to amplify braking force.

Vacuum brake booster a device that uses vacuum to amplify the braking force.

Vacuum canister a tank-like enclosure to serve the requirements of vacuum-actuated accessories.

Vacuum chamber A term used for vacuum canister or vacuum reserve tank.

Vacuum delay valve An orifice-controlled valve which delays a vacuum signal to a diaphragm, such as in the distributor vacuum-advance unit.

Vacuum diverter valve A term used for vacuum pot or vacuum motor.

Vacuum gauge An analogue or digital instrument used for measuring vacuum.

Vacuum hose A rubber-like, plastic or nylon hose used to transmit vacuum.

Vacuum hydraulic power unit A unit consisting of a vacuum brake cylinder or chamber, hydraulic cylinder(s), and control valve in which driver effort is combined with force from the cylinder piston or chamber diaphragm to displace fluid under pressure for actuation of the brake (s).

Vacuum kicker A computer-controlled throttle device used to increase idle RPM during certain operating conditions, such as when the air-conditioning system is in operation.

Vacuum line A rubber-like, plastic or nylon tube used to transmit a vacuum reading from one place to another.

Vacuum modulator A component of the air-temperature sensor that regulates the amount of vacuum signal that is applied to the servo motor.

Vacuum motor 1. A device having a chamber containing a vacuum diaphragm used to create movement in a component, such as a heated air-intake blend door, when engine vacuum is routed to the chamber.

2. A device designed to provide mechanical control by the use of a vacuum.

Vacuum pot A term used for vacuum motor.

Vacuum power unit A device for operating the doors and valves of an air-conditioner control or duct system using vacuum as a source of power.

Vacuum programmer A device with a bleed valve that changes vacuum pressure by bleeding more or less air, thereby controlling the vacuum signal.

Vacuum pump An electro-mechanical device used to rid a system of excess moisture and air.

Vacuum Regulator / Solenoid Regulator valve provides vacuum feedback to carburettor.

Vacuum reserve tank A container used to store and provide reserve engine vacuum.

Vacuum Retard Delay Valve Delays the loss of vacuum signal during some vehicle deceleration modes.

Vacuum secondaries The secondary bores of a four-barrel carburettor that are opened by a vacuum.

Vacuum selector valve A vacuum valve that controls the vacuum motors, which, in turn, operate the airflow-control doors.

Vacuum switch An electrical switch controlled by a vacuum.

Vacuum tank A term used for vacuum canister, vacuum chamber, or vacuum reserve tank.

Vacuum tap A point at which a vacuum can be released from an enclosure.

Vacuum technology A Unaxis core competence; essential in numerous surface treatment processes. Vacuum assures dust-free environments and the precisely controlled conditions essential to uniform surface coatings and thermal surface hardening. Unaxis is changing this area of manufacturing with its innovative production processes and equipment.

Vacuum test A test to determine the presence of a vacuum.

Vacuum Testing A test on the intake manifold pressure to determine the condition of the piston rings, valves, ignition timing and valve timing or to test for a restricted exhaust system

Vacuum-advance control An emissions-control system component that allows vacuum advance only during certain modes of engine operation.

Vacuum-advance solenoid An electrically operated two-position valve that provides or prevents a vacuum signal to the distributor vacuum-advance unit.

Vacuum-check relay a mechanical, air-operated device that closes off a vacuum line to a pot whenever the manifold vacuum pressure falls below the applied vacuum pressure.

Vacuum-check valve 1. An air-operated mechanical device that closes a vacuum line to the vacuum reserve tank whenever the manifold vacuum pressure falls below the reserve vacuum pressure.

2. In a power-brake system, a one-way valve on the power cylinder vacuum-supply lose connection to maintain system vacuum under certain conditions.

Vacuum-controlled temperature sensing valve A valve that connects the manifold vacuum to the distributor advance under hot-idle conditions.

Vacuum-suspended power brake A type of power brake in which both sides of the piston are subjects to a vacuum.

Valence bond The extent to which an atom combines or "bonds" with other atoms.

Valence electron The electron in the outermost shell or orbit of an atom.

Valence Ring The outermost orbital path of electrons around the nucleus of the atom.

Valence Shell The outermost shell or orbital path of an atom.

Validation The process of determining that the requirements are the correct requirements and that they are complete. The system life cycle process may use requirements that are derived requirements in system validation. Validation is the process of ensuring that the product conforms to defined user needs, requirements, and/or specifications under defined operating conditions. Design validation is performed on the final product design with parts that meet design intent. Production validation is performed on the final product design with parts that meet design intent produced with production processes intended for normal production.

Validity list A list of valid bulletins supplied by the manufacturer.

Valleys The annular grooves or recessed areas between the lands of a valve spool.

Value added Activities or actions taken that add real value to the product or service.

Value Engineering 1. Value Engineering is a structured methodology for applying value analysis or function analysis to increase customer or user value.

2. A formal technique to eliminate, without impairing essential functions or characteristics, anything that unnecessarily increases the cost of a product. It is a disciplined system for accomplishing the functions that the customer needs and wants at the lowest cost.

Value guides A used-car price guide.

Value stream The set of specific actions required to bring a specific product through three critical management tasks of any business: *Problem-solving, Information management* and *physical transformation*.

Valve A circular-stemmed device used to control the flow or air/fuel mixture in and the flow of burned gasses out of the engine.

Valve angle The angle at which the valve face is machined.

Valve arrangement The way valves are placed in an engine, such as valves-in-head or valves-in-block.

Valve assembly 1. A tire valve stem that fits through the wheel to allow the tire to be filled with air. A cap seals the valve and prevents foreign debris from entering and damaging the pin.

2. A term used for valve body assembly.

Valve body assembly An assembly containing the parts that act as the "brain" of an automatic transmission, receiving messages on gear selected, vehicle speed, throttle opening and engine load, then operating automatically by sending fluid to apply the appropriate bands and/or clutches.

Valve clearance A term used for valve lash.

Valve face The tapered section of the valve head making contact with the valve seat.

Valve float 1. A valve that does not close completely.

2. A valve that does not close at the proper time.

Valve guide The bore in the cylinder head through which the valve stem passes.

Valve head A term designating the enlarged end of the valve.

Valve job The reconditioning of a cylinder head, including the valves.

Valve keeper A small part that fits into the retainer groove located near the tip of the valve stem to secure the valve and valve spring.

Valve key A term used for valve keeper.

Valve lash The specified clearance between a valve stem end and a rocker arm on an OHV engine or valve stem end and the camshaft on an OHV engine; necessary to allow for heat expansion.

Valve lift Distance that the valve moves from the closed to the open position.

Valve lifter the cylindrical component that presses against the lobe of a camshaft and moves up and down as the cam lobe rotates, opening and closing an intake or exhaust valve. Virtually all modern valve lifters are of an hydraulic design that uses a cushion foil to promote quiet operation.

Valve overlap The time that the closing of the exhaust valve overlaps the opening of the intake valve at the end of the exhaust stroke and at the beginning of the intake stroke, when the intake and exhaust valves are partially open at the same time depending on the spacing of the lobe centres on the camshaft and the cam's duration.

Valve plate A plate containing suction and/ or discharge valves located under the compressor heads.

Valve pocket An area cut into the top of a piston to provide valve clearance.

Valve position sensor A sensor that relays information, in electronic form, to a process relative to the position of a valve.

Valve rack A holder that keeps valves in proper order when removed, to ensure proper reassembly.

Valve relief A term used for valve pocket.

Valve retainer A term used for valve keeper or valve key.

Valve rotator A device that rotates the valve while the engine is running.

Valve seat The ring of hard metal to which the valve seals.

Valve spool A spool-shaped valve such as that in a power-steering unit.

Valve spring A small coil spring that closes the valve and keeps the lifter in contact with the camshaft.

Valve stem 1. A device mounted in the rim or inner tube to provide a method of increasing or decreasing air pressure inside a tire.

2. The long, slim, round part of a valve.

Valve tappet A term used for valve lifter.

Valve temperature The operational temperature of a valve that may reach 1,400°F at the head and be as low as 100°F at the stem.

Valve timing The actual opening and closing of the valves in relationship to the number of degrees of crankshaft rotation.

Valve train The valves and camshaft(s) within an engine, and any parts attached to the valves, such as rockers and pushrods, to move them up and down.

Vane A flat, extended surface such as that on an impeller of a pump.

Vane passing syndrome A type of cavitation caused by the impeller/ cutwater clearance being too small.

Vane pump a pump having small vanes in an elliptically shaped housing.

Vanes The scientifically designed and positioned fins that direct or redirect fluid flow inside the torque converter.

Vapour A gaseous form of matter.

Vapour lines Lines used to carry gas or vapour.

Vapour liquid ratio A measurement of the ratio of vapour to liquid at a given temperature used to help determine a gasoline's tendency to contribute to vapour lock in an automotive fuel system.

Vapour lock A condition in carburetted engines where excessive heat has caused the fuel in the fuel line or fuel pump to boil, blocking the flow of fuel to the carburettor and preventing the engine from starting.

Vapour pressure Below this pressure, the liquid at this temperature will vaporize.

Vapour recovery system 1. That part of the evaporative-emissions control system that prevents gasoline vapours from escaping to the atmosphere by trapping them in the charcoal canister to be drawn into the engine and burned when the engine is started.

2. The provisions of a fuel nozzle that prevent gasoline vapours from entering the atmosphere when a vehicle is being fueled.

Vapour return line A hose used to carry fuel vapour from the fuel pump or the liquid vapour separator back to the fuel tank.

Vapour saver system A term used for vapour recovery system.

Vapour/fuel separator a term used for vapour/ liquid separator.

Vapour/liquid separator Part of the evaporative-emissions control system that prevents liquid (gasoline) from flowing through the vent line to the charcoal canister.

Vaporization The process of tuning a liquid, such as gasoline, into a vapour which is often accomplished after the atomized fuel leaves the carburettor.

Vaporize The fluid passes from a liquid to a gaseous state. If this happens at the seal faces the seal faces will be blown open.

Varactor A semiconductor diode whose capacitance varies with the applied voltage, used extensively for automatic tuning in radio receivers.

Variability reduction A multi-part strategy to reduce product variation and make a product more robust or fit to use through design of experiments, design within process capabilities, and process improvement.

Variable resistor A resistor that provides for an infinite number of resistance values within a specific range, such as a rheostat or potentiometer.

Variable speed motor Used to control flow in a system by varying the frequency of the motor. Recommended for circulating systems and any other system where the main head is friction losses in the piping system.

Variable venturi carburettor A carburettor having an aperture that increases or decreases in size as controlled by intake manifold vacuum.

Variable-aperture carburettor A term used for variable venturi carburettor.

Variable-assist steering A power-steering system that varies the amount of assistance it provides according to driving conditions. It provides maximum assistance at low speeds for manoeuvres such as turning into a parking space or turning a corner after leaving a stop light. It provides minimum assistance at cruising or highway speeds to provide greater vehicle stability.

Variable-capacitor A capacitor whose capacitance can be adjusted by rotating a shaft or crew.

Variable-displacement engine An engine designed so that displacement can be altered by activating and deactivating cylinders.

Variable-displacement pump A pump that automatically regulates output volume of fluid based on the needs of the transmission.

Variable-orifice steering system High powered steering assistance is provided during low speed cornering and parking; normal power steering assistance is provided at higher speeds.

Variable-pitch spring A valve spring with unevenly spaced coils.

Variable-pitch stator A torque-converter design for transmissions used in off-highway, special-equipment applications.

Variable-rate coil spring a spring having an average spring rate based on load at a predetermined deflection in a variety of wire sizes and shapes.

Variable-rate spring A spring that increasingly resists deflection as the load is increased, such as a coil spring having its coils closer together at one end, or in a single-leaf spring that is thinner near the ends.

Variable-ratio power steering A power-steering system in which the response of the vehicle wheels varies according to how much the steering wheel is turned.

Variable-ratio steering A power-steering system that varies the steering ratio at different vehicle speeds.

Variable-speed fan A fan having a fluid-filled or thermostatic-actuated clutch that senses the temperature of the air coming through the radiator and adjusts its speed accordingly.

Variational geometry A capability of 2d and 3d modelling systems in which the user defines a model by dimensions and constraints, which are then solved by a series of simultaneous equations to create and modify geometry.

Varistor A semiconductor whose resistance decreases as the voltage is increased.

Varistor A two-electrode semiconductor device having a voltage-dependent nonlinear resistance.

Varnish 1. A coating of deposits that collect on parts; caused by oxidation and contamination of liquids; can cause binding or restriction

2. Baked on oil and grease.

Varnish dip The process of dipping a transformer or coil in varnish to bind or protect materials.

Vault A product data management (pdm) system data storage areas or databases. Information stored in pdm system vaults is controlled by system rules and processes.

Vehicle 1. A means of conveyance or transport, sometimes motorized, that moves on wheels, runners, or tracks.

2. A generic term for a car or truck.

3. The combination of binder and solvents or diluents, which are used to put the binder in a liquid, usable form.

Vehicle body clearance The distance from the inside of the inner tire to the spring or other nearest frame or body structure.

Vehicle classification Any of several standard methods of classifying vehicles such as engine type, body/frame construction, fuel consumption structure, or type of drive. The classifications most common to consumers are the use of body shape, seat arrangement, and number of doors.

Vehicle construction types Passenger cars of today use one of two types of construction: conventional body-over-frame and unitized or unibody.

Vehicle emission-control information The emission-control information shown directly on a label on each vehicle.

Vehicle engine tests a test involving engine rigs or vehicles being placed into cold rooms at -22 degrees F to -40 degrees F (-30 degrees C to -40 degrees C) that are started after a cold-soak period to test all engine components in a cold-start situation.

Velocity A measurement of the speed of the liquid in the system. Measured in feet or meters per second. The pump is a constant velocity device.

Velocity head Part of the total head calculation. Derived from the formula h = V2/2g

Velocity stacks The short tubes attached to carburettors that are tuned to a resonate frequency that forces more air into the carburettors. Also referred to as stacks.

Velocity The rate of change of displacement; dx/dt.

Velocity vector Magnitude of speed plus direction.

Velvetouch A brand of sintered-metal brake linings.

Vendor A manufacturer or supplier of an item.

V-engine An engine having two banks or rows of cylinders set at an angle to form a V.

Vent 1.A controlled opening through which air or other vapours can escape.

2. A condition whereby fresh outside air may be introduced into the vehicle

Vent port The forward port in the master cylinder bore.

Vent To remove air or gas from the system. It is important to vent the stuffing box in vertical pumps to prevent the seal faces from running dry.

Vented disc brakes A brake disc that has cooling passages between the friction surfaces.

Ventilate the block To throw a rod through the block

Ventilated rotor A disc-brake rotor whose friction surfaces are separated by cooling fins having open, ventilated passages between each face for heat dissipation.

Ventilation gates Ventilation apertures that penetrate the exterior walls of the carousel. The gates incorporate a system of closable louvers doors that are used to attenuate the amount of wind entering and exiting the enclosure.

Ventilation To circulate fresh air through a space or system.

Venturi 1. A narrow area in a pipe through which a liquid or a gas is permitted to flow.

2. The construction of the air passageway between the choke and throttle plates in a carburettor, so as air flows past this constriction, its velocity increases and a partial vacuum is produced, thereby promoting a flow of gas from the main gas nozzle.

Venturi principle the condition of a gas or liquid flowing through a pipe and the pipe diameter is narrow in one place, the flow speeds up in the narrow area, causing a lower pressure in the narrow area, which may be defined as venturi.

Venturi vacuum amplifier An exhaust-gas recirculation system having a device that uses the weak venturi vacuum signal to regulate the application of strong manifold vacuum to the EGR valve, usually including a reservoir that supplies sufficient vacuum when the engine is producing too little for proper operation.

Venus Second planet from the Sun, a terrestrial planet.

Verification 1. Confirmation by examination and provision of objective evidence that specified requirements have been fulfilled.

2. Verification is the process of evaluating a system or component to determine whether the products of a given phase satisfy the conditions imposed at the start of that phase.

3. The process of verifying the functional and performance requirements of a design, be it a chip, board, or system. Many different kinds of verification tools are in use today, including simulation, formal verification, various types of physical analysis tools, emulation, and rapid prototyping. Most design verification strategies employ many or all of these approaches to assure the reliability of the final product prior to its manufacture.

Verification function Tasks, actions, and activities to be performed to evaluate progress and effectiveness of the evolving system (people, product, and process) solutions and to measure compliance with requirements. Analysis, (including simulation) demonstration, test, and inspection are verification approaches used to evaluate; risk; people, product, and process capabilities, compliance with requirements; and proof of concept.

Vernier callipers A precision measuring instrument with both a fixed and a sliding jaw, and a depth gauge; used to measure thickness and depth.

Vernier Rocket engine of small thrust used for fine adjustments in velocity and trajectory.

Version An identified and documented body of software. Modifications to a version of software (resulting in a new version) require configuration management actions by the supplier, acquirer or his agent, or both.

Version The version of an object or product structure is used to distinguish between the changes made to the different object or structure as it changes during its lifecycle

Vert A term used for convertible.

Vertical load The weight on a tire that serves as the input for a tire's performance.

Vertical position The position of welding in which the axis is approximately vertical.

Vertical Sensitivity An indication of how much the vertical amplifier can amplify a week signal

Vertical-load capacity The maximum recommended vertical downward force that can be safely applied to a coupling device.

Very high frequencies (VHF) The band of signal frequencies between 30 and 300 mhz.

Very low frequencies Those radio frequencies between 3 and 30 kHz.

Vetronix A manufacturer of Automotive tools and equipment such as the Mastertech scantool.

Vibration A rapid back and forth motion.

Vibration damper A device designed to dampen or reduce crankshaft torsional (twisting) vibrations that might otherwise cause crankshaft cracks or breakage.

Vibrator coil An induction coil constructed so that the magnetism of the core operates the make and break or vibrator of the primary circuit.

Video monitor A CRT that does not have an RF amplification, only video and deflection circuitry. Commonly used as a computer display.

Vidicon An imaging device consisting of a vacuum tube, in which an electron beam is swept across a phosphor coating on the glass where the image is focused, and its electrical potential varies in proportion to the level of light it encounters. This varying potential becomes the basis of the video signal produced.

Virtual customer The term virtual customer refers to the use of technology and, more-specifically, web-based tools to gather customer input and feedback throughout the product development process to better understand and address customer needs.

Virtual instrument A collection of Components that provides the instrumentation required to perform an experiment.

Virtual prototyping Virtual prototyping refers to the use of numerical analysis tools to analyze a design instead of building and testing a physical prototype.

Virtual reality Technology that enables users to "enter" and navigate through a computer-generated 3d environment. It allows users to change their viewpoint and interact with objects created in the environment in a way that mimics the real world.

Viscose friction Friction between moving parts,, such as in a compressor resulting from using an oil that is too heavy or thick.

Viscosity grade The numerical rating of a fluid's resistance to flow.

Viscosity index A number to indicate the change in viscosity of an oil when heated.

Viscosity rating A numerical indicator of the viscosity of an engine oil established by the American Petroleum Institute.

Viscosity Resistance to pouring. Higher viscosities can restrict seal movement. Centrifugal pumps can handle a maximum viscosity similar to 30 weight oil at room temperature. Above this viscosity a positive displacement pump should be used.

Viscous 1. Thick.

2. Tendency to resist flowing.

Viscous coupling A coupling device having input and output shafts fitted with thin, alternating discs in a closed chamber filled with fluid.

Viscous damper A type of crankshaft harmonic balancer that contains a liquid to control engine vibration.

Viscous fan A cooling fan attached to a viscous coupling.

Viscous friction The friction between layers of a liquid.

Viscous-drive fan clutch A belt-driven, water-pump mounted, liquid-filled or thermostatically controlled cooling fan clutch that drives the fan fast at high temperatures and slow at low temperatures.

Viscus converter clutch (VCC) A torque converter clutch that uses a silicone fluid to transfer and multiply torque, the silicon absorbs shock and dampens engine vibration during clutch application.

Visible light Electromagnetic radiation that can be seen by the human eye.

Visual controls Displaying the status of an activity so every employee can see it and take appropriate action.

Visual inspection An inspection by sight as opposed to smell, hearing, touch, or instrumentation.

Viton An E.I. Dupont Dow manufactured elastomer widely used in the sealing industry. The generic name is fluorocarbon. Many of these compounds are attacked by water and steam.

Voice of the customer Customer input and feedback, both positive and negative, including needs, wants, comparisons, relative importance, likes, dislikes, problems, and suggestions. The voc is used to drive the product definition and support techniques such as qfd.

Voice synthesizer A computer-controlled phoneme generator capable of reproducing the phonemes used for basic speech, putting them into the right combination to create words and sentences.

Voice warning system A warning system that uses a voice synthesizer to alert the driver of monitored conditions.

Void A term used to describe a policy contract that is free of all legal effect.

Voiture omnibus A French term for carriage for all, loosely translated to bus.

Volatile 1. Evaporating rapidly.

2. Easily vaporizes or explodes

Volatile memory Memory that does not retain data when power is interrupted to the computer.

Volatile organic compound (VOC) A term that relates to chemicals that are negligibly photochemically reactive.

Volatility 1. Term used to describe a gasoline's tendency to change from a liquid to a vapour.

2. The measure of how rapidly a liquid evaporates.

Volcanism Volcanic force or activity.

Volcano An opening in a planet's crust that allows magma to reach the surface.

Volksrod A Volkswagen hot rod.

Volt (V) Unit of measurement of electromotive force or potential difference. Symbol E, in electricity; symbol V in semiconductor circuits.

Voltage A derivative electrical quantity, E, measured in the unit Volts and defined in terms of the independently obtained Ampere, I, and the unit of resistance, Ohm (R) by Ohm's Law E = IR.

Voltage Cupling A setting that enables an oscilloscope to read a type of voltage. AC coupling allows the scope to filter out the DC and read only AV voltage, DC coupling allows the scope to read AC and DC voltages, Ground coupling displays the ground trace

Voltage divider Tapped or series resistance or impedance across a source voltage to produce multiple voltages.

Voltage drop Difference in potential between two points in a passive component or circuit.

Voltage efficiency The ratio of average voltage during discharge to average voltage during recharge under specified conditions of charge and discharge.

Voltage limit Maximum or minimum value in a voltage range.

Voltage Limiter Connected through the resistor network of a voltage regulator It controls the field voltage for the required amount of charging

Voltage limiting Bounding circuit used to set specified maximum or minimum voltage levels.

Voltage mode The functioning of a power supply so as to produce a stabilized output voltage.

Voltage monitor A circuit or device that determines whether or not an output voltage is within some specified limits.

Voltage potential The electrical pressure at a particular point.

Voltage regulation The process of holding voltage constant between selected parameters, the extent of which is expressed as a percent.

Voltage regulator An electrical device to control the output of a generator or alternator.

Voltage Scale The voltage defined by each vertical division on the scope display

Voltage source A power source that tends to deliver constant voltage.

Voltage Spike A momentary increase or decrease in voltage

Voltage stabilization The use of a circuit or device to hold constant an output voltage within given limits.

Volt-ampere (VA) Measurement unit of apparent power. Whenever a power supply output is AC (alternating current), the potential exists for the voltage and current waveforms to be out-of-phase or non-sinusiodal. In these cases, the VA product can be much larger than the real power delivered. In an extreme case, the voltage and current sine waveforms may be 90 degrees out-of-phase. The power supply providing this output is delivering no real output power but could be well over its maximum voltage or current output rating.

Voltmeter An instrument used to read the pressure behind the flow of electrons.

Volumetric efficiency (VE) 1. A percentage ratio that varies with the engine's rpm measuring the difference between the air-fuel mixture actually entering a cylinder and the amount that can enter under ideal conditions.

2. The ratio between the amount of air-fuel mixture that actually enters the cylinder and the amount that could enter under ideal conditions expressed as a percentage.

Volumetric Energy Density Potential energy in a given volume of fuel.

Volute casing Derives is name from a spiral shaped casing surrounding the pump impeller. It converts velocity energy to pressure energy.

Vortex oil flow 1. The circular oil-flow path as it leaves the impeller, travels to the turbine, then through the stator and back into the impeller.

2. A mass of whirling fluid within a torque converter when impeller speed is high and turbine speed is low.

Vortex pump A type of pump used for excessive solids. The impeller is recessed into the volute. A very low efficiency design, but practical in many applications.

Vortexing liquid Creating a "whirlpool affect" that can draw air into the suction of the pump.

V-pulley Used in automotive applications to drive the accessories, such as a water pump, generator, alternator, power steering, and air-conditioner compressor.

W

Waddle A term used for lateral runout waddle.

Wafer A polished slice cut from a drawn cylinder of monocrystalline silicon, and the starting point in chip manufacture. A wafer is less than 1 mm thick and may have a diameter between 25 and 300 mm, depending on the material involved.

Wafer processing A Unaxis division, and the term for wafer handling and production.

Wagon A station wagon.

Wail To perform at peak efficiency.

Waiver A written authorization to accept a configuration item, which during production of after having been submitted for inspection, is found to depart from specified requirements, but nevertheless is considered suitable for use "as is" or after re-work by an approved method.

Waiver The agreement to forego premium payment during a period of disability.

Wake up signal An input signal used to notify the body computer that an engine start is about to happen.

Walk-away lease Another name for a closed-end lease. A leasing contract in which the residual value, or final value, of the vehicle at the end of the term, has been specified. The lessee is not required to buy the vehicle, or make up any shortfall in its residual value.

Walking orbit An orbit in which gravitational influences are used to induce a precession in a satellite's orbital plane.

Wander The abnormal tracking or steering of a vehicle.

Wankel engine An engine concept developed in Germany in the 1950's having a three-sided rotor in a slightly hourglass-shaped oval chamber.

Warehouse management systems Software that integrates mechanical and human performed activities with an information system to effectively manage warehouse business processes and direct warehouse activities.

Warm up To allow an engine to reach its normal operating temperature.

Warm-up cycle The PCM uses warm-up cycles instead of key cycles to clear DTC's. A warm-up cycle is achieved when the engine coolant rises at least 22o C (40°F) from start-up and achieves a minimum temperature of 71°C (160°F)

Warm-up regulator A device in a fuel injection system to adjust the air/fuel mixture while the engine is warming up.

Warning blinker A term used for hazard system.

Warning flasher A device found in the turn signal and hazard flasher circuit that causes the warning lamps to flash on and off .

Warning light A light on the dash to warn of a problem.

Warp A slight twist or curve in a surface.

Warpage Bending or twisting in a part that was originally straight.

Washboard The corrugated surface of an unpaved road.

Washer A round, metal device with a hole in the middle to help secure a nut or a bolt.

Waste Anything that uses resources, but does not add real value to the product or service.

Waste spark A spark occurring during the exhaust stroke on a computerized ignition system.

Wastegate A turbocharger relief valve to prevent the buildup of too much pressure.

Wastegate Control Solenoid A solenoid used to control boost output on turbocharged modles of Chrysler vehicles

Water brake A type of absorption unit found on some dynamometers.

Water burnout The application of bleach or water to the rear wheels prior to a burnout to clean, scuff, and heat the tire surfaces for better traction immediately before a drag race.

Water column A manometer.

Water control valve A mechanically operated or vacuum-operated shutoff valve that stops the flow of hot water to the heater core.

Water cooled Using water as a heat transfer medium.

Water diverter A device used to direct the flow of coolant in a head or block.

Water fade A brake fade that occurs when water is trapped between the brake linings and the drum or rotor and the coefficient of friction is reduced.

Water filter A replaceable filter used to remove impurities from an engine-cooling system.

Water glass A common term for sodium silicate.

Water horse power The calculated horse power coming out of the pump using the formula WHP = head x gpm/3960

Water jacket 1. The hollow passages inside the cylinder head and engine block through which coolant flows to carry away the heat.

2. The open spaces within the cylinder clock and cylinder head where coolant flows.

Water pump A device, usually located on the front of the engine and driven by one of the accessory drive belts, that circulates the coolant by causing it to move from the lower radiator-outlet section into the engine by centrifugal action of a finned impeller on the pump shaft.

Water soluble Any material that will dissolve in water.

Water valve An electrical-, mechanical-, or vacuum-operated device that controls the flow of coolant to the heater core.

Water wash The forcing of exhaust air and fumes from a spray booth through water so that the vented air is free of thermal-sprayed particles or fumes.

Water-cooled system A term used to identify a liquid cooling system such as one that uses water and antifreeze.

Waterfall A name given to a particular form of presentation of the project life cycle. Rather than being broken into distinct periods of controlled phases, all activities appear as one long hierarchical succession.

Waterfall development model Waterfall development model undertakes the development of the entire system in a series of development phases and activities. This approach assumes the following: a) the requirements are knowable in advance of implementation; b) the requirements have no unresolved, high-risk implications; c) the nature of the requirements will not change very much during development; d) the right architecture for implementing the requirements is well understood; and e) there is enough calendar time to proceed sequentially. The projects using the waterfall development model are checked for proper execution and quality through validation of entry requirements and exit criteria at each phase. This model contrasts with the spiral development model.

Water-heated choke A bimetallic spring in the choke assembly that opens the throttle valve during warm up when the engine coolant reaches a specific temperature.

Watt (W) A unit of power equal to one joule of work performed per second; 746 watts are the equivalent of one horsepower. The watt is named for James Watt, Scottish engineer (1736-1819) and pioneer in steam engine design.

Wattage A specific quantity of electrical power.

Watt-Hour (Wh) Unit of energy measurement, equal to one watt per hour (3600 joules).

Watt's linkage A three-bar arrangement of a live or de Dion rear axle to prevent lateral movement.

Wave scavenging Internal exhaust resonating that increases the extraction of exhaust gases.

Wave spring A disc washer used when axial space is at a premium. Does not allow very much axial travel of the seal.

Wave The generic term for a pattern that repeats over time

Waveform A graphical representation of the change in voltage on the line over a specific period of time

Waveform Capture rate Refers to how quickly an oscilloscope acquires waveforms, expressed as waveforms per second

Waveform point A digital value that represents the voltage of a signal at a specific point in time. Waveform points are calculated from sample points and stored in memory

Wavefront correction control system (WCCS) The Wavefront Correction Control System is responsible for determining and transmitting corrective values for observed errors in the wavefront delivered by the telescope.

Wavelength The distance from crest to crest, or trough to trough, of an electromagnetic or other wave. Wavelengths are related to frequency: The longer the wavelength, the lower the frequency.

Wax 1. A compound to shine the painted surface.

2. To beat a competitor in a race.

Wax pellet-type thermostat A term used for pellet thermostat.

Ways Machined surface on the calliper support on which a sliding calliper slides.

Wear and use Normal depreciation of a vehicle under average daily use.

Wear compensator A device mounted on the clutch cover having an actuator arm that fits in a hole in the release-sleeve retainer.

Wear limit A manufacturer's specifications as to the durability of a part in terms of serviceability.

Wear mated parts A condition that exists when two parts rub against each other.

Wear pattern Visible wear.

Wear ring Used with closed impeller pumps to restrict leakage from the high pressure side of the pump to the low pressure side. Should be replaced when the recommended clearance is doubled.

Wear sensor A projection on an inboard brake pad that causes a squealing sound when the brake pads are worn thin.

Weather modulator A term used for cold-weather modulator.

Weather pack Connector A type of connector that seals the terminal ends aganst water pennatration

Weatstone Bridge A series parallel arrangement of resistors between an input terminal and ground

Weave bead A type of weld bead made with transverse oscillation.

Web 1. The inner part of a brake shoe that is perpendicular to the table and which all of the springs and other linkage parts (hardware) attach.

2. In computer terms web is short for the world wide web.

Wedge combustion chamber A combustion chamber resembling a wedge.

Wedge head A cylinder head with wedge-shaped combustion chambers to provide a large quench area.

Wedge To shift vehicle weight in oval-track racing by raising or lowering the springs with small blocks or wedges.

Wedge-actuated brakes A braking system that uses air pressure and air brake chambers to push a wedge and roller assembly into an actuator that is located between the adjusting and anchor pistons.

Wedge-shaped combustion chamber a type of combustion chamber that is shaped similar to a wedge or "V," designed to increase the movement of air and fuel to aid in mixing.

Wedging factor a factor that takes into account the multiplication of force between the V-belt and pulley-groove surfaces that occurs because of the wedging action of the belt in the groove.

Weedburner The exhaust headers on a drag racer that swoop close to the ground and to the rear.

Weekend warrior A part-time, weekend racer having a regular weekday job.

Weenie A tire.

Weibull distribution A failure distribution that is very useful in reliability activities because it can be used to model many other life distributions. By adjusting the beta factor, or shape parameter, of the weibull distribution, it can be made to model a decreasing, constant, or increasing hazard rate. The weibull distribution provides reasonably accurate failure analysis and failure forecasts with extremely small samples.

Weight The measure of the earth's gravitational force or pull on an object.

Weight transfer The momentary shift of a vehicle's weight forward or rearward.

Weightlessness A state experienced in a ballistic trajectory (i.e. in orbit or free fall) when, because the gravitational attraction is opposed by equal and opposite inertial forces, a body experiences no mechanical stress.

Weight-saver spare tire A compact spare tire.

Weight-to-power ratio The ratio of a vehicle's weight to its horsepower.

Weld a localized merging of metals or nonmetals produced wither by heating the materials to the welding temperature, with or without the application of pressure or by the application of pressure alone and with or without the use of filler material.

Weld axis A line through the length of the weld, perpendicular to and at the geometric center of its cross section.

Weld bead A weld resulting from a pass.

Weld crack A crack located in the weld metal or heat-affected zone.

Weld metal The portion of a fusion weld that has been completely melted during welding.

Weld pool The localized volume of molten metal in a weld prior to its solidification as weld metal.

Weldability The capacity of material to be welded under the imposed fabrication conditions into a specific, suitably designed structure and to perform satisfactorily in the intended service.

Welded metal bellows A seal design used to eliminate the use of elastomers. Excellent for cryogenic and hot applications. Not as effective for hot petroleum applications because of "coking" problems

Welding A joining process that produces merging of materials by heating them to the welding temperature, with or without the application of pressure or by the application of pressure alone, and with or without the use of filler material.

Welding arc a controlled electrical discharge between the electrode and the work piece that is formed and sustained by the establishment of a gaseous, conductive medium.

Welding filler metal The metal or alloy added in making a weld joint that alloys with the base metal to form weld metal in a fusion-welded joint.

Welding rod A form of welding filler metal, normally packaged in straight lengths, that does not conduct the welding current.

Welding tip That part of an oxyfuel gas welding torch from which the gases issue.

Weldment An assembly of two or more metal parts that have been joined by fusing metal with the application of heat.

Western Union splice The electrical connection made by paralleling the bared ends of two conductors and then twisting these bared ends, each around the other.

Wet bulb A device, such as a thermometer, having a wet sock over its sensing element.

Wet Compression Test A test of the engines compression with two table spoons of oil added to the cylinder to determine if the low compression from the initial compression test is caused by leaking piston rings, leaking valves, or from a hole in the head gasket or piston.

Wet Compression Testing A test of the engines compression pressure with two tablespoons of engine oil added through the spark plug hole to seal the rings and determine if the compression leak is from the rings or the valves

Wet end The part of the pump that gets wet from the pumping fluid. Includes the volute, stuffing box, impeller wear rings, and shaft or sleeve.

Wet liner A cylinder liner that is in contact with the coolant.

Wet sleeve A term used for wet liner.

Wet tank a supply reservoir.

Wet-bulb temperature The ambient temperature measured with a wet-bulb thermometer.

Wet-disk clutch A clutch having a friction disk that operates in a bath of oil.

Whale tale A large, horizontal spoiler at the rear of a vehicle.

Wheatstone Bridge A network of four resistances, an emf (voltage) source, and an indicator connected such that when the four resistances are matched, the indicator will show a zero deflection or "null" reading. Prototype of most other bridge circuits.

Wheel A circular frame or hub of an axle to which a tire is attached.

Wheel adapter A metal plate that permits the use of a wheel having a different bolt pattern.

Wheel alignment A condition where the wheels and tires are in proper position on the vehicle.

Wheel balance The equal distribution of weight of a wheel with a mounted tire.

Wheel base The distance from the centre of the front wheels to the centre of the rear wheels.

Wheel cans The wheel wells on a race car having a full width body.

Wheel centerline An imaginary line through the center of the tire, a vertical line if the tire is exactly in an upright position.

Wheel chocks Blocks to prevent the wheels of a vehicle from rolling during a repair or service procedure.

Wheel cylinder piston The device in a wheel cylinder that expands, causing the brake shoe to contact the drum.

Wheel cylinder The hydraulic slave cylinder mounted on the backing plate of a drum brake assembly. The wheel cylinder convert hydraulic pressure from the master cylinder to mechanical force that applies the brake shoes.

Wheel estate A mobile home.

Wheel meter A chassis dynamometer.

Wheel nut Threaded nuts used to retain the wheel to the studs on the hub assembly.

Wheel offset the distance between the centreline of the rim and the mounting plate of the wheel

Wheel rebound Downward wheel and suspension movement.

Wheel rim Circular steel, aluminium, or magnesium components on which the tires are mounted, manufactured from stamped, or pressed steel discs that are riveted or welded together to form the circular rim.

Wheel shimmy 1. The wobbling motion during rotation of a dynamically unbalanced tire.

2. Rapid inward and outward oscillations of the front wheels.

Wheel size Determined by the diameter and width of the wheel on which the tire is mounted. A 15-inch wheel has a diameter of 15 inches. A 15 x 7 wheel has a 15-inch diameter and a 7-inch width.

Wheel spindle The spindle on which a wheel is mounted.

Wheel tramp The wheel-lifting action or hopping motion caused by static unbalance. Motion may be up-and-down or forward-and-backward, caused by centrifugal force acting on a heavy tire section located near the thread-face center.

Wheel tubs Wheel wells installed to accommodate oversized tires.

Wheel weights Weights used during the balancing process to equalize the weight mass around and across the tire; attached to the rim with clips or special adhesive.

Wheel-and-axle speed sensors Electromagnetic devices used to provide wheel speed information for an anti-lock brake system.

Wheelbase distance, center to center, from front axle to rear axle. Wheelbase is important because it indicates available body length and weight distribution between front and rear axles.

Wheelie A wheelstand, lifting the front wheels off the pavement.

Wheelie bars A pair of long bars with wheels extending from the rear of the vehicle to prevent wheelstanding.

Wheels A car.

Wheel-slip brake-control system A system which automatically controls rotational wheel slip during braking.

Wheel-slip sensor a device used in combination with the wheel slip brake-control system to send the rate of angular rotation of the wheel(s) and transmit signals to the logic controller.

Wheel-speed sensor A sensor on each wheel used to monitor speed for anti-lock braking systems.

Wheelstand A mainstay in drag races, this refers to when the front end of car lifts up during a race. It is also known as a wheelie.

White flag A signal to a closed-course race driver that she or he is about to begin the last lap.

White grease A multi-porpoise lithium based grease.

Wide ratios A transmission setup with wide spreads between speeds.

Wide-open throttle(WOT) Having the throttle wide open, at full speed.

Wide-open-throttle cut-out relay A relay that cuts power to the compressor clutch or other accessories during heavy acceleration.

Wi-Fi Wireless Fidelity. A wireless communications network conforming to IEEE 802.11 specifications.

Wild Great in performance and/or appearance.

Winch A power-driven spool having a wire cable.

Wind resistance A term used for air resistance.

Wind up 1. The tendency of the rear axle to rotate with the wheels during hard acceleration.

2. To rev an engine.

Windage tray A metal meshed screen in the oil pan to deflect oil away from the crankshaft.

Winded time A quarter-mile drag racing time recorded with the advantage of a tail wind.

Winding A conductor wrapped onto a magnetic core or core form, e.g., a transformer primary or secondary. Change from the nominal DC output voltage.

Winding area The cross-sectional area of a bobbin or magnetic core that can be filled with a winding.

Winding length The allowable length on a core form or bobbin that can be occupied by a single layer winding. One dimension of the core window.

Winding machine A machine designed to put wire onto a bobbin or core form, or onto a toroid.

Windings 1. The turns of wire around a core, as in a relay.

2. The three separate bundles in which wires are grouped in a stator.

Windmill A supercharger.

Window net A net in the window of the driver's door that serves as a restraining device to keep the driver's arm inside.

Window regulator In a power-window system, the device that converts rotary motion of the motor into the linear, vertical movement of the window.

Windows When spelled with a capital, Windows is short of the Microsoft Windows Operating System.

Windscreens The British term for windshield.

Windshield the forward-facing window in a vehicle.

Windshield wiper Mechanical arms that sweep back and forth across the windshield to remove water, snow, or dirt.

Windsor experiment Chrysler Canada, in partnership with government and educational institutions, has launched a number of initiatives collectively referred to as the Windsor Experiment. The Windsor Experiment conducts studies to benchmark "best practice" training and education in Europe, and applied the results in various skills development programs.

Wing A wing-shaped spoiler.

Wing nut A nut having wing-like flanges for ease of installation and removal.

Wing/mudgard English term for Fender

Winston cup The creme de la creme of nascar racing, the drivers are going for all the marbles in this race that spans the entire season as drivers accumulate points at each of 31 events. The driver that accumulates the most points, not the most wins, becomes the eventual winston cup winner.

WIP turn rate A measure of the speed at which work-in-process moves through a plant. Typically calculated by dividing the value of total annual shipments at plant cost (for the most recent full year) by the average WIP value at plant cost.

Wipe To defeat a competitor in a race.

Wiped out 1. To be defeated in a race.

2. To crash.

Wiper A term used for windshield wiper.

Wire bonding A market sector served by the Unaxis Assembly & Packaging division (ESEC). Wire bonding equipment makes electrical connections using fine gold, aluminium or copper wires to bridge the pads on the bare semiconductor (chip) and corresponding contacts on the substrate (lead frame).

Wire cloth A material that may be cut to size and used for filtration.

Wire placement The position and routing of wires on an automobile, determined by factory engineers to provide a safe, practical, and economical route.

Wireframe A geometric model that describes 3d geometry by outlining its edges, similar to a "stick figure".

Wiring diagram A map-like drawing that shows the wiring arrangement and colours of a vehicle wiring harness.

Wiring harness The major assembly of vehicles electrical wiring system.

Wiring protective devices A device used to prevent damage to the wiring system by maintaining proper wire routing and retention with the use of special clips, retainers, straps, and supplementary insulators to provide additional protection to the conductor over what the insulation itself is capable of providing.

Wishbone A term used for A-arm or A-frame.

With tracer A term that indicates a solid or dashed line of a contrasting colour on wire insulation, used for identification purposes where many wires are involved.

Withstand voltage The maximum voltage that can be applied between separate circuits without causing failure.

Witness lines Lines scribed on the surface of adjacent parts prior to disassembly to ensure proper realignment when they are reassembled.

Wobble An unsteady movement off a normal axis.

Wobble plate An offset plate that is secured to the main shaft and moves the pistons to and fro.

Wobble-plate compressor 1. An offset plate that is secured to the main shaft and moves the piston to and fro

2. A term used for swash-plate compressor.

Woodruff key A half-moon-shaped key used to prevent a pulley or gear from turning on a shaft. Also known as half-moon key.

Woody A vehicle having wood panel or simulated wood panel bodywork.

Work 1. The change of position of an object against an opposing force.

2. The product of a force and the distance through which it acts.

Work bench Any bench on which work is done; usually having a vise and one or more drawers.

Work breakdown structure (WBS) A product-oriented listing, in family tree order, of the hardware, software, services and other work tasks which completely defines a product or program. The listing results from project engineering during the development and production of a materiel item. A WBS relates the elements of work to be accomplished to each other and to the end product.

Work breakdown structure Work breakdown structure is a hierarchical tree structure decomposing a project into activities and sub-activities to help define and control the project and its elements of work.

Work hardening The brittleness of a metal part due to stress from bending, hammering, or fatigue.

Workflow systems / workflow management systems Workflow Systems are systems to support the coordination, communication and control of business processes by means of information technology for the purpose of improving and better managing these processes. Workflow Systems automate a business process, in whole or part, during which documents, information or tasks are passed from one participant to another for action, according to a set of procedural rules.

Working chamber An area in a shock absorber where the pressure and vacuum are produced.

Working Voltage The specified operating voltage, or range of voltages, of a component, device or cell.

Work-in-process inventory (WIP) The amount or value of all materials, components, and subassemblies representing partially completed production; anything between the raw material/purchased component stage and finished-goods stage.

Works team A factory-supported racing team.

World trade organization (WTO) Created by the Uruguay Round and successor to the GATT, this new organization began operations on January 1, 1995.

World-class manufacturer A somewhat arbitrary designation that can be supported by performance results related to various manufacturing metrics. (World-class metrics may vary from one industry to another.) Typically, it denotes "best in class" producers on a worldwide basis. In the broadest sense, world-class manufacturers are those perceived to deliver the greatest value at a given price level.

Worm A type of gear on which the teeth resemble large threads.

Worm and sector A type of steering-gear assembly that imparts a rotary motion to a straight-line motion.

Worm bearing preload The adjustment of a work gear bearing to prevent or reduce backlash.

Worm gear 1. The gear at the end of the steering column in a worm-and-sector steering system.

2. A gear that is shaped like a shaft with gear teeth cut in a continuous spiral around its outer surface. A worm gear changes the axis of rotation when it turns another gear.

Worm hole A galvanic reaction that erodes metal for the outer walls of wet cylinder sleeves.

Worm shaft A steering gearbox component having spiral grooves that resemble a coiled worm, to transfer motion from the steering-wheel shaft to the pitman shaft.

Worm tracks Oil flow circuits cast within a valve body.

Wraparound headers Exhaust headers that fit closely to the engine block.

Wraparound seat A body-conforming bucket seat.

Wrench 1. To twist.

2. A hand tool for twisting and / or holding bolt heads and nuts.

Wrist pin A pin used to attach the connecting rod to the piston.

Wrist-pin bushing A bushing that supports the wrist pin in a connecting rod.

Writing Speed The ability of an analogue oscilloscope to provide a visible trace of the movement of a signal from one point to another

Wye A mechanical or electrical assembly connected in the form of a y.

X

X-axis 1.The longitudinal axis around which a vehicle structure rolls from side to side.

2. Represents the total time across the face of the oscilloscope.

X-bar chart x-bar chart A quality control chart that monitors the mean of the process. A sample of *n* parts is collected from the process every so many parts or time periods. The mean of the sample is plotted on the control chart and a determination is made if the process is "under control" or not.

X-chassis A conventional chassis design, used until the late 1960's which narrows in the center, giving the vehicle a rigid structure that is designed to withstand a high degree of twist having a heavy front cross member to support the upper and lower suspension control arms and coil springs.

X-rays A band of electromagnetic radiation intermediate in wavelength between ultraviolet radiation and gamma rays. Because x-rays are absorbed by the atmosphere, x-ray astronomy is performed in space.

Xylene 1.A chemical solvent (C_8H_{11}) used to remove grease and paint.

Y

Y axis The Y axis represents the voltage displayed on the face of the oscilloscope

Yield Produced product related to scheduled product.

Yield point elongation Strain at yield point of a material. It is an indication of ductility.

Yield strength elongation Strain corresponding to yield strength of material. It is an indication of ductility.

Yield strength Indication of maximum stress that can be developed in a material with-out causing plastic deformation. It is the stress at which a material exhibits a specified permanent deformation and is a practical approximation of elastic limit. Offset yield strength is determined from a stress-strain diagram. It is the stress corresponding to the intersection of the stress-strain curve, and a line parallel to its straight line portion offset by a specified strain. Offset for metals is usually specified as 0.2%, i.e., the intersection of the offset line and the 0-stress axis is at 0.2% strain. Offset for plastics is usually 2%.

Yield value Stress in an adhesive joint at which a marked increase in deformation occurs without an increase in load.

Young's modulus Alternate term for modulus of elasticity in tension or compression.

Y-pipe A Y-shaped pipe, such as an exhaust pipe, that merges two passages into one.

Z

Zap The vertical axis around which a vehicle structure's front and rear ends swing back and forth.

Z-axis 1. Represents the intensity or brightness of the display on an Oscilloscope

2. In a Cartesian coordinate system, the axis which is perpendicular to the x-y plain

Zeeman effect The splitting of energy levels of an atom, ion or molecule because of a magnetic field.

Zener region The region current enters into when it flows through a diode in the reverse direction

Zener voltage The reverse voltage at which breakdown occurs in a zener diode.

Zenith angle The angle between vertical and the optical axis of the Telescope. It is the complement of the altitude.

Zerk Fitting A nipple-like lubrication fitting through which grease is applied to a chassis or suspension joint with a grease gun.

Zero Balance With transducers like strain gauges, the output is large compared to the changes caused by the strain. Setting a zero balance subtracts an offset (actually a fraction of the bridge excitation voltage), so the changes caused by the strain can be accurately measured.

Zero Emissions Vehicle (ZEV) An electric, Hydrogen, or other vehicles that produce zero levels of HC, CO, and NOx

Zero gravity A condition in which gravity appears to be absent. Zero gravity occurs when gravitational forces are balanced by the acceleration of a body in orbit or free fall.

Zero lash No clearance between valve lifter and camshaft lobe.

Zero lift trajectory A trajectory in which the control system acts to maintain a condition of no aerodynamic lift on the rocket.

Zero toe Adjusting the wheels so they point straight ahead.

Zero-gap ring A piston ring that does not have end clearance.

Zinc inner liner A thin zinc-metal leaf between the longer steel leaves of a leaf-spring suspension system to control sliding friction between the leaves, and prevent corrosion on certain models.

Zing To unintentionally over rev an engine.

Zoomies Exhaust headers that sweep back and up toward the top of the rear tires on an open-wheeled drag racer.

Zyglow A non-destructive testing system using a dye penetrant and ultraviolet light to check non-magnetic parts for faults and cracks.